General Systems Theory 2.0

-- General Architectural Theory Using the SBC Architecture --

William S. Chao

CONTENTS

4

8

PREFACE

Human beings have employed the notion of systems so widely in all kinds of scientific studies. Systems definition is an artifact created by humans to describe what a system is. A system has been defined, by general systems theory 1.0, hopefully to be an integrated whole, embodied in its components, their interrelationships with each other and the environment, and the principles and guidelines governing its design and evolution. This general systems theory 1.0 definition of a system possesses one cardinal deficiency. The deficiency comes from that it does not describe the integration of systems structure and systems behavior.

Systems structure and systems behavior are the two most significant views of a system. In order to achieve a truly integrated whole of a system, we first need to integrate the systems structure and behavior together. In other words, integration of the systems structure and systems behavior results in the integration of a whole system. Since general systems theory 1.0 does not describe the integration of systems structure and systems behavior, very likely it only hopes and will never be able to actually form an integrated whole of a system. In this situation, general systems theory 1.0 is powerless in defining a system suitably.

SBC (i.e. structure-behavior coalescence) architecture provides an elegant way to integrate the structure and behavior of a system. A system is therefore redefined, by general systems theory 2.0 (general architectural theory), truly to be an integrated whole, using the SBC architecture, embodied in its assembled components, their interactions (or handshakes) with each other and the environment, and the principles and guidelines governing its design and evolution. Since general systems theory 2.0 describes the integration of systems structure and systems behavior, definitely it is able to form an integrated whole of a system. In this situation, general systems theory 2.0 is fully capable of defining a system.

In this book, we shall detail the general systems theory 2.0 defining a system through the application of SBC architecture. By this book's introduction and elaboration of SBC architecture which covers the: a) evolution&motivation view, b) multi-level (hierarchical) view and c) systemic view of a system, all readers will understand clearly how the general systems theory 2.0 helps us define a truly integrated whole of a system.

ABOUT THE AUTHOR

Dr. William S. Chao is the CEO & founder of SBC Architecture International®. SBC (Structure-Behavior Coalescence) architecture is a systems architecture which demands the integration of systems structure and systems behavior of a system. SBC architecture applies to hardware architecture, software architecture, enterprise architecture, knowledge architecture and thinking architecture. The core theme of SBC architecture is: "Architecture = Structure + Behavior."

William S. Chao received his bachelor degree (1976) in telecommunication engineering and master degree (1981) in information engineering, both from the National Chiao-Tung University, Taiwan. From 1976 till 1983, he worked as an engineer at Chung-Hwa Telecommunication Company, Taiwan.

William S. Chao received his master degree (1985) in information science and Ph.D. degree (1988) in information science, both from the University of Alabama at Birmingham, USA. From 1988 till 1991, he worked as a computer scientist at GE Research and Development Center, Schenectady, New York, USA.

Dr. William S. Chao has been teaching at National Sun Yat-Sen University, Taiwan since 1992 and now serves as the president of Association of Enterprise Architects, Taiwan Chapter. His research covers: systems architecture, hardware architecture, software architecture, enterprise architecture, knowledge architecture and thinking architecture.

PART I: GENERAL SYSTEMS THEORY 1.0 VERSUS GENERAL SYSTEMS THEORY 2.0

Chapter 1: Introduction to Systems

The word "system" originates from the Greek term, systēma, meaning "composition" or "whole". The notion of systems has been so widely used in all kinds of scientific studies such as systems analysis and design [Hoff10, Shel11], systems architecting [Maie09, Mull11], systems architecture [Burd10, Roza11], systems bible [Gall03, Kill09], systems biology [Klip09, Voit12], system dynamics [Forr61, Ogat03, Palm09], systems ecology [Jorg12, Odum94], systems engineering [Beam90, Kass07, Koss11], systems medicine [Pork78, Weil00, Weil04], systems modeling [Frie11], systems physiology [Raff11, Sher09], systems requirement [Bere09, Grad06], systems science [Warf06], systems theory [Bert69, Luhm12], systems thinking [Chec99, Ghar11, Mead08], systems view [Bert81, Lasz96].

In this chapter, we first introduce the general systems theory 1.0. We then introduce the physical and virtual systems. A physical system exists in the physical, concrete, or real world. A virtual system exists in the virtual, abstract, or notional world. A system has a boundary. The system itself is inside the boundary and the environment is outside the boundary. After that, we then introduce the high order systems. A system evolves when it changes and the final section of this chapter will introduce the evolution of a system.

1-1 General Systems Theory 1.0

All things that strike us as something independent are essentially parts of a system. We usually call the parts of a system its components. Components are sometimes labeled as parts, entities, objects, building blocks and non-aggregated systems [Chao09, Chao14a, Chao14b].

In the 1920s, Ludwig von Bertalanffy wrote: there exist models, principles, and laws that apply to generalized systems or their subclasses, irrespective of their particular kind, the nature of their constituent elements, and the relationships or "forces" between them [Bert69]. In this book, we refer what Ludwig von Bertalanffy proposed and developed as the general systems theory 1.0.

The need for defining a system arises because any real-life system is inherently complicated. It is impossible to comprehend fully the intricate interrelationships of any system of the real world with its environment, or to describe all its components and each of its details. Systems definition is an artifact created by humans to describe what a system is [Kapo94].

Every system is something the whole. Systems emphasize the holistic vision. General systems theory 1.0 defines a system, in Figure 1-1, hopefully to be an integrated whole, embodied in its components, their interrelationships with each other and the environment, and the principles and guidelines governing its design and evolution [Chec99, Ghar11, Mead08].

A system, hopefully is an integrated whole,
embodied in its components,
their interrelationships with each other and the environment,
and the principles and guidelines governing its design and evolution.

Figure 1-1   General Systems Theory 1.0 Defining a System

A system defined by the general systems theory 1.0 has the following characteristics: 1) hopefully, it is an integrated whole; 2) it is embodied in its assembled components; 3) components are interrelated with each other and the environment; 4) it evolves; and 5) it uses structural decomposition [Chao12, Ghar11] rather than functional decomposition [Scho10].

Systems definition is used to describe what a system is. Without a systems definition, everybody has his own saying about a system and never be able to reach a consensus. For example, John Irving thinks the *Wardrobe_A* is embodied in its assembled components of *Drawer_1* and *Drawer_2*, their interrelationships with each other and the environment; Sandra Woods thinks the *Wardrobe_A* is embodied in its assembled components of *Drawer_1*, *Drawer_2*, *Drawer_3* and *Drawer_4*. It is impossible for John Irving and Sandra Woods to work together on the *Wardrobe_A* if they can not reach a common definition. To solve the conflict between John Irving and Sandra Woods, here comes the general systems theory 1.0 defining the *Wardrobe_A*, shown in Figure 1-2, hopefully to be an integrated whole embodied in its assembled components of *Drawer_1*, *Drawer_2* and *Drawer_3*, their interrelationships with each other and the environment, and the principles and guidelines governing its design and evolution.

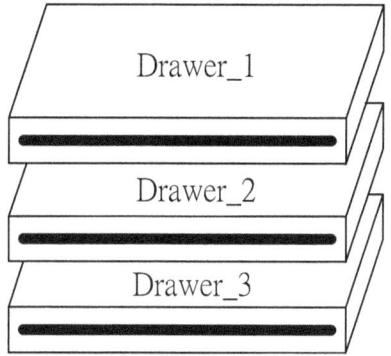

Figure 1-2   General Systems Theory 1.0 Defining the *Wardrobe_A*

As a second example, general systems theory 1.0 defines an *Eyeglasses*, shown in Figure 1-3, hopefully to be an integrated whole embodied in its assembled components of *Frames* and *Lenses*, their interrelationships with each other and the environment, and the principles and guidelines governing its design and evolution.

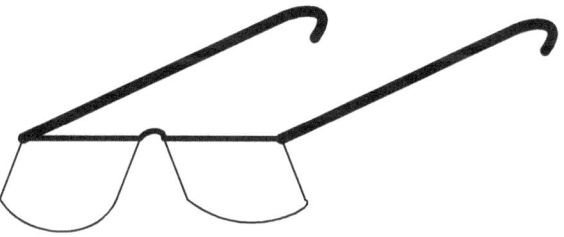

Figure 1-3   General Systems Theory 1.0 Defining an *Eyeglasses*

As the third example, general systems theory 1.0 defines a *Swing*, shown in Figure 1-4, to be hopefully an integrated whole embodied in its assembled components of *Ropes* and *Seat*, their interrelationships with each other and the environment, and the principles and guidelines governing its design and evolution.

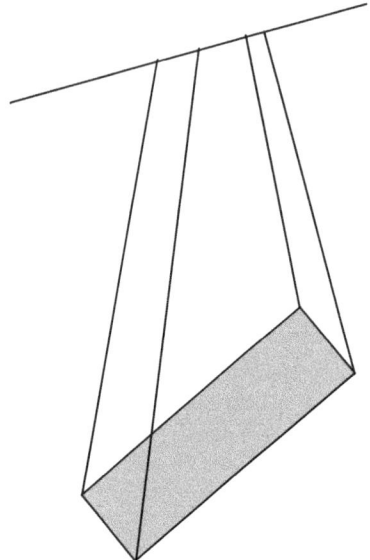

Figure 1-4   General Systems Theory 1.0 Defining a *Swing*

1-2 Physical and Virtual Systems

In general, the systems are divided into two categories: 1) physical systems and 2) virtual systems.

A physical system exists in the physical world [Acko68]. A physical system is also called a concrete or real system. For example, a *Bicycle* composed of *Wheels*, *Frame* and *Pedal*, shown in Figure 1-5, is a physical, concrete, or real system.

Figure 1-5   A *Bicycle* is a Physical System

As a second example, a *Chair* composed of *Seat*, *Back* and *Legs*, shown in Figure 1-6, is a physical, concrete, or real system.

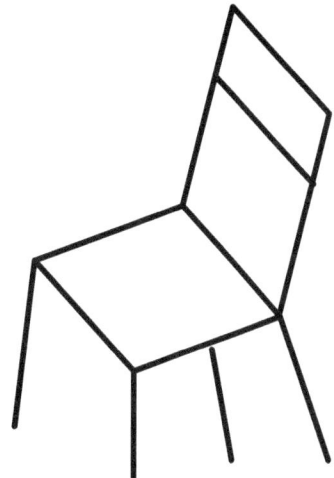

Figure 1-6   A *Chair* is a Physical System

A virtual system is a system that is composed of non-physical components, i.e., ideas, thoughts, or notions. A virtual system exists in the virtual, abstract, or notional world. For example, a fairy tale *"Jack and the Beanstalk"* composed of *"Jack"* and *"the Giant,"* shown in Figure 1-7, is a virtual, abstract, or notional system.

Figure 1-7   *Jack and the Beanstalk* is a Virtual System

As a second example, For example, a software *Multi-Tier Personal Data System* composed of *MTPDS_GUI*, *Age_Logic*, *Overweight_Logic* and *Personal_Database*, shown in Figure 1-8, is a virtual, abstract, or notional system.

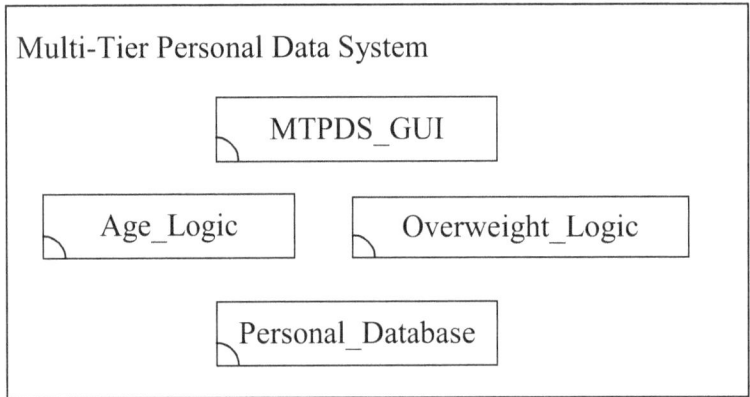

Figure 1-8   *Multi-Tier Personal Data System* is a Virtual System

1-3 Boundary and Environment of a System

We scope a system by defining its boundary as shown in Figure 1-9. All components of the system are inside the boundary while the environment is outside the boundary.

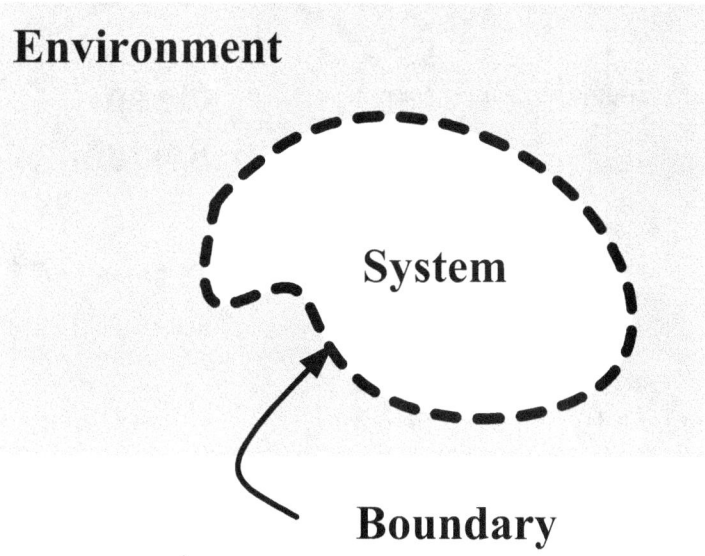

Figure 1-9   Boundary and Environment of a System

The environment is also known as the surroundings. A system may or may not interrelate with the environment. An open system shall interrelate with the environment through the exchange of matter, energy, data, information, or message as shown in Figure 1-10.

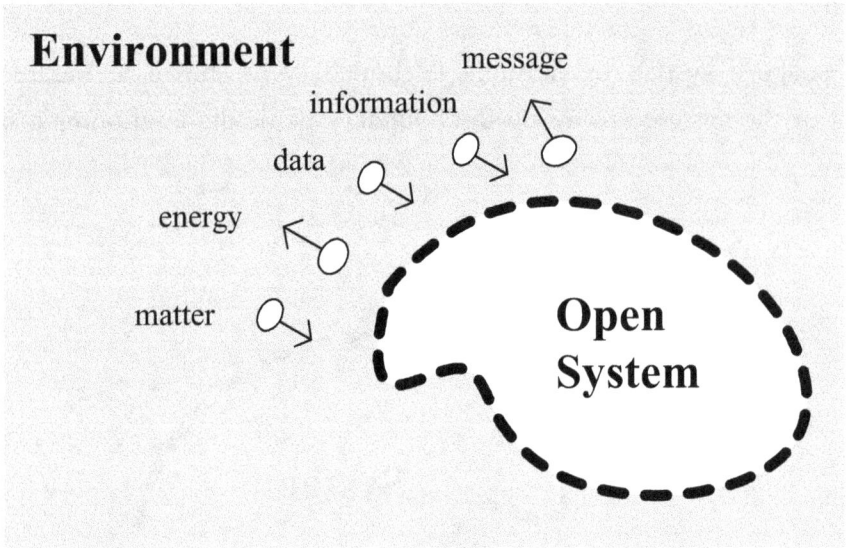

Figure 1-10   Open System Interrelates with the Environment

An isolated system does not interrelate with the environment at all. There is no exchange of matter, energy, data, information, or message between the isolated system and the environment as shown in Figure 1-11.

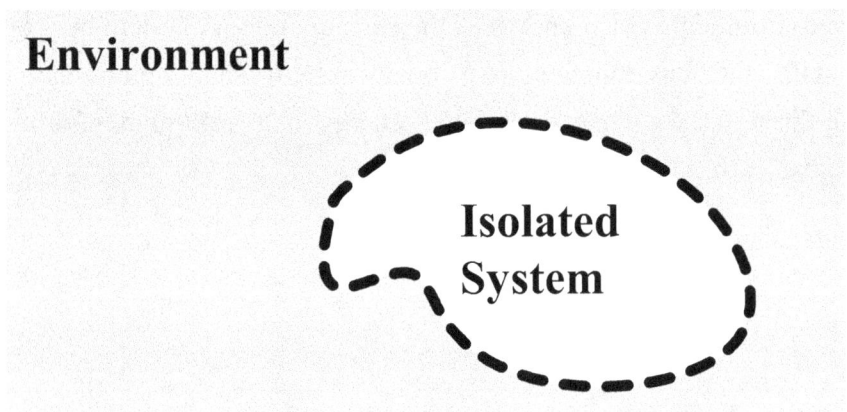

Figure 1-11   Isolated System Does Not Interrelate with the Environment

1-4 Higher-Order Systems

Higher-order systems interrelate with the environment through the exchange of not only matter, energy, data, information, or message but also systems as shown in Figure 1-12.

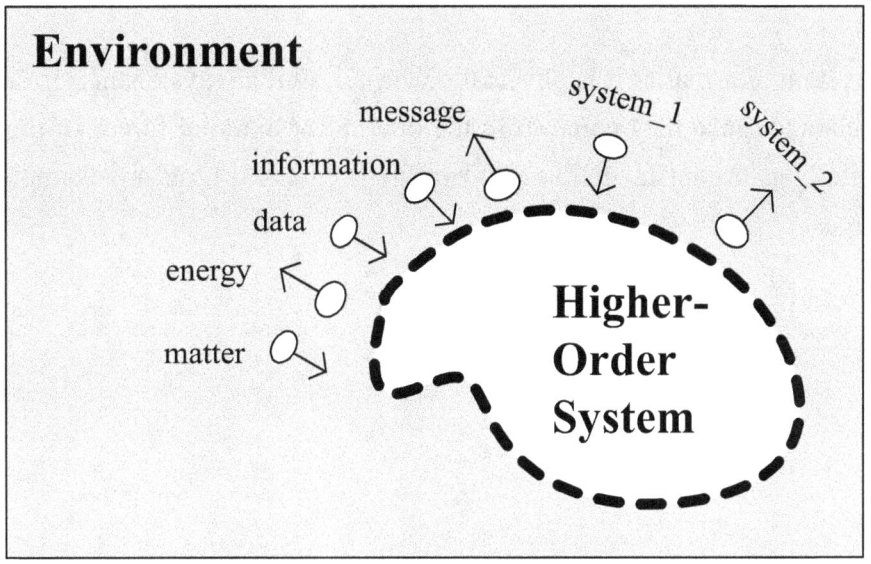

Figure 1-12　Higher-Order System

Human brain is regarded as higher-order systems. The human brain is a higher-order system, because it is able to produce a large number of systems, as shown in Figure 1-13. In the figure, *system_1*, *system_2* and *system_n* are the output of the human brain.

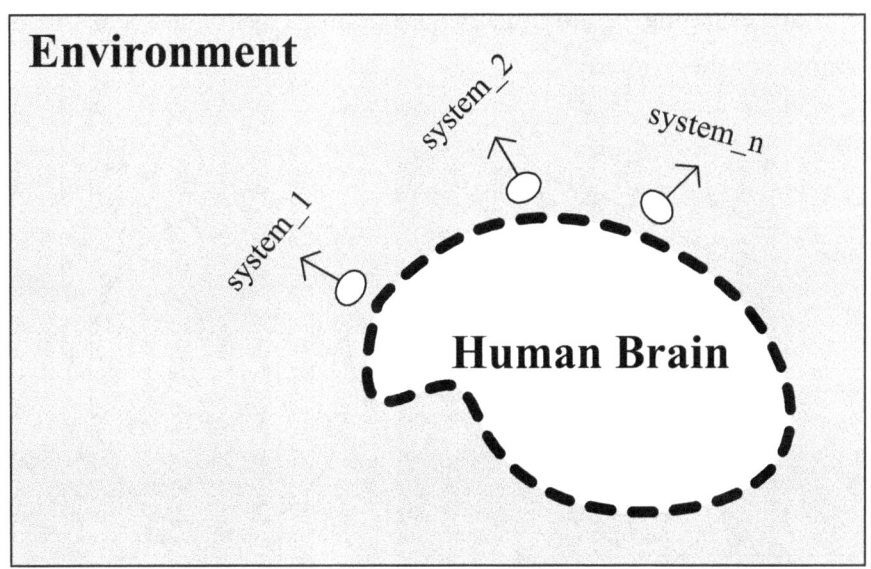

Figure 1-13　Human Brain is a Higher-Order System

1-5 Evolution of a System

A system, not matter it is physical or virtual, will always change from time to time. The change cause may come from the internal or external forces of the system. A self-replicating organism cell, as shown in Figure 1-14, is an example of the internal forces.

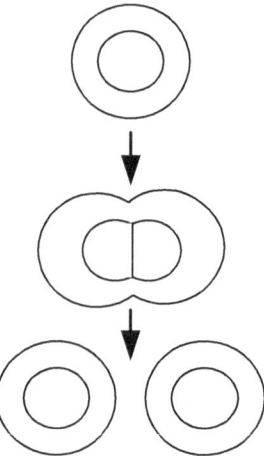

Figure 1-14   Self-Replicating Organism Cell

A worker reshaping, rebuilding, or remodeling a system, as shown in Figure 1-15, is an example of the external forces.

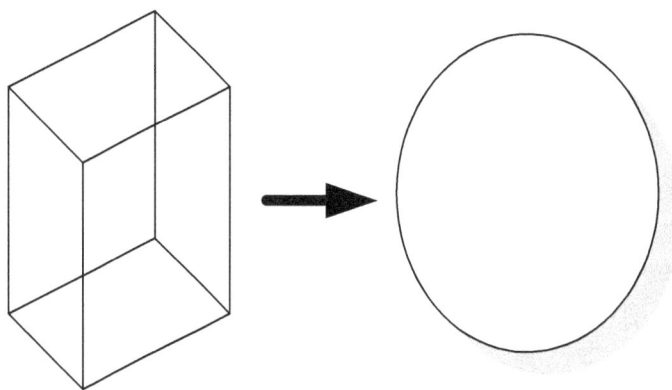

Figure 1-15   Reshape a System

A system evolves when it changes. Evolution of a system is shown in Figure 1-16.

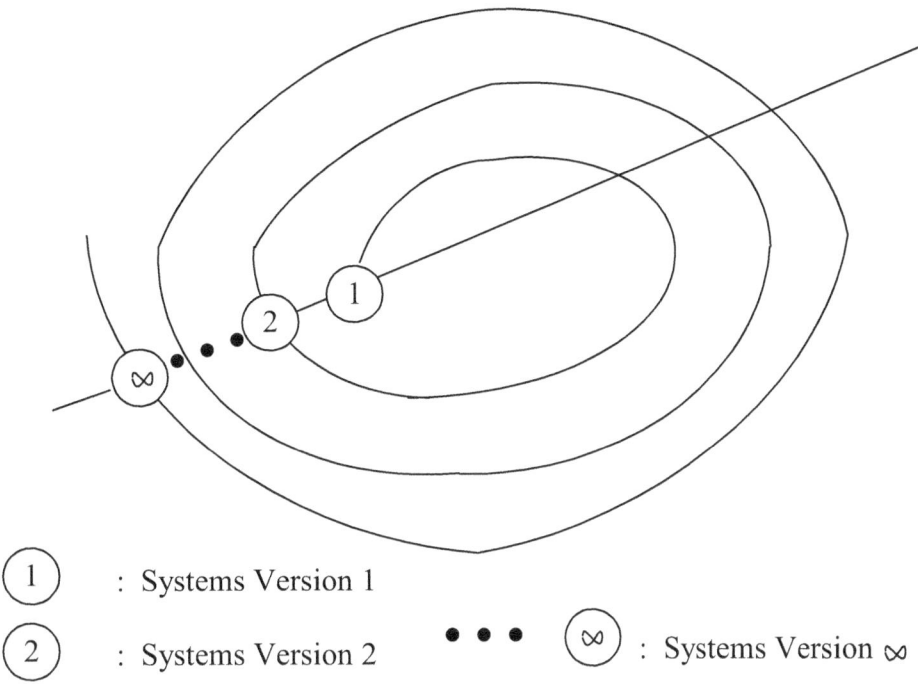

1 : Systems Version 1

2 : Systems Version 2   • • •   ∞ : Systems Version ∞

Figure 1-16   Evolution of a System

For example, Figure 1-17 shows the general systems theory 1.0 systems definition *version 1* defining the *House_B* to be hopefully an integrated whole embodied in its assembled components of *Roof_1*, *Window_1* and *Door_1*, their interrelationships with each other and the environment, and the principles and guidelines governing its design and evolution.

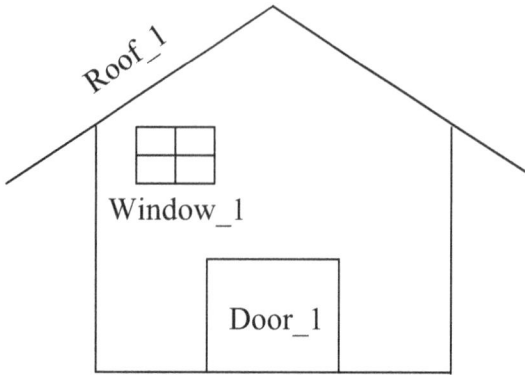

Figure 1-17   General Systems Theory 1.0 Systems Definition *Version 1*
Defining the *House_B*

After the *house_B* changes and evolves, Figure 1-18 shows the general systems theory 1.0 systems definition *version 2* defining the *House_B* to be hopefully an integrated whole embodied in its assembled components of *Roof_1*, *Window_1*, *Window_2* and *Door_1*, their interrelationships with each other and the environment, and the principles and guidelines governing its design and evolution.

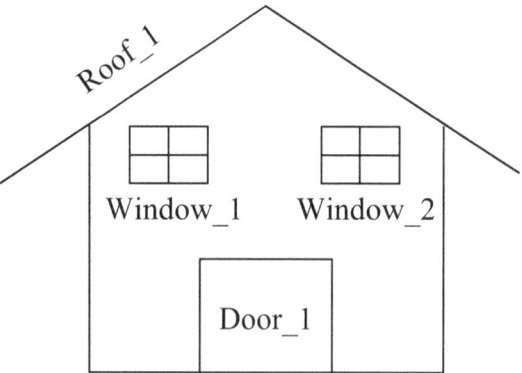

Figure 1-18   General Systems Theory 1.0 Systems Definition *Version 2*
Defining the *House_B*

Chapter 2: Introduction to Systems Architecture

A system comprises multiple views such as strategy/version n, strategy/version n+1, concept, analysis, design, implementation, structure, behavior and input/output data views. A systems model is required to describe and represent all these multiple views.

The systems model describes and represents the system multiple views possibly using two different approaches. The first one is the non-architectural approach and the second one is the architectural approach. The non-architectural approach respectively picks a model for each view. The architectural approach, instead of picking many heterogeneous and separated models, will use only one single coalescence model.

When used as a knowledge repository of a system, systems architecture becomes a communicating tool for comprehension enhancement, internal collaboration and interworking with partners. Systems architecture also supplies documented systems structures and systems behaviors.

Systems architecture should not be constructed in one step. On the contrary, systems architects will iteratively and evolutionally construct each version of the systems architecture. Iterations and evolutions allow systems architects to demonstrate incremental value of their works and obtain early feedback of the systems architecture.

2-1 Multiple Views of a System

In general, a system is extremely complex that it consists of several evolution&motivation views such as strategy/version n and strategy/version n+1 views; it also consists of various multi-level (hierarchical) views such as concept, analysis, design and implementation views; it also consists of many systemic views such as structure, behavior and input/output data views [Kend10, Pres09, Somm06].

Figure 2-1 shows that in a system all these strategy/version n, strategy/version n+1, concept, analysis, design, implementation, structure, behavior and input/output data views represent the multiple views of a system.

30

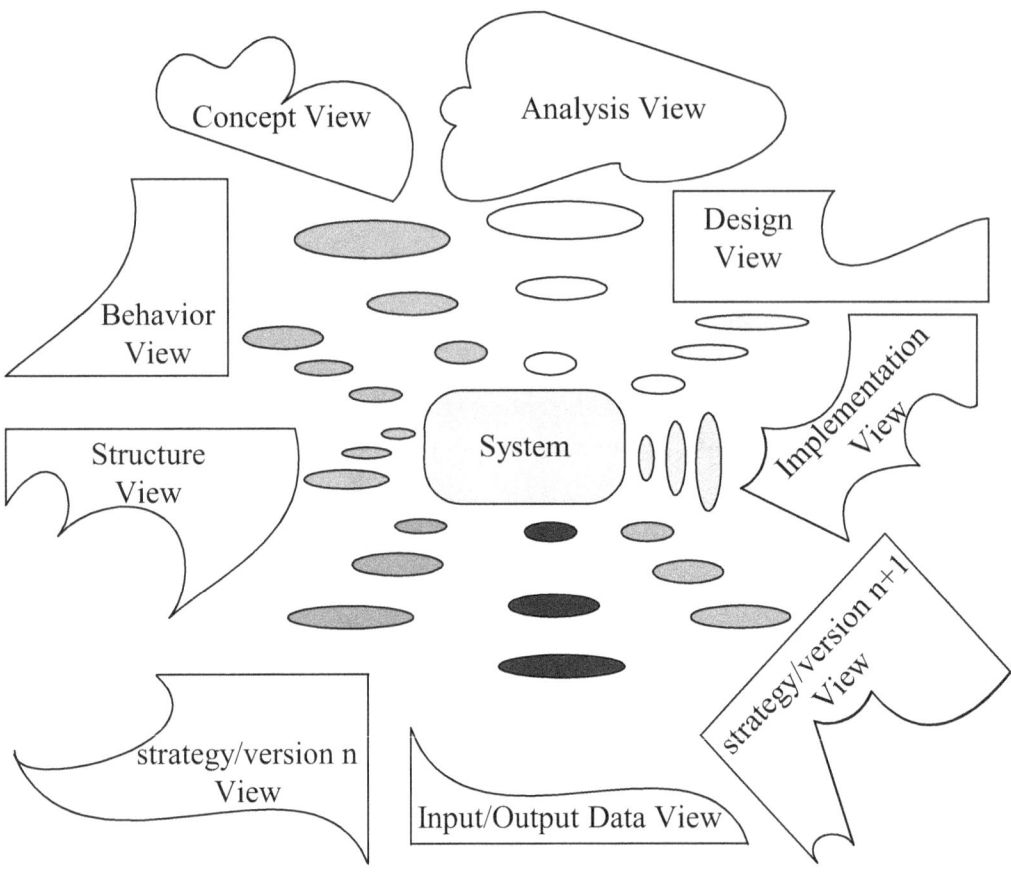

Figure 2-1   Multiple Views of a System

Among the above multiple views, the structure and behavior views are perceived as the two prominent ones. The structure view focuses on the systems structure which is described by components and their composition while the behavior view concentrates on the systems behavior which involves interactions (or handshakes) among the external environment's actors and components. Strategy/version n, strategy/version n+1, concept, analysis, design, implementation and input/output data views are considered to be other views as shown in Figure 2-2.

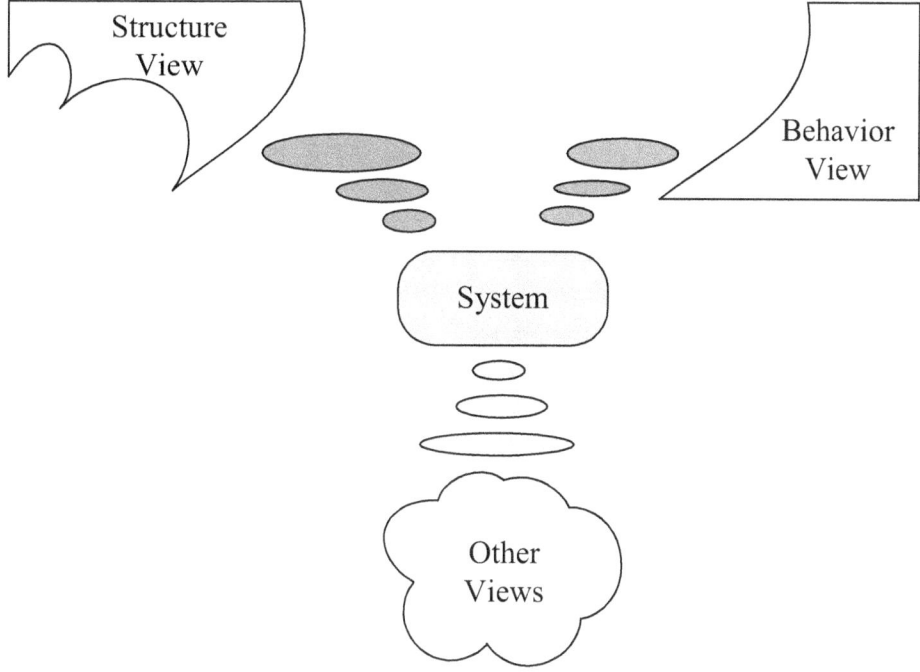

Figure 2-2   Structure, Behavior, and Other Views

Either Figure 2-1 or Figure 2-2 represents the multiple views of a system. In some situations Figure 2-1 is used and in other situations Figure 2-2 is used.

2-2 Systems Model

A systems model (SM) is a virtual system, distinguished from a physical system, used to describe and represent either the physical or virtual systems.

Figure 2-3 shows a physical system in which there are two buildings located in the upper left side and right underneath. The upper left building is Jackson Hotel and the right underneath building is Clinton Theater.

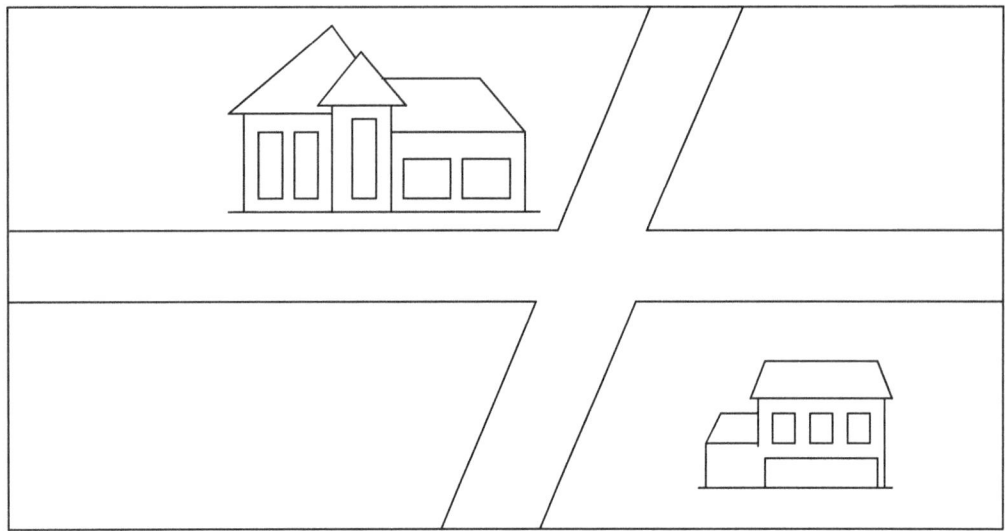

Figure 2-3   A Physical System

To model the physical system in Figure 2-3 we may then obtain a map as shown in Figure 2-4. The map is a kind of systems model used to describe and represent the physical system.

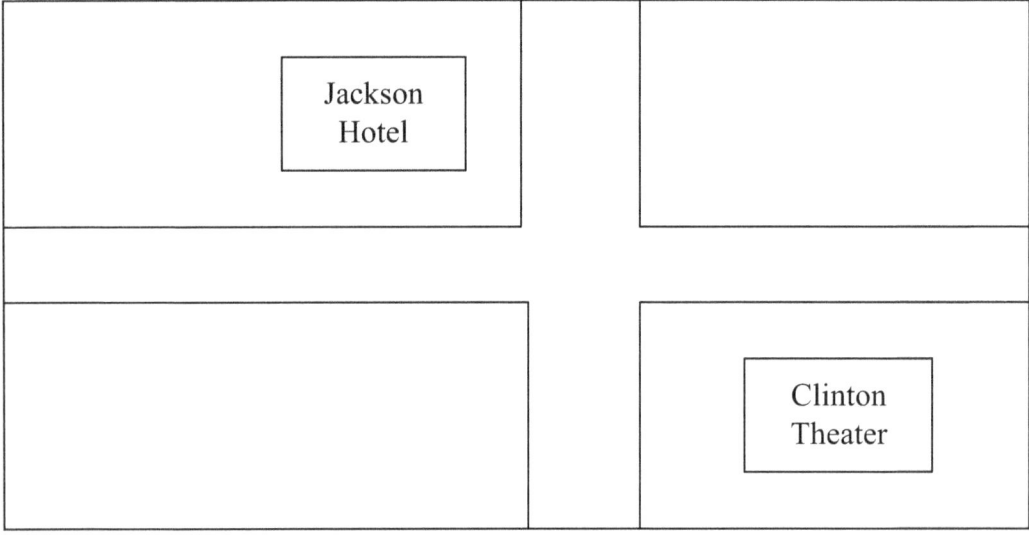

Figure 2-4   Map as a Systems Model

Besides describing and representing systems in the physical world, a systems model can also describe and represent systems in the virtual world. The virtual world includes a software system, a virtual reality, or a thought within a person's mind, etc. Figure 2-5 shows that a fashion designer is designing a new suit of clothes. Designing a suit of clothes, being a thought inside a person's mind, belongs to the virtual world.

Figure 2-5   Thought inside a Person's Mind

To model the thought within a person's mind in Figure 2-5, we may then use a clothes design diagram as shown in Figure 2-6. The clothes design diagram is a kind of systems model used to describe and represent a person's thought.

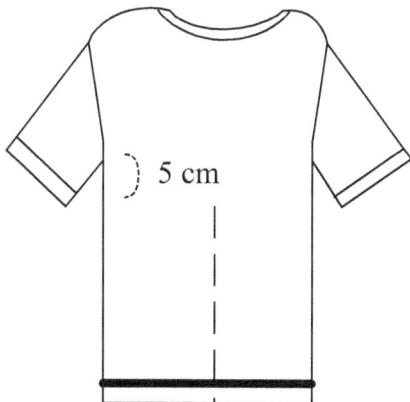

Figure 2-6   Clothes Design Diagram as a System Model

2-3 Non-Architectural Approaches Versus Architectural Approaches

A system is exceptionally complex that it includes multiple views such as strategy/version n, strategy/version n+1, concept, analysis, design, implementation, structure, behavior and input/output data views.

The systems model describes and represents the system multiple views possibly using two different approaches. The first one is the non-architectural approach and the second one is the architectural approach.

The non-architectural approach, also known as the model multiplicity approach [Dori95, Dori02, Dori16], respectively picks a model for each view as shown in Figure 2-7, the strategy/version n view has the strategy/version n model, the strategy/version n+1 view has the strategy/version n+1 model, the concept view has the concept model, the analysis view has the analysis model, the design view has the design model, the implementation view has the implementation model, the structure view has the structure model, the behavior view has the behavior model, and the input/output data view has the input/output data model. These multiple models, are heterogeneous and not related to each other, and thus become the primary cause of model multiplicity problems [Dori95, Dori02, Dori16, Pele02, Sode03].

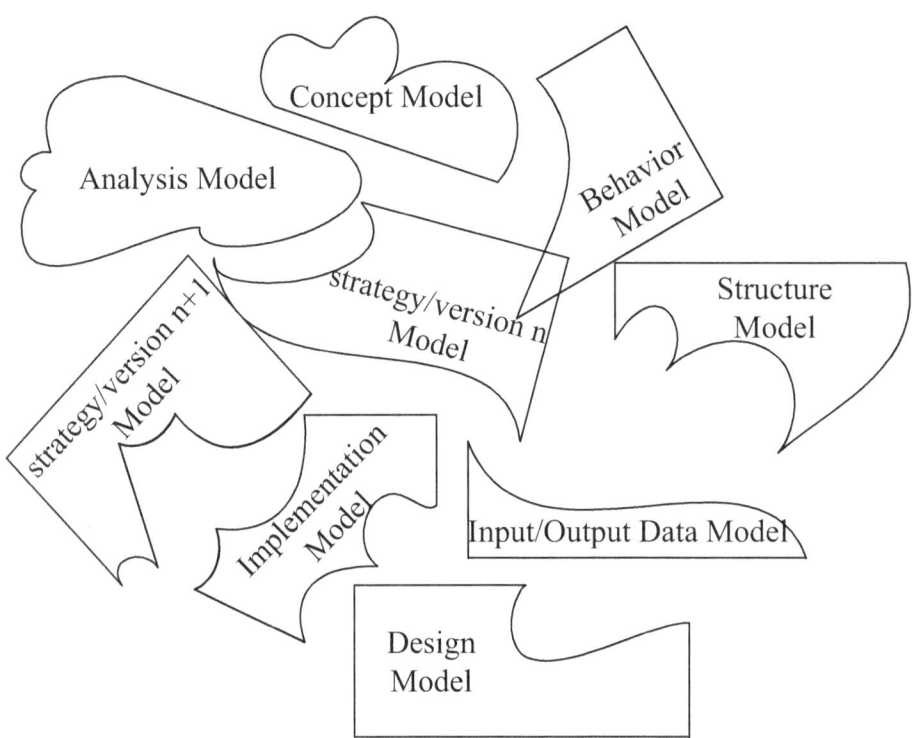

Figure 2-7   The Non-architectural Approach Picks a Model for Each View

The architectural approach, also known as the model singularity approach [Dori95, Dori02, Dori16, Pele02, Sode03], instead of picking many different models, will use only one single model as shown in Figure 2-8. The strategy/version n, strategy/version n+1, concept, analysis, design, implementation, structure, behavior, and input/output data views are all integrated in this multiple views coalescence (MVC) model of systems architecture [Chao14a, Chao14b, Clem02, Clem10, Dike01, Gort06, Putm00, Roza05, Shaw96, Tayl09, Wang99].

Figure 2-8   Systems Architecture Uses a Single Model

Figure 2-7 has many models. Figure 2-8 has only one model. Comparing Figure 2-7 with Figure 2-8, we unquestionably conclude that an integrated, holistic, united, coordinated, coherent and coalescence model is more favorable than a collection of many heterogeneous and separated models.

2-4 Definition of Systems Architecture

Involved systems are extremely complex in every aspect so that each stakeholder needs a blueprint or model to capture their essential structures and behaviors. Systems architecture is such a blueprint or model.

There are several well-know definitions of systems architecture [Dam06, Mino08, O'Rou03, Roza05]. ANSI/IEEE 1471-2000 defines systems architecture as: "the fundamental organization of a system, embodied in its components, their relationships to each other and the environment, and the principles governing its design and evolution." The Open Group defines systems architecture as either "a formal description of a system, or a detailed plan of the system at component level to guide its implementation," or as "the structure of components, their interrelationships,

36

and the principles and guidelines governing their design and evolution over time"
[Rayn09, Toga08].

Concluding the above definitions, we now give systems architecture a definition of our own as shown in Figure 2-9.

Systems architecture is an integrated whole of a system's multiple views, i.e., structure, behavior, and other views, embodied in its components, their interactions with each other and the environment, and the principles and guidelines governing its design and evolution.

Figure 2-9   Definition of Systems Architecture

From the above definition, we find out that systems architecture is an integrated whole of a system's multiple views, i.e., structure, behavior and other views, embodied in its assembled components, their interactions (or handshakes) with each other and the environment, and the principles and guidelines governing its design and evolution. That is, systems architecture is an integrated and coalescence model of multiple views. In this coalescence model, structure, behavior and other views are all included in it as shown in Figure 2-10. We do not supply each view a respective model in this systems architecture coalescence model.

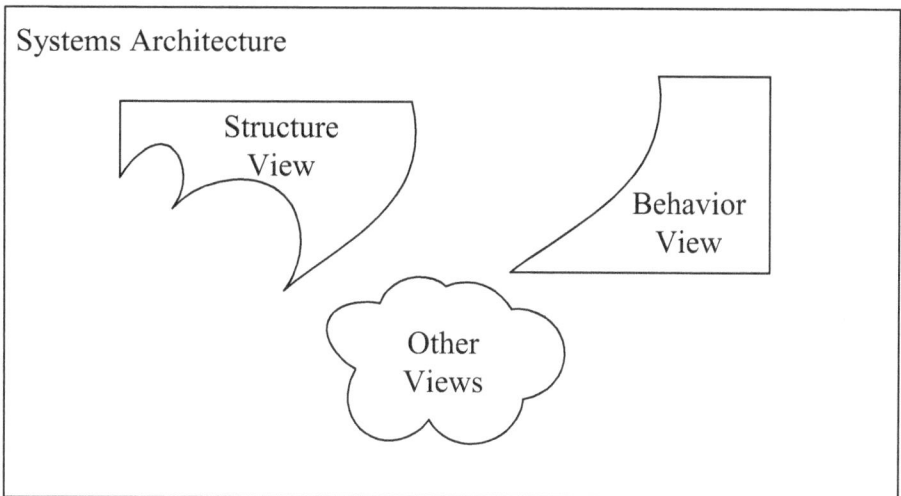

Figure 2-10   All Multiple Views are Included in This Systems Architecture

Since multiple views are embodied in a system's assembled components which belong to the structure view, they shall not exist alone. Multiple views must be loaded on the structure view just like a cargo is loaded on a ship as shown in Figure 2-11. There will be no multiple views if there is no structure view. Stand-alone multiple views are not meaningful.

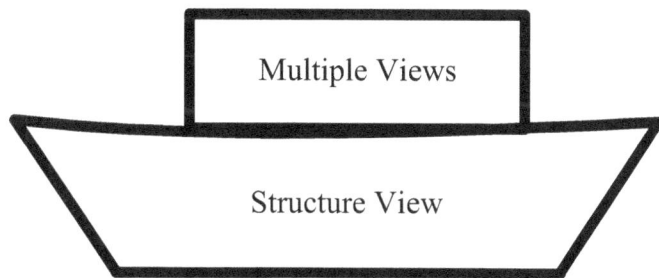

Figure 2-11   Multiple Views Must be Loaded on the Structure View

2-5 Architecture Description Language

An architecture description is a formal description and representation of a system. A description of the systems architecture has to grasp the essence of the system and its details at the same time. In other words, an architecture description not only provides an overall picture that summarizes the whole system, but also contains enough detail that the system can be constructed and validated.

The language for architecture description is called the architecture description language (ADL) [Clem02, Clem10, Dike01, Roza05, Shaw96, Tayl09]. An ADL is a special kind of language used in describing the architecture of a system.

Since the architectural approach uses a coalescence model for all multiple views of a system, the foremost duty of ADL is to make the strategy/version n, strategy/version n+1, concept, analysis, design, implementation, structure, behavior and input/output data views all integrated and coalesced within this architecture description.

2-6 Systems Architecture as a Knowledge Repository

Based on its definition, systems architecture can be regarded as a knowledge repository of a system. Each stakeholder, through structure, behavior and other views, submits his own knowledge and expertise to this repository when the systems architecture is built up, as shown in Figure 2-12.

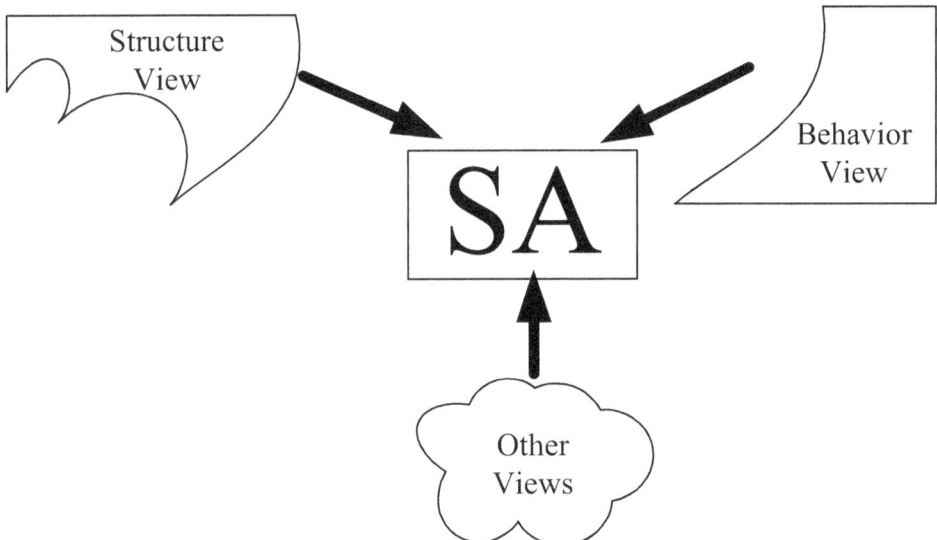

Figure 2-12   Each Stakeholder Submits His Own Knowledge and Expertise

On the other hand, any stakeholder, if there is any request then he would query the system architecture. The result of the query is gathered into a view for stakeholders to see or read, as shown in Figure 2-13.

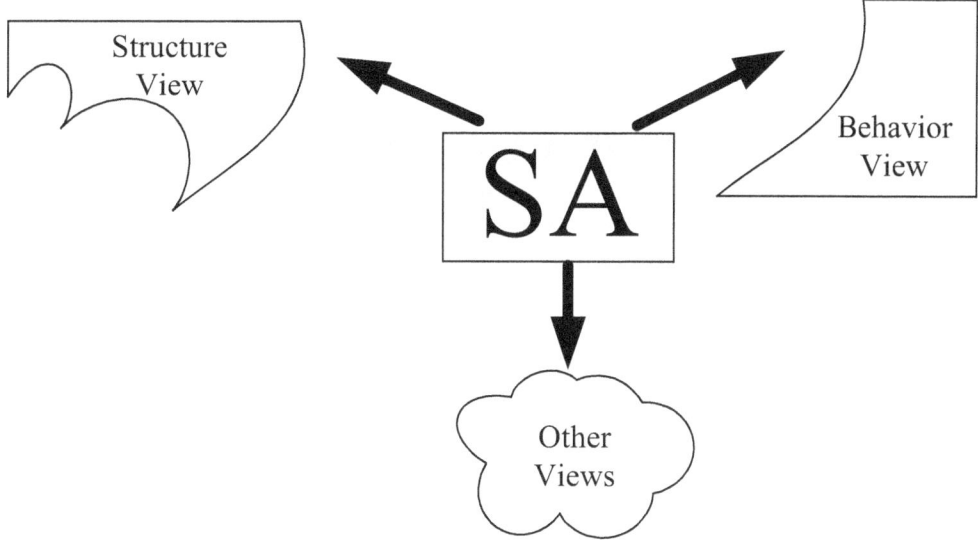

Figure 2-13   Views for Stakeholders to See or Read

Combining the above two figures, Figure 2-14 tells us that systems architecture is exactly a knowledge repository of a system. Stakeholders can submit and acquire knowledge to and from the systems architecture.

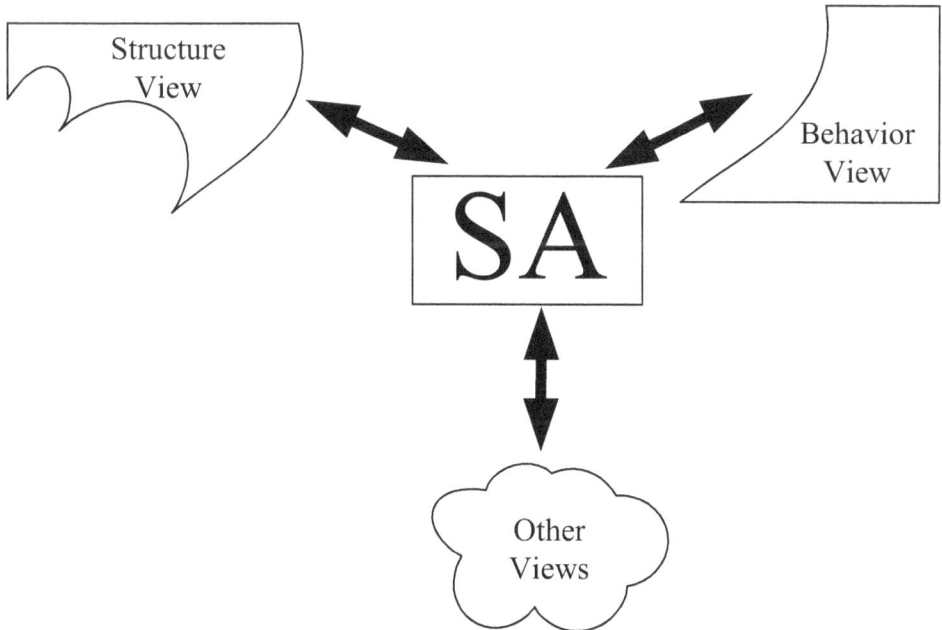

Figure 2-14   SA as a Knowledge Repository of a System

When used as a knowledge repository of a system, systems architecture becomes a communicating tool for comprehension enhancement, internal collaboration and interworking with partners. The systems architecture also supplies documented systems structures and systems behaviors.

2-7 Constructing the Systems Architecture Iteratively and Evolutionally

Systems architecture shall not be constructed in one step. On the contrary, a systems architect must construct the systems architecture iteratively and evolutionally. Iterations and evolutions allow systems architects to demonstrate incremental values of their works and obtain early feedback of the systems architecture.

Figure 2-15 shows that the systems architecture *version 1*, *version 2*, *version 3*, *version 4,...*, and *version* ∞ are constructed iteratively and evolutionally by a systems architect.

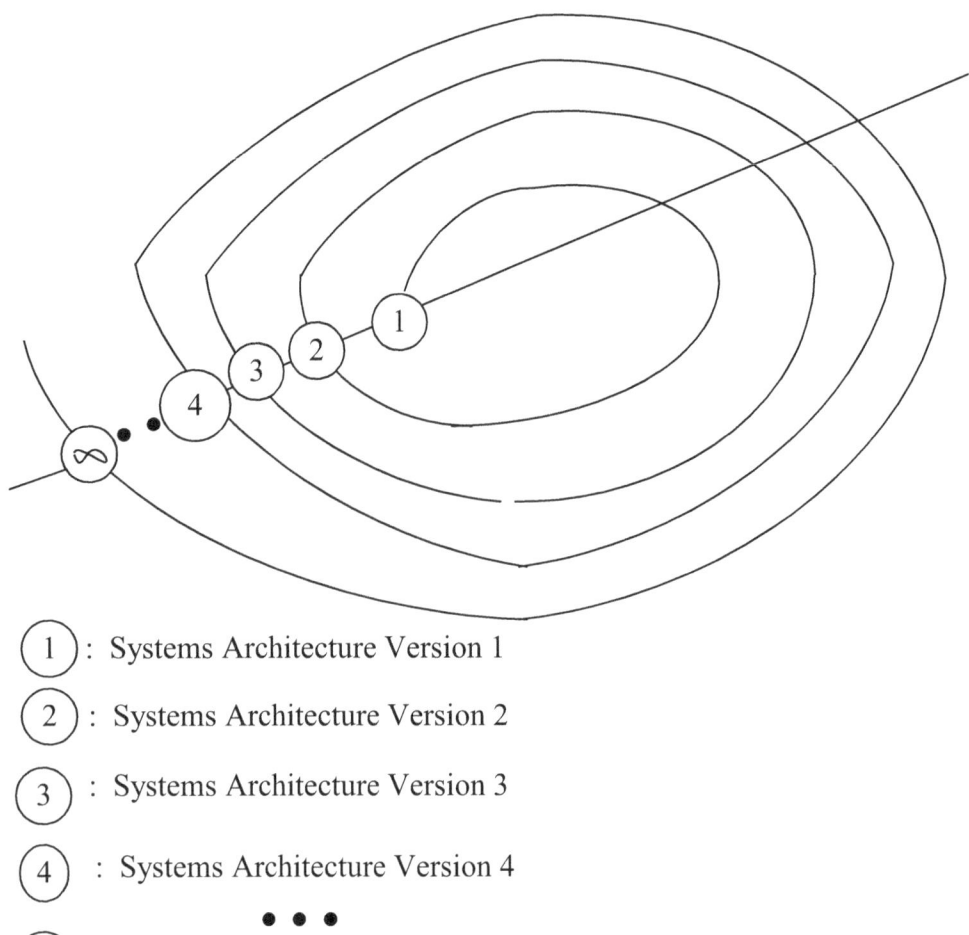

① : Systems Architecture Version 1

② : Systems Architecture Version 2

③ : Systems Architecture Version 3

④ : Systems Architecture Version 4

• • •

∞ : Systems Architecture Version ∞

Figure 2-15   Systems Architecture is Constructed Iteratively and Evolutionally

Systems architecture *version n* is sometimes referred to as the baseline (As-Is) architecture which represents the current system that has been formally reviewed and agreed upon. On the other hand, systems architecture *version n+1* is sometimes referred to as the target (To-Be) architecture which represents the goal system that will be formally constructed.

2-8 Architecture Development Method

If we adopt the iterative and evolutional construction of systems architecture approach, then we would obtain the architecture development method (ADM) [Toga08] as shown in Figure 2-16.

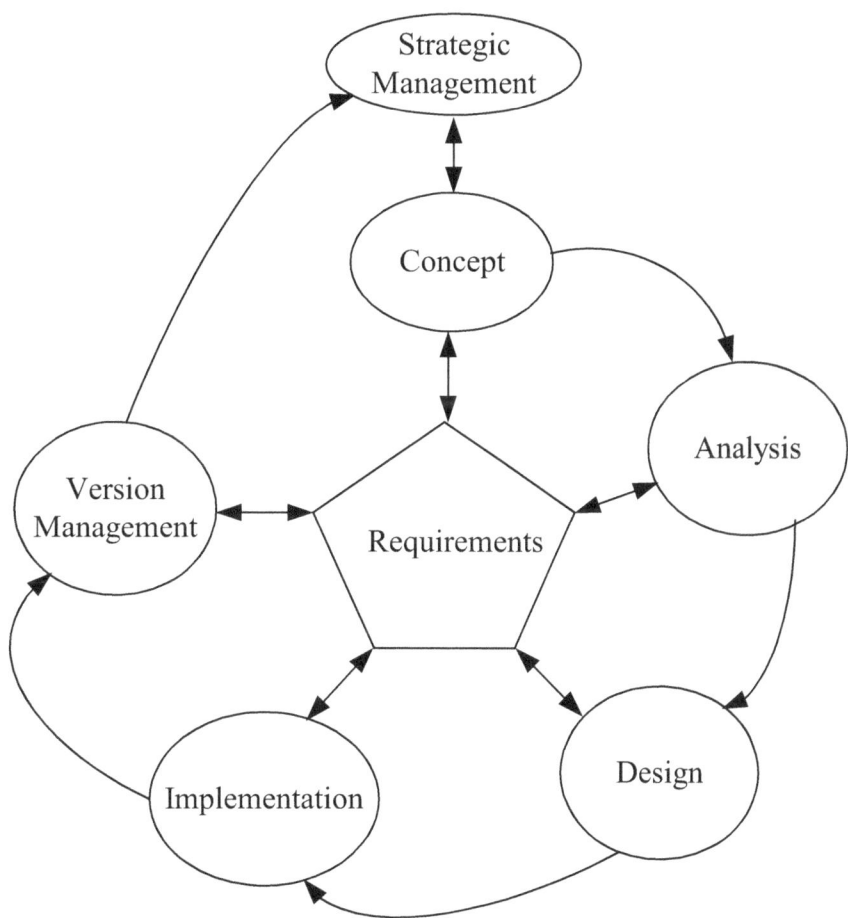

Figure 2-16   Architecture Development Method

The iterative and cyclic ADM, being utilized by a systems architect to accomplish each version management of systems architecture, shall do the strategic management first and then go through the concept, analysis, design and implementation phases of systems architecture construction. Every phase checks with

42

the requirements to make sure that each version of the constructed systems architecture is what the users want.

The output of strategic management is a strategy and the output of version management is a version of systems architecture. Accordingly, each strategy is mapped to a version of systems architecture as shown in Figure 2-17.

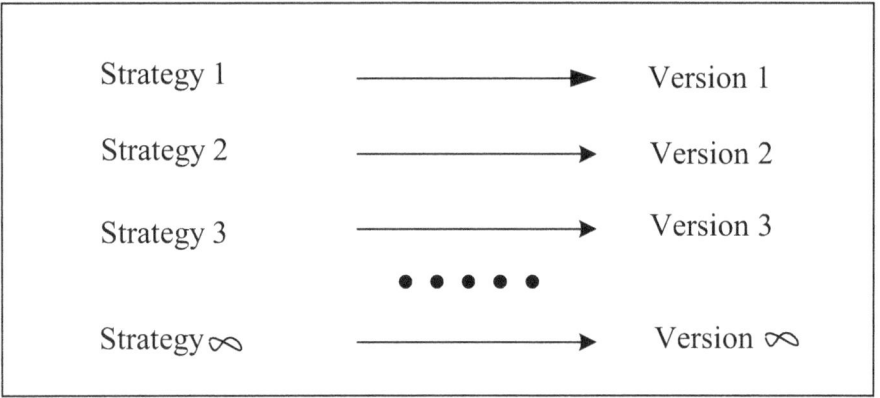

Figure 2-17   Each Strategy is Mapped to a Systems Architecture Version

2-9 View Model

A system comprises multiple views such as strategy/version n, strategy/version n+1, concept, analysis, design, implementation, structure, behavior and input/output data views. We can represent all these multiple views in a one-dimensional array as shown in Figure 2-18.

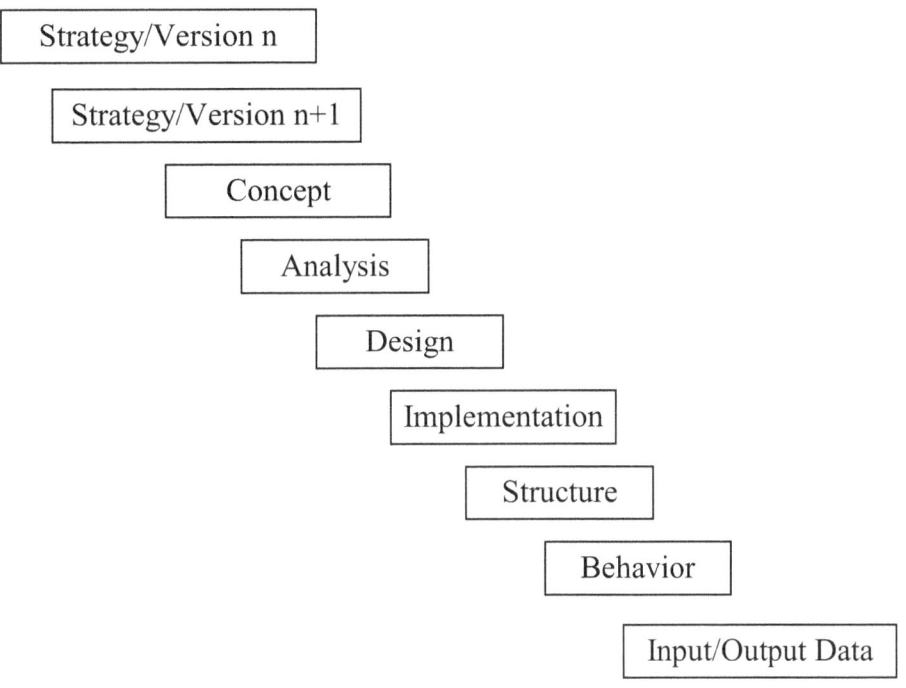

Figure 2-18   Array Representation of Multiple Views

We can also describe and represent all these multiple views in a three-dimensional matrix as shown in Figure 2-19.

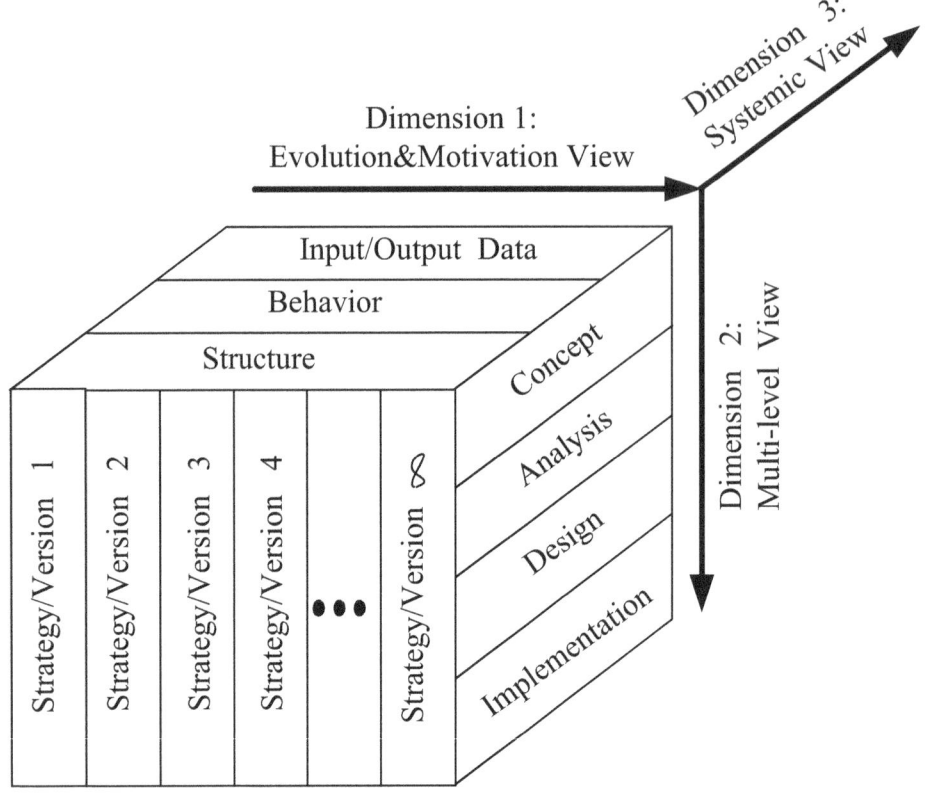

Figure 2-19   Three-dimensional Matrix Representation of Multiple Views

In the matrix representation of multiple views, dimension 1 stands for the evolution&motivation view which contains the strategy/version 1, strategy/version 2, strategy/version 3, strategy/version 4,…, and strategy/version ∞ views; dimension 2 stands for the multi-level (hierarchical) view which contains the concept, analysis, design and implementation views; dimension 3 stands for the systemic view which contains the structure, behavior, input/output data views. The matrix representation of multiple views is also called a view model (VM) or architecture framework (AF) [Chao09, Dam06, Mino08, O'Rou03].

Chapter 3: Structure-Behavior Coalescence for Systems Architecture

In general, multiple views coalescence (MVC) architecture is synonymous with the systems architecture. Since structure and behavior views are the two most prominent ones among multiple views, integrating the structure and behavior views becomes a superb approach for integrating multiple views of a system. In other words, structure-behavior coalescence (SBC) leads to the coalescence of multiple views. Therefore, we conclude that SBC architecture is also synonymous with the systems architecture.

3-1 Multiple Views Coalescence to Achieve the Systems Architecture

Systems architecture has been defined as a coalescence model of multiple views. Multiple views coalescence uses only a single coalescence model as shown in Figure 3-1. Strategy/version n, strategy/version n+1, concept, analysis, design, implementation, structure, behavior and input/output data views are all integrated in this MVC architecture.

Figure 3-1   MVC Architecture

Generally, MVC architecture is synonymous with the systems architecture. In other words, multiple views coalescence sets a path to achieve the systems architecture as shown in Figure 3-2.

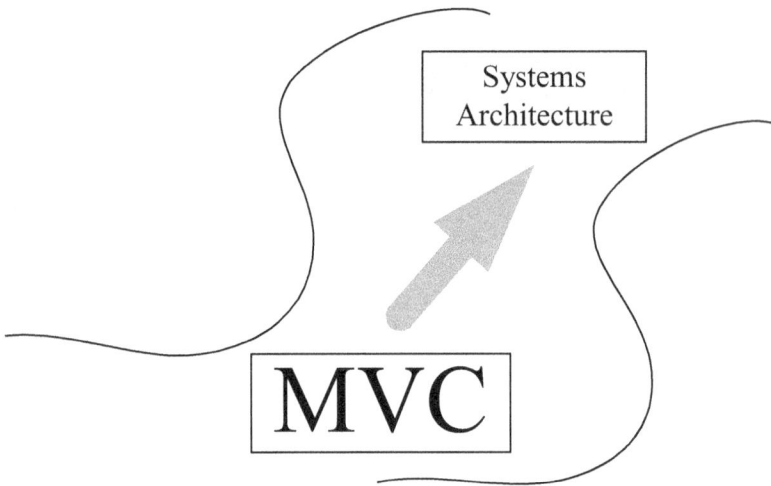

Figure 3-2   MVC to Achieve the Systems Architecture

In the MVC architecture, multiple views must be attached to or built on the systems structure. In other words, multiple views shall not exist alone; they must be loaded on the systems structure just like a cargo is loaded on a ship as shown in Figure 3-3. There will be no multiple views if there is no systems structure. Stand-alone multiple views are not meaningful.

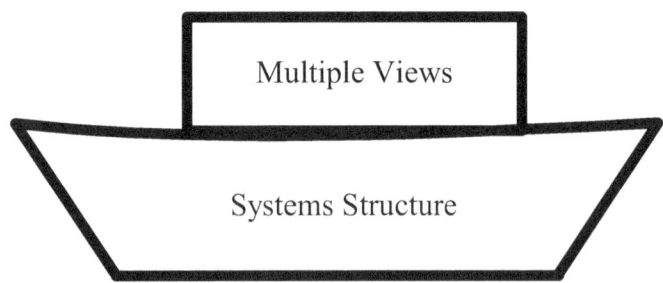

Figure 3-3   Multiple Views are Loaded on the Systems Structure

3-2 Integrating the Systems Structures and Systems Behaviors

By integrating the systems structure and systems behavior, we obtain structure-behavior coalescence (SBC) within the system as shown in Figure 3-4.

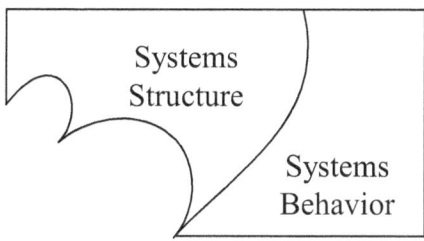

Figure 3-4   Structure-Behavior Coalescence

Structure-behavior coalescence has never been used in any systems model (SM) for systems development except the SBC architecture. There are many advantages to use the structure-behavior coalescence approach to integrate the systems structure and systems behavior.

SBC architecture uses a single model as shown in Figure 3-5. Systems structures and systems behaviors are integrated in this SBC architecture.

Figure 3-5   SBC Architecture

Since systems structures and systems behaviors are so tightly integrated, we sometimes claim that the core theme of SBC architecture is: "Systems Architecture = Systems Structure + Systems Behavior," as shown in Figure 3-6.

48

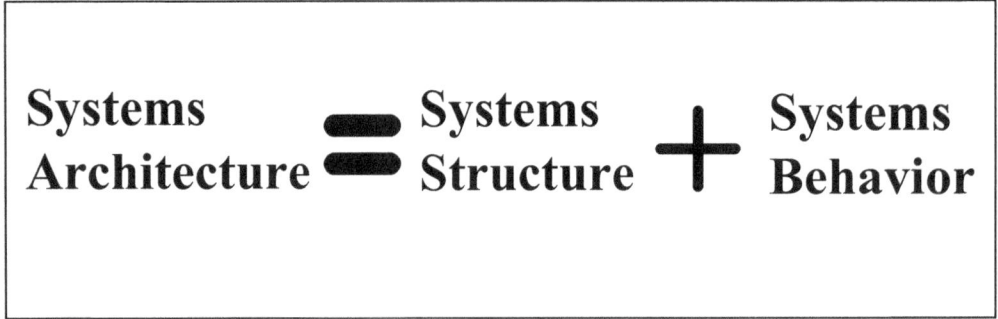

Figure 3-6   Core Theme of SBC Architecture

So far, systems behaviors are separated from systems structures in most cases [Pres09, Roza05, Somm06]. For example, the well-known structured systems analysis and design (SSA&D) approach uses structure charts (SC) to represent the systems structure and data flow diagrams (DFD) to represent the systems behavior [Denn08, Kend10, Your99]. SC and DFD are two different models. They are so separated like that there is "Pacific Ocean" between them, as shown in Figure 3-7.

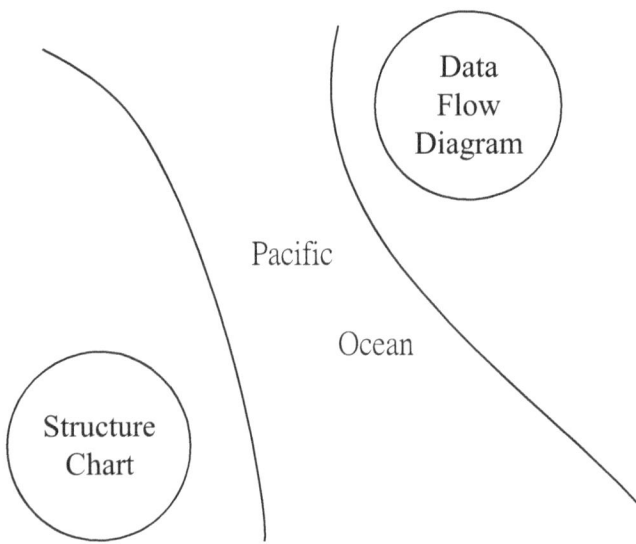

Figure 3-7   Two Heterogeneous and Separated Models

3-3 Structure-Behavior Coalescence to Facilitate Multiple Views Coalescence

Since structure and behavior views are the two most prominent ones among multiple views, integrating the structure and behavior views is clearly the best way to integrate multiple views of a system. In other words, structure-behavior coalescence facilitates multiple views coalescence as shown in Figure 3-8.

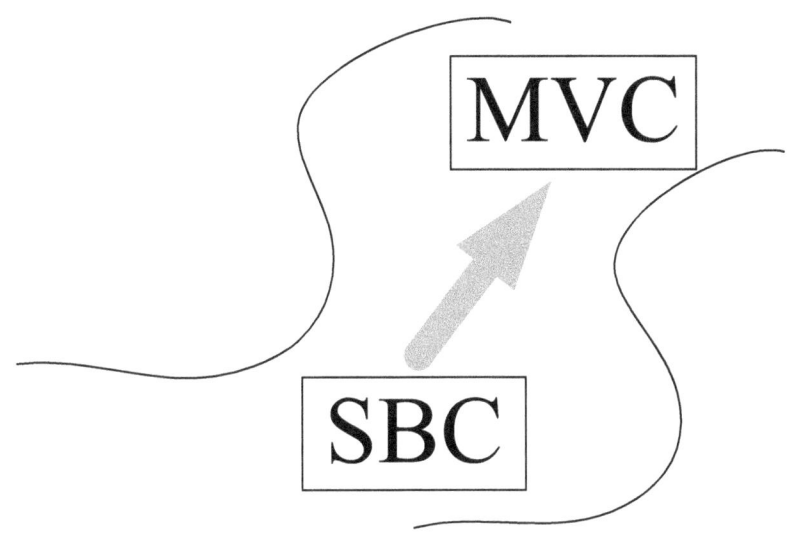

Figure 3-8   SBC Facilitates MVC

3-4 Structure-Behavior Coalescence to Achieve the Systems Architecture

Figure 3-2 declares that multiple views coalescence sets a path to achieve the desired systems architecture with the most efficient approach. Figure 3-8 declares that structure-behavior coalescence facilitates multiple views coalescence.

Combining the above two declarations, we conclude that structure-behavior coalescence sets a path to achieve the systems architecture as shown in Figure 3-9. In this case, SBC architecture is also synonymous with the systems architecture.

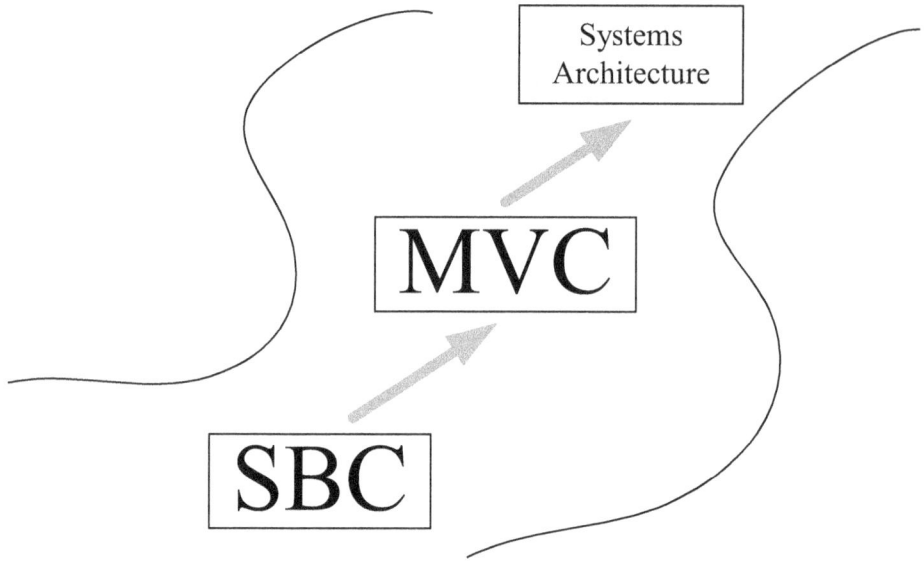

Figure 3-9   SBC to Achieve the Systems Architecture

SBC architecture strongly demands that the structure and behavior views must be coalesced and integrated. This never happens in other architectural approaches such as Zachman Framework [O'Rou03], The Open Group Architecture Framework (TOGAF) [Rayn09, Toga08], Department of Defense Architecture Framework (DoDAF) [Dam06] and Unified Modeling Language (UML) [Rumb91]. Zachman Framework does not offer any mechanism to integrate the structure and behavior views. TOGAF, DoDAF and UML do not, either.

In the SBC architecture, a systems behavior must be attached to or built on a systems structure. In other words, a systems behavior can not exist alone; it must be loaded on a systems structure just like a cargo is loaded on a ship as shown in Figure 3-10. There will be no systems behavior if there is no systems structure. A stand-alone systems behavior is not meaningful.

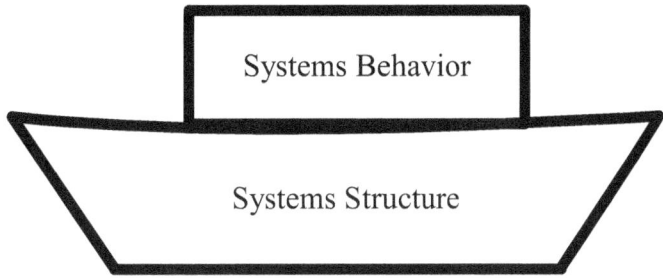

Figure 3-10   A Systems Behavior Must be Loaded on a Systems Structure

Chapter 4: Introduction to SBC Architecture

As discussed in the previous chapter, SBC (i.e. structure-behavior coalescence) architecture is the recommended systems architecture approach. SBC architecture includes: a) SBC process algebra (SBC-PA) or SBC architecture description language (SBC-ADL), b) SBC architecture development method (SBC-ADM) and c) SBC view model (SBC-VM).

An architecture description language (ADL) is a special kind of language used in describing the architecture of a system. SBC-ADL uses six fundamental diagrams to formally grasp the essence of a system and its details at the same time.

Through the iterative and cyclic architecture development method (ADM), a systems architect is able to construct the systems architecture smoothly. SBC-ADM, being utilized by a systems architect to accomplish each version management of the systems architecture, shall do the strategic management first and then go through the concept, analysis, design and implementation phases of systems architecture construction. Every phase checks with the requirements to make sure that each version of the constructed systems architecture is what the users want.

View model (VM) is a three-dimensional matrix representation of multiple views of a system. In the SBC view model, dimension 1 stands for the evolution&motivation view which contains the strategy/version 1, strategy/version 2, strategy/version 3, strategy/version 4, strategy/version ∞ views; dimension 2 stands for the multi-level (hierarchical) view which contains the concept, analysis, design and implementation views; dimension 3A stands for the SBC-ADL view which contains the architecture hierarchy diagram (AHD), framework diagram (FD), component operation diagram (COD), component connection diagram (CCD), structure-behavior coalescence diagram (SBCD) and interaction flow diagram (IFD) views; dimension 3B stands for the multi-layer view which contains the business layer, application layer, data layer and technology layer views.

4-1 Definition of SBC Architecture

Here, let us first give the SBC architecture a definition as shown in Figure 4-1.

SBC architecture,
through structure-behavior coalescence,
truly is an integrated whole of a system's multiple views, i.e., structure, behavior, and other views, embodied in its components, their interactions with each other and the environment, and the principles and guidelines governing its design and evolution.

Figure 4-1   Definition of SBC Architecture

From the above definition, we find out that SBC architecture, through structure-behavior coalescence, is a truly integrated whole of a system's multiple views, i.e., structure, behavior and other views, embodied in its assembled components, their interactions (or handshakes) with each other and the environment, and the principles and guidelines governing its design and evolution.

SBC architecture includes: a) SBC process algebra (SBC-PA) [Chao17a, Chao17b, Chao17c, Chao17d, Chao17e, Chao17f] or SBC architecture description language (SBC-ADL), b) SBC architecture development method (SBC-ADM) and c) SBC view model (SBC-VM).

4-2 SBC Architecture Description Language

An architecture description language is a special kind of language used in describing the architecture of a system [Shaw96, Tayl09].

A description of the systems architecture has to grasp the essence of a system and its details at the same time. In other words, a systems architecture description not only provides an overall picture that summarizes the system, but also contains enough detail that the system can be constructed and validated.

SBC-ADL uses six fundamental diagrams to formally describe the integration of systems structure and systems behavior of a system. These diagrams, as shown in Figure 4-2, are: a) architecture hierarchy diagram (AHD), b) framework diagram (FD), c) component operation diagram (COD), d) component connection diagram (CCD), e) structure-behavior coalescence diagram (SBCD) and f) interaction flow diagram (IFD).

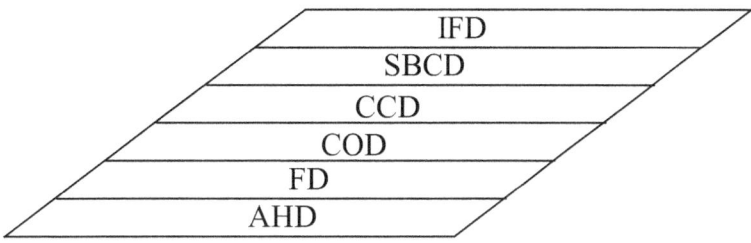

Figure 4-2   Six Fundamental Diagrams of SBC-ADL

SBC-ADL uses AHD, FD, COD, CCD, SBCD and IFD to depict the systems structure and systems behavior of a system as shown in Figure 4-3.

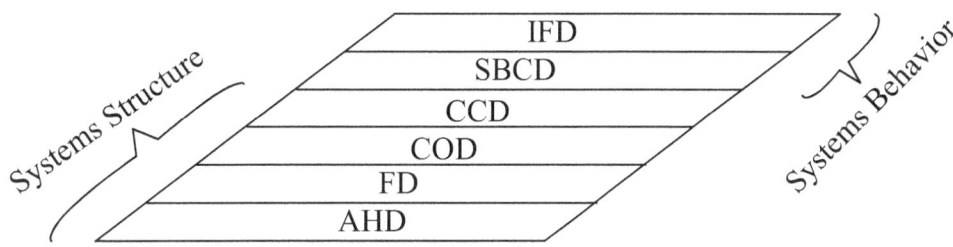

Figure 4-3   Systems Structure and Systems Behavior of a System

Examining the SBC-ADL approach, we find out that it depicts the systems structure first and then depicts the systems behavior later, not the other way around. The reason SBC-ADL does so lies in that the systems behavior must be attached to or built on the systems structure. With the systems structure and attached systems behavior, then, we can smoothly get the systems architecture as shown in Figure 4-4.

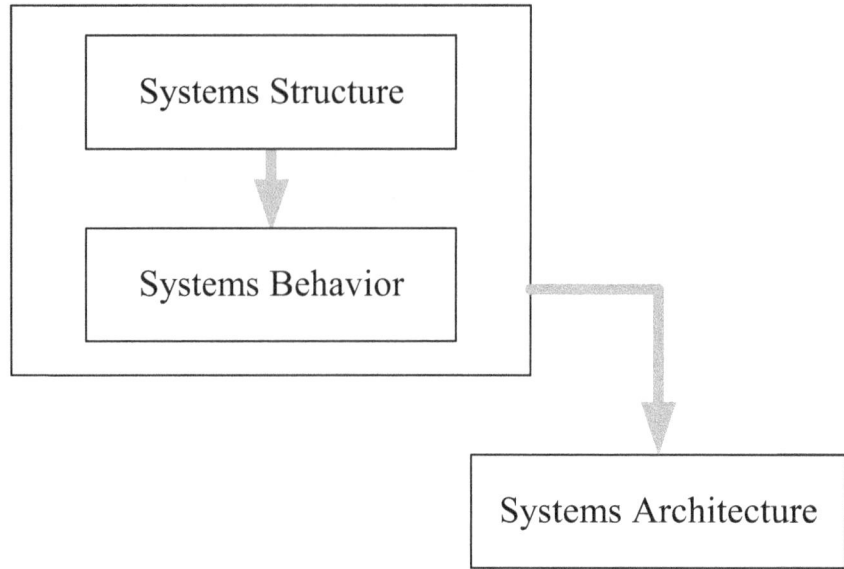

Figure 4-4   Systems Behavior is Attached to the Systems Structure

Let us ask the opposite question. Can the systems structure be attached to or built on the systems behavior? The answer is "no" as shown in Figure 4-5.

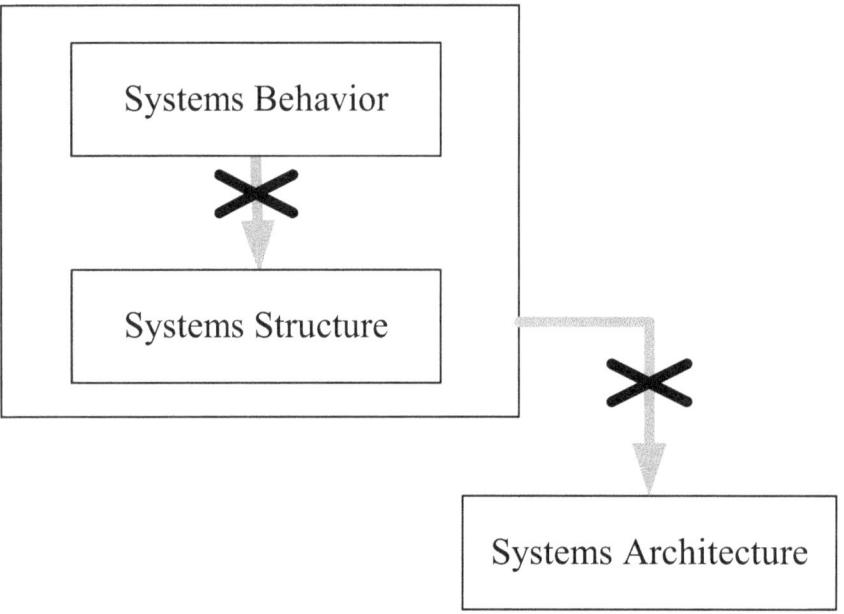

Figure 4-5   Systems Structure is not Attached to the Systems Behavior

In the SBC-ADL, the systems behavior must be attached to or built on the systems structure. In other words, a systems behavior shall not exist alone; it must be loaded on a systems structure just like a cargo is loaded on a ship as shown in Figure 4-6. There will be no systems behavior if there is no systems structure. A stand-alone systems behavior is not meaningful.

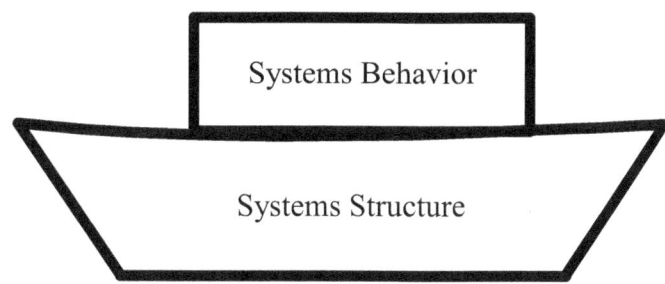

Figure 4-6   A Systems Behavior is Loaded on a Systems Structure

AHD, FD, COD and CCD belong to a system's systems structure. SBCD and IFD belong to a system's systems behavior. Concluding the above discussion, we perceive that SBC-ADL will describe AHD, FD, COD and CCD first then describe SBCD and IFD later when it constructs a systems architecture.

4-2-1 Systems Structure Is Composed of Components

Through architecture hierarchy diagram (AHD), systems architects shall clearly observe the multi-level (hierarchical) decomposition and composition of a system. AHD is the first fundamental diagram to achieve structure-behavior coalescence.

As an example, Figure 4-7 shows that the *QQQ* system is composed of *B1*, *B2*, *A1*, *D1*, *D2* and *T1*; *A1* is composed of *A11*; *T1* is composed of *T11* and *T12*; *A11* is composed of *A111*; *A111* is composed of *A1111* and *A1112*. Among them, *QQQ*, *A1*, *T1*, *A11* and *A111* are aggregated systems while *B1*, *B2*, *A1111*, *A1112*, *D1*, *D2*, *T11* and *T12* are non-aggregated systems.

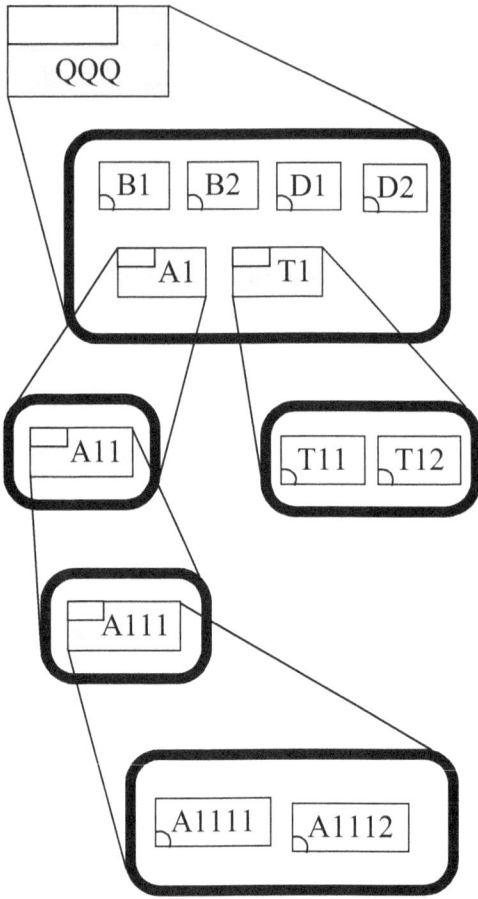

Figure 4-7   AHD of the *QQQ* System

There are aggregated and non-aggregated systems in an architecture hierarchy diagram. Non-aggregated systems are sometimes labeled as components, parts, entities, objects and building blocks [Chao09, Chao12, Chao14a, Chao14b].

Framework diagram (FD) represents the decomposition and composition of a system in a multi-layer (also referred to as multi-tier) manner. Only components will appear in a FD. FD is the second fundamental diagram to achieve structure-behavior coalescence.

As an example, Figure 4-8 shows a FD of the *QQQ* system. In the figure, *Business_Layer* contains the *B1* and *B2* components; *Application_Layer* contains the *A1111* and *A1112* components; *Data_Layer* contains the *D1* and *D2* components; *Technology_Layer* contains the *T11* and *T12* components.

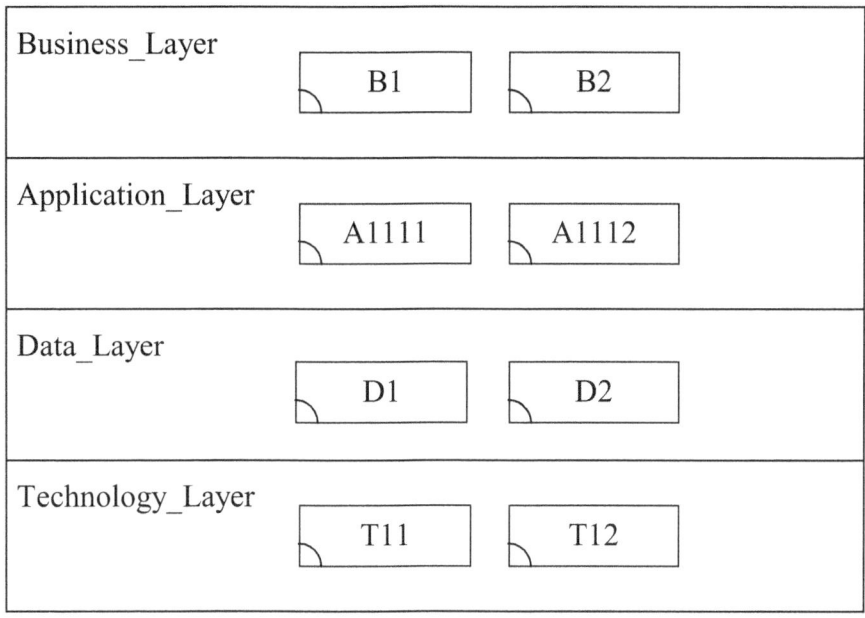

Figure 4-8   FD of the *QQQ* System

Systems structure, apart from outlining the components, shall also describe all operations for each component and all connections among components and actors in the external environment.

For a system, we use a component operation diagram (COD) to demonstrate all components' operations. COD is the third fundamental diagram to achieve structure-behavior coalescence. Figure 4-9 shows a COD of the *QQQ* system. In the figure, component *B1* has two operations: *op_01* and *op_02*; component *B2* has one operation: *op_03*; component *A1111* has one operation: *op_04*; component *A1112* has two operations: *op_05* and *op_06*; component *D1* has two operations: *op_07* and *op_08*; component *D2* has one operation: *op_9*; component *T11* has one operation: *op_10*; component *T12* has one operation: *op_11*.

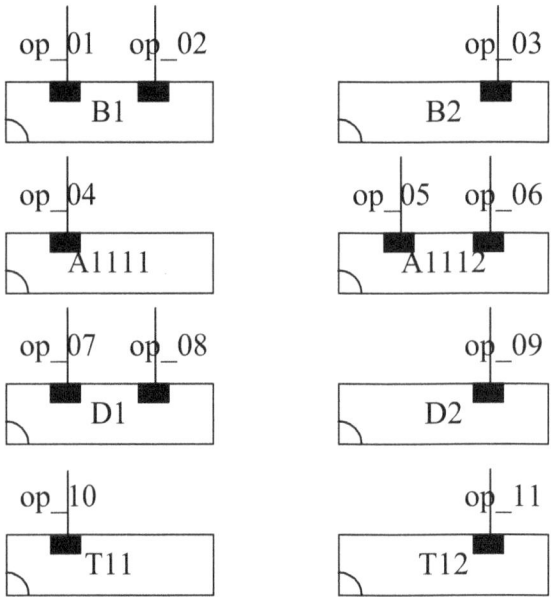

Figure 4-9   COD of the *QQQ* System

We use a component connection diagram (CCD) to describe how the components and actors (in the external environment) are connected within a system. CCD is the fourth fundamental diagram to achieve structure-behavior coalescence. Figure 4-10 exhibits a CCD of the *QQQ* system.

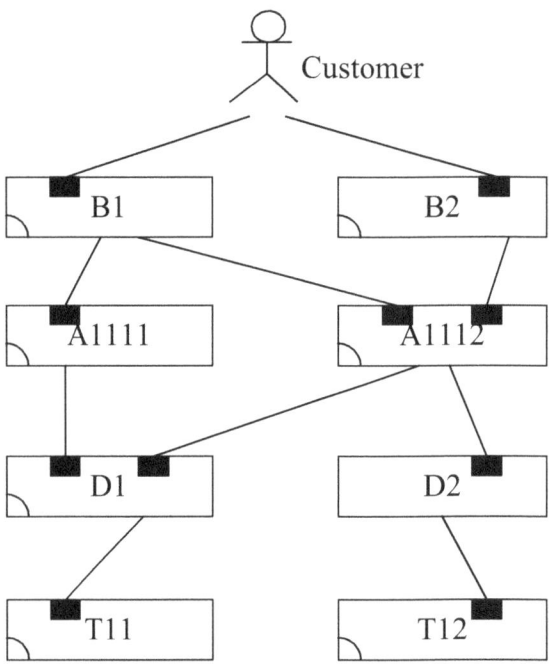

Figure 4-10   CCD of the *QQQ* System

In Figure 4-10, we see that actor *Customer* has a connection with each one of the *B1* and *B2* components; component *B1* has a connection with each one of the *A1111* and *A1112* components; component *B2* has a connection with the *A1112* component; component *A1111* has a connection with the *D1* component; component *A1112* has a connection with each one of the *D1* and *D2* components; component *D1* has a connection with the *T11* component; component *D2* has a connection with the *T12* component.

4-2-2 Interactions among Components and Actors to Draw forth the Systems Behavior

In a system, if the components, and among them and the external environment's actors to interact (or handshake), these interactions will draw forth the systems behavior. That is, "interaction" plays an important factor in integrating the systems structure and systems behavior for a system.

We use a structure-behavior coalescence diagram (SBCD) to describe how the systems structure and systems behavior are integrated within a system. SBCD is the fifth fundamental diagram to achieve structure-behavior coalescence. Figure 4-11 exhibits a SBCD of the *QQQ* system. In this example, an actor interacting with eight components shall describe the overall systems behavior. Interactions among the *Customer* actor and the *B1*, *A1111*, *D1* components draw forth the *qqq_1* behavior. Interactions among the *Customer* actor and the *B1*, *A1112*, *D1*, *T11* components draw forth the *qqq_2* behavior. Interactions among the *Customer* actor and the *B2*, *A1112*, *D2*, *T12* components draw forth the *qqq_3* behavior.

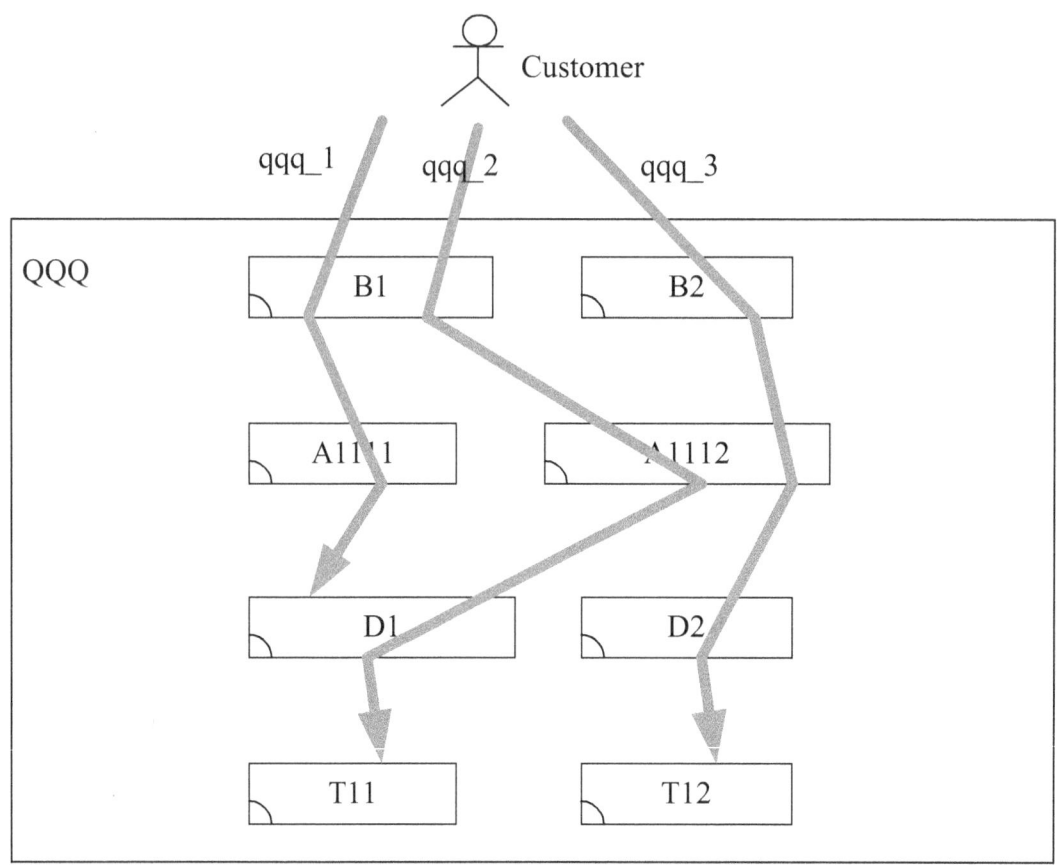

Figure 4-11   SBCD of the *QQQ* System

The overall behavior of a system is a collection of all its individual behaviors. All individual behaviors are mutually independent of each other. They tend to be executed concurrently [Hoar85, Miln89, Miln99]. For example, the overall behavior of the *QQQ* system includes the *qqq_1*, *qqq_2* and *qqq_3* behaviors. In other words, the *qqq_1*, *qqq_2*, and *qqq_3* behaviors are combined to produce the overall behavior of the *QQQ* system.

The major purpose of using the architectural approach, instead of separating the structure model from the behavior model, is to achieve one single coalesced model. In Figure 4-11, systems architects are able to see that the systems structure and systems behavior coexist in the SBCD. That is, in the SBCD of the *QQQ* system, architects not only see its systems structure but also see (at the same time) its systems behavior.

The overall behavior of a system consists of many individual behaviors. Each individual behavior represents an execution path. We use an interaction flow diagram (IFD) to demonstrate this individual behavior. IFD is the sixth fundamental diagram utilized to achieve structure-behavior coalescence. The overall behavior of the *QQQ* system includes three behaviors: *qqq_1*, *qqq_2* and *qqq_3*. Each of them is described by an individual IFD.

Figure 4-12 shows an IFD of the *qqq_1* behavior. First, actor *Customer* interacts with the *B1* component through the *op_01* operation call interaction. Next, component *B1* interacts with the *A1111* component through the *op_04* operation call interaction. Finally, component *A1111* interacts with the *D1* component through the *op_07* operation call interaction.

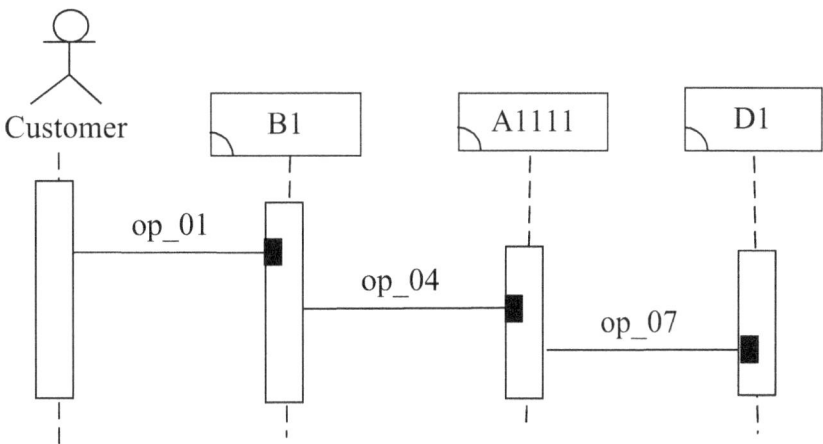

Figure 4-12   IFD of the *qqq_1* Behavior

Figure 4-13 shows an IFD of the *qqq_2* behavior. First, actor *Customer* interacts with the *B1* component through the *op_02* operation call interaction. Next, component *B1* interacts with the *A1112* component through the *op_05* operation call interaction. Continuingly, component *A1112* interacts with the *D1* component through the *op_08* operation call interaction. Finally, component *D1* interacts with the *T11* component through the *op_10* operation call interaction.

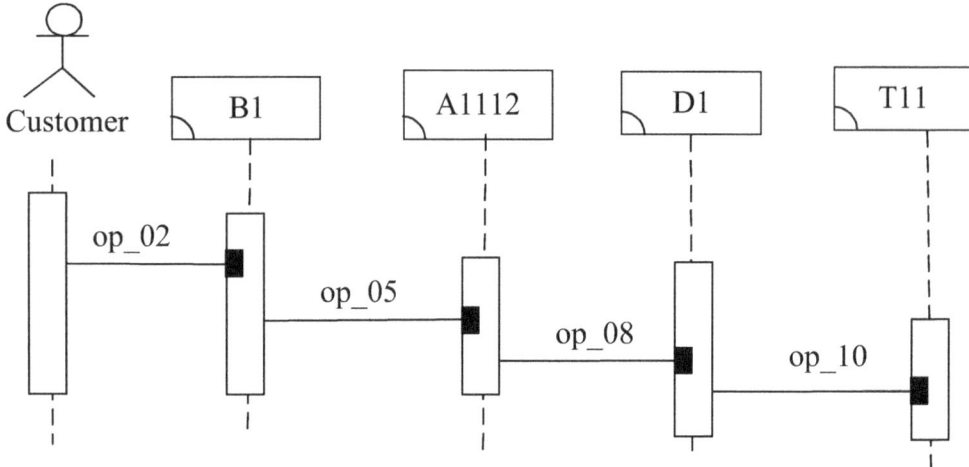

Figure 4-13   IFD of the *qqq_2* Behavior

Figure 4-14 shows an IFD of the *qqq_3* behavior. First, actor *Customer* interacts with the *B2* component through the *op_03* operation call interaction. Next, component *B2* interacts with the *A1112* component through the *op_06* operation call interaction. Continuingly, component *A1112* interacts with the *D2* component through the *op_09* operation call interaction. Finally, component *D2* interacts with the *T12* component through the *op_11* operation call interaction.

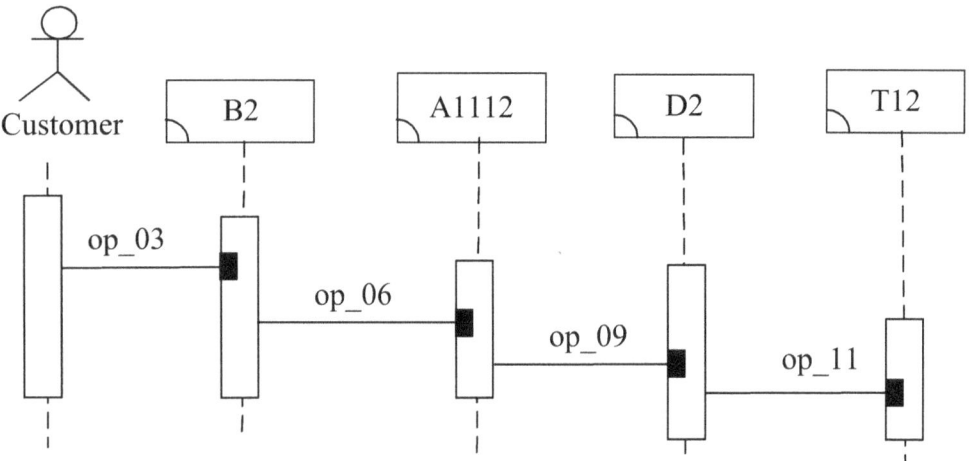

Figure 4-14   IFD of the *qqq_3* Behavior

4-3 SBC Architecture Development Method

The term of architecture development method (ADM) is first used in the open group architecture framework [Toga08]. Through the iterative and cyclic ADM, a systems architect is able to construct the systems architecture smoothly.

SBC architecture development method (SBC-ADM), being utilized by a systems architect to accomplish each version management of the systems architecture, shall do the strategic management first and then go through the concept, analysis, design and implementation phases of systems architecture construction. Every phase shall check with the requirements to make sure that each version of the constructed systems architecture is what the users want as shown in Figure 4-15.

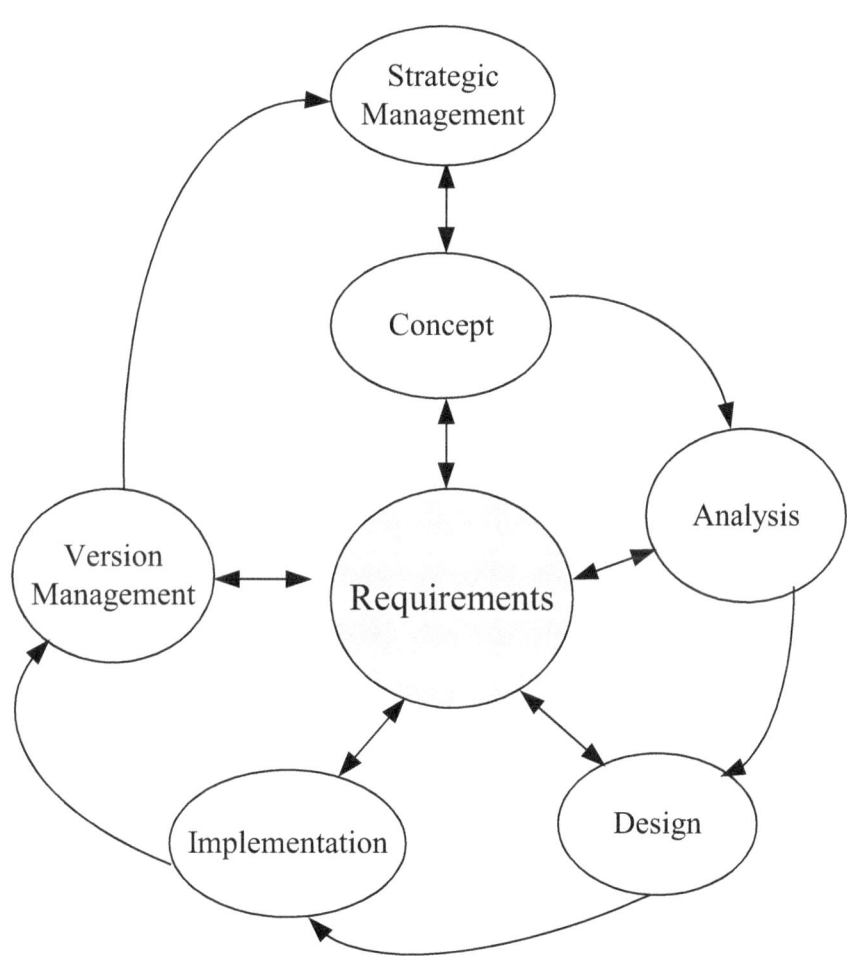

Figure 4-15   SBC Architecture Development Method

The output of strategic management is a strategy and the output of version management is a version of systems architecture. Accordingly, each strategy is mapped to a version of systems architecture as shown in Figure 4-16.

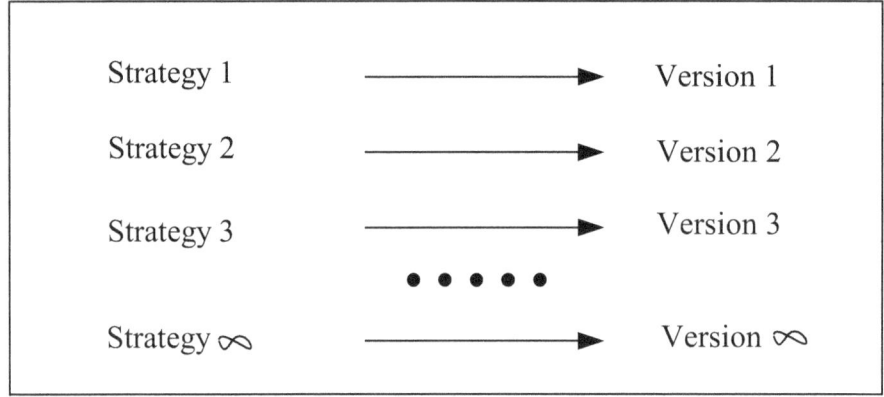

Figure 4-16   Each Strategy is Mapped to a Version of
Systems Architecture

4-4 SBC View Model

View model [Chao09, Dam06, Mino08, O'Rou03] is a three-dimensional matrix representation of a system's multiple views as shown in Figure 4-17. In the figure, dimension 1 stands for the evolution&motivation view which contains the strategy/version 1, strategy/version 2, strategy/version 3, strategy/version 4,..., and strategy/version ∞ views; dimension 2 stands for the multi-level (hierarchical) view which contains the concept, analysis, design and implementation views; dimension 3 stands for the systemic view which contains the structure, behavior, input/output data views.

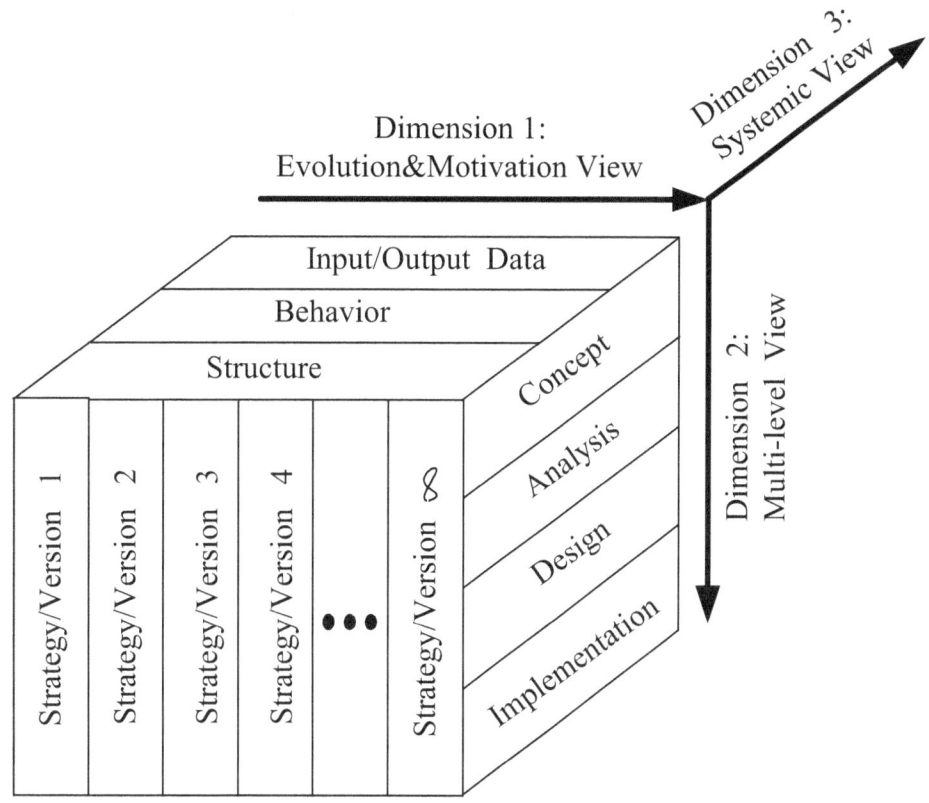

Figure 4-17   View Model

SBC view model (SBC-VM) [Chao09, Dam06, Mino08, O'Rou03] adopts a simplifier representation of multiple views as shown in Figure 4-18.

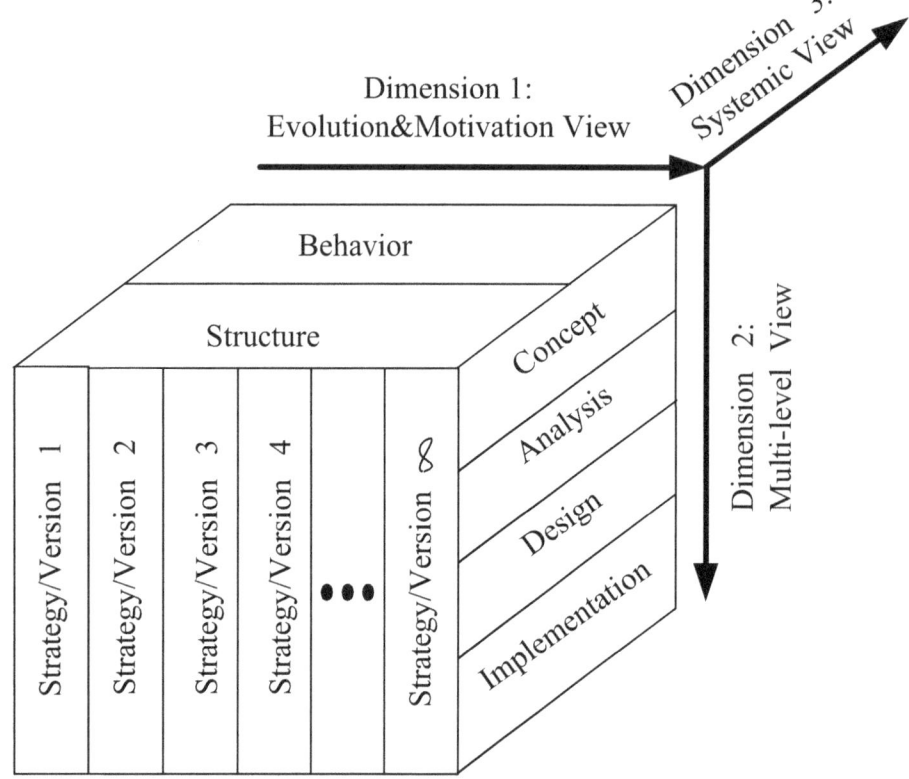

Figure 4-18   SBC View Model

In the SBC view model, dimension 1 stands for the evolution&motivation view which contains the strategy/version 1, strategy/version 2, strategy/version 3, strategy/version 4,…, and strategy/version ∞ views; dimension 2 stands for the multi-level (hierarchical) view which contains the concept, analysis, design and implementation views; dimension 3 stands for the systemic view which contains the structure and behavior views. According to the approach of SBC-ADL, the structure view consists of AHD, FD, COD and CCD; the behavior view consists of SBCD and IFD. Also, FD consists of business layer, application layer, data layer and technology layer. Adding these ideas to Figure 4-18, we then get the complete SBC view model as shown in Figure 4-19.

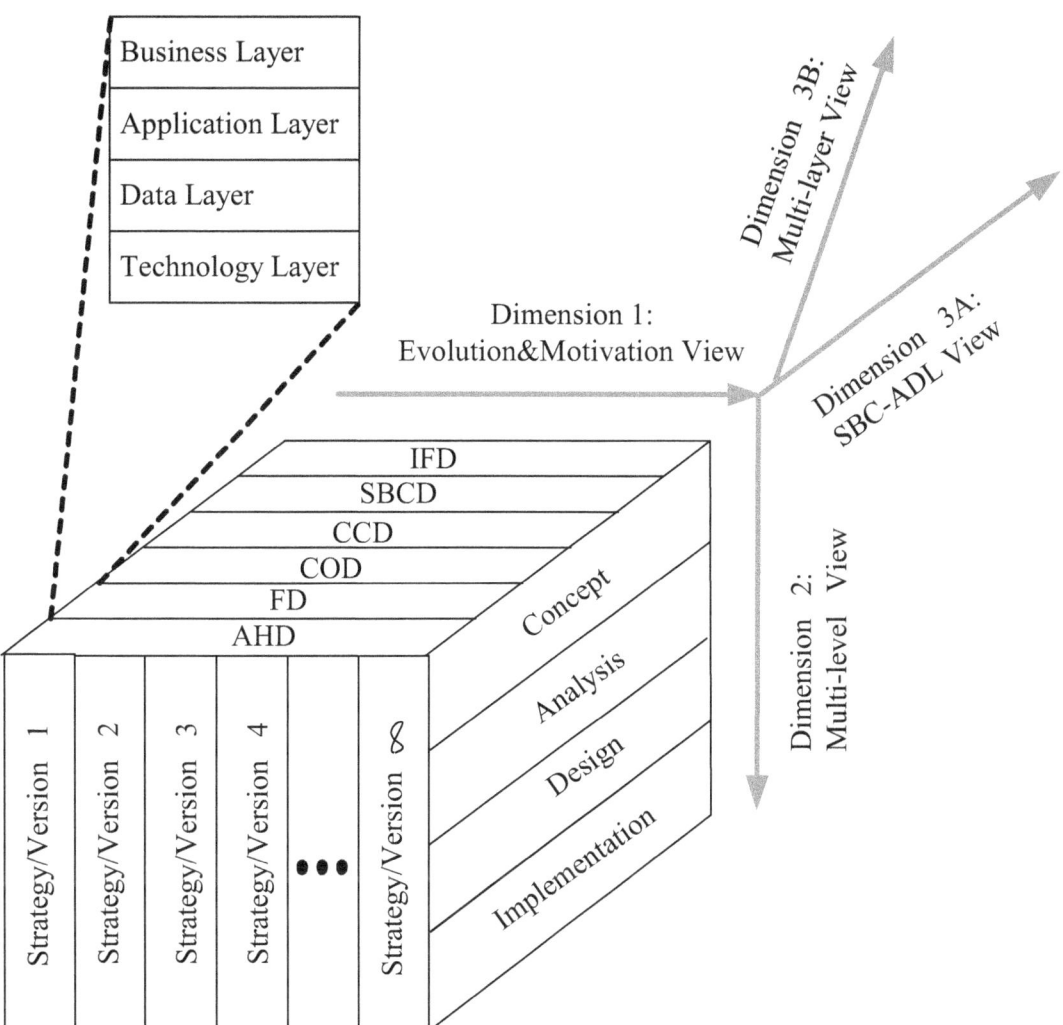

Figure 4-19   Complete SBC View Model

Chapter 5: Introduction to General Systems Theory 2.0

Because general systems theory 1.0 defining a system does not describe the integration of systems structure and systems behavior, very likely it will never be able to form a truly integrated whole of a system. In this situation, general systems theory 1.0 is powerless in defining a system appropriately.

SBC architecture provides an elegant way to integrate the structure and behavior of a system. Therefore, general systems theory 2.0 (general architectural theory) shall use the SBC architecture to define a system. A system is redefined, by general systems theory 2.0, shown in Figure 5-1.

A system,
through the SBC architecture,
truly is an integrated whole,
embodied in its components,
their interactions with each other and the environment,
and the principles and guidelines governing its design and evolution.

Figure 5-1   General Systems Theory 2.0 Defining a System

According to the above definition, general systems theory 2.0 uses the SBC architecture to define a system. Based on the SBC view model shown in Figure 5-2, there are three significant dimensions for general systems theory 2.0 (general architectural theory) to define a system.

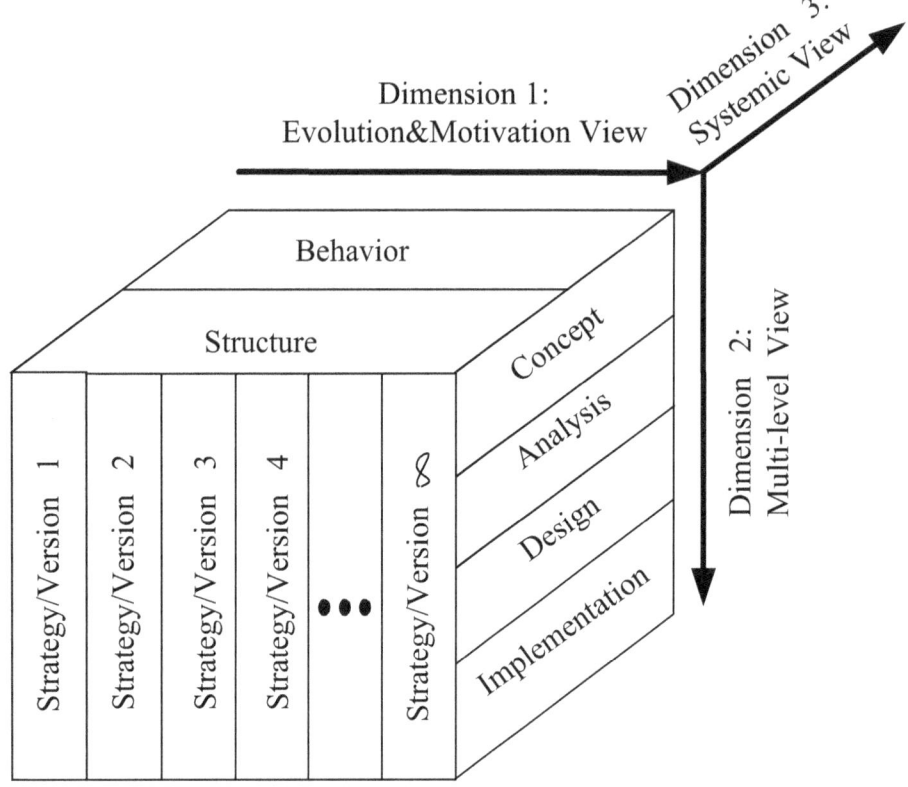

Figure 5-2   SBC View Model

Dimension 1 stands for the evolution&motivation view which contains the strategy/version 1, strategy/version 2, strategy/version 3, strategy/version 4,…, and strategy/version ∞ views. Dimension 2 stands for the multi-level (hierarchical) view which contains the concept, analysis, design and implementation views. Dimension 3 stands for the systemic view which contains the structure and behavior views.

5-1 Evolution&Motivation View

A system, not matter it is physical or virtual, will always change from time to time. A system evolves when it changes. Evolution of a system is shown in Figure 5-3.

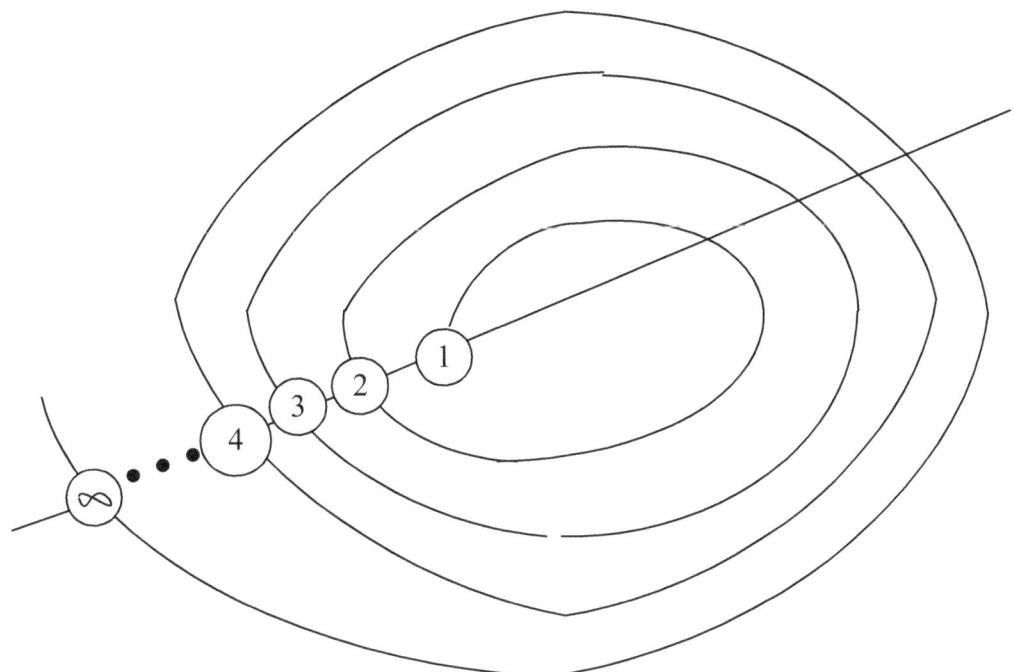

(1) : Systems Definition Version 1

(2) : Systems Definition Version 2

(3) : Systems Definition Version 3

(4) : Systems Definition Version 4

• • •

(∞) : Systems Definition Version ∞

Figure 5-3   Evolution of a System

Each time when a system changes or evolves, we shall get a new version of its systems definition. In the above figure, *version 1* stands for the original systems definition of a system and evolves into *version 2*, *version 3*, *version 4*,…, and *version ∞* gradually.

Evolution of a system is represented by the SBC architecture, as the evolution&motivation view shown in Figure 5-4.

Dimension 1:
Evolution&Motivation View

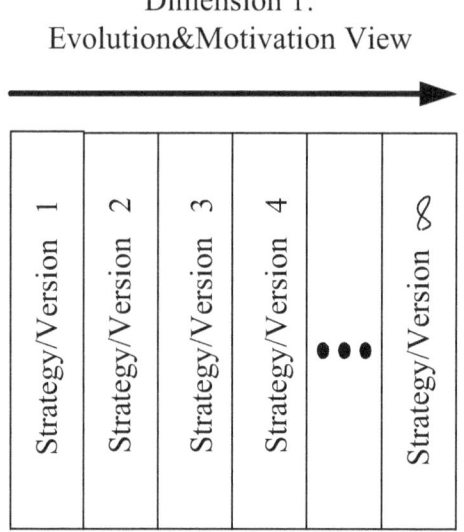

Figure 5-4   Evolution&Motivation View

In the evolution&motivation view, we see that each strategy is mapped to a version of systems definition as shown in Figure 5-5.

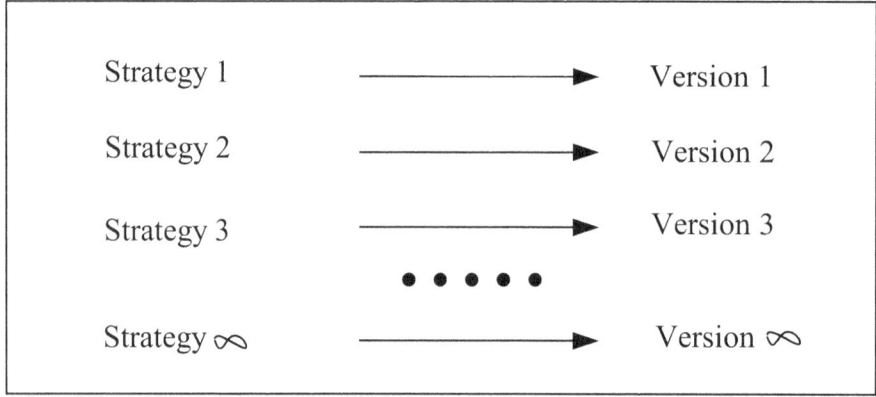

Figure 5-5   Each Strategy is Mapped to a Version of
Systems Definition

5-2 Multi-Level View

In the SBC architecture, multi-level (hierarchical) view contains the concept, analysis, design and implementation views as shown in Figure 5-6.

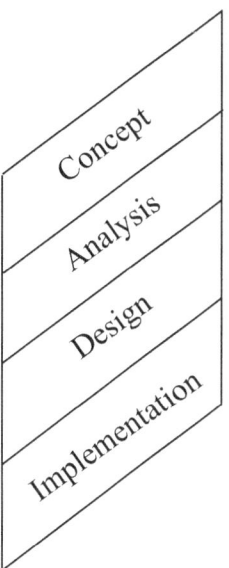

Figure 5-6   Multi-Level (Hierarchical) View

Architectural construction involves multi-level decisions (i.e. views) made regarding how a systems architecture should be constructed. These multi-level decisions may lead us to the following multi-level architecture diagram, as shown in the example of Figure 5-7. In the figure, *QQQ* is composed of *B1*, *B2*, *A1*, *D1*, *D2* and *T1*; *A1* is composed of *A11*; *T1* is composed of *T11* and *T12*; *A11* is composed of *A111*; *A111* is composed of *A1111* and *A1112*.

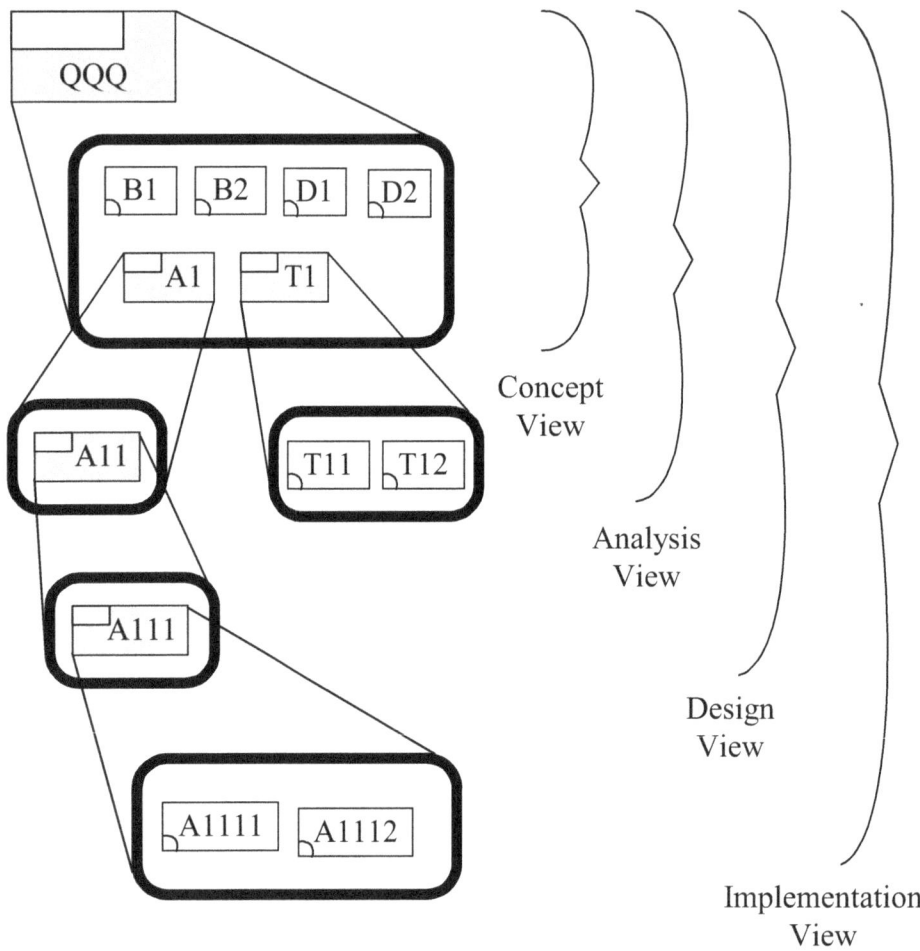

Figure 5-7   Multi-Level Architecture Diagram

There are aggregated and non-aggregated systems in Figure 5-7. Systems *QQQ*, *A1*, *T1*, *A11* and *A111* are categorized into aggregated systems while systems *B1*, *B2*, *A1111*, *A1112*, *D1*, *D2*, *T11* and *T12* are categorized into non-aggregated systems.

Concept view corresponds to an executive summary for an administrator who wants an estimate of the scope of the system, what it would cost, and how it would relate to the general environment in which it will operate. Analysis view corresponds to a summary for an analyzer who works on the analysis of a system. Analysis view is one level down structural decomposition (with observation congruence verification) of the concept view [Chao15a, Chao15b, Chao15c, Chao15d, Chao15e]. Design view describes what a designer has accomplished for his task. Design view is one level down structural decomposition (with observation congruence verification) of the analysis view. Implementation view shows what an implementer has done for his work. Implementation view is one level down structural decomposition (with observation congruence verification) of the design view.

Figures 5-7 gives us some ideas of architecture construction in brief, but is short of systemic views. Figure 5-8 demonstrates a multi-level architecture relating to systemic views.

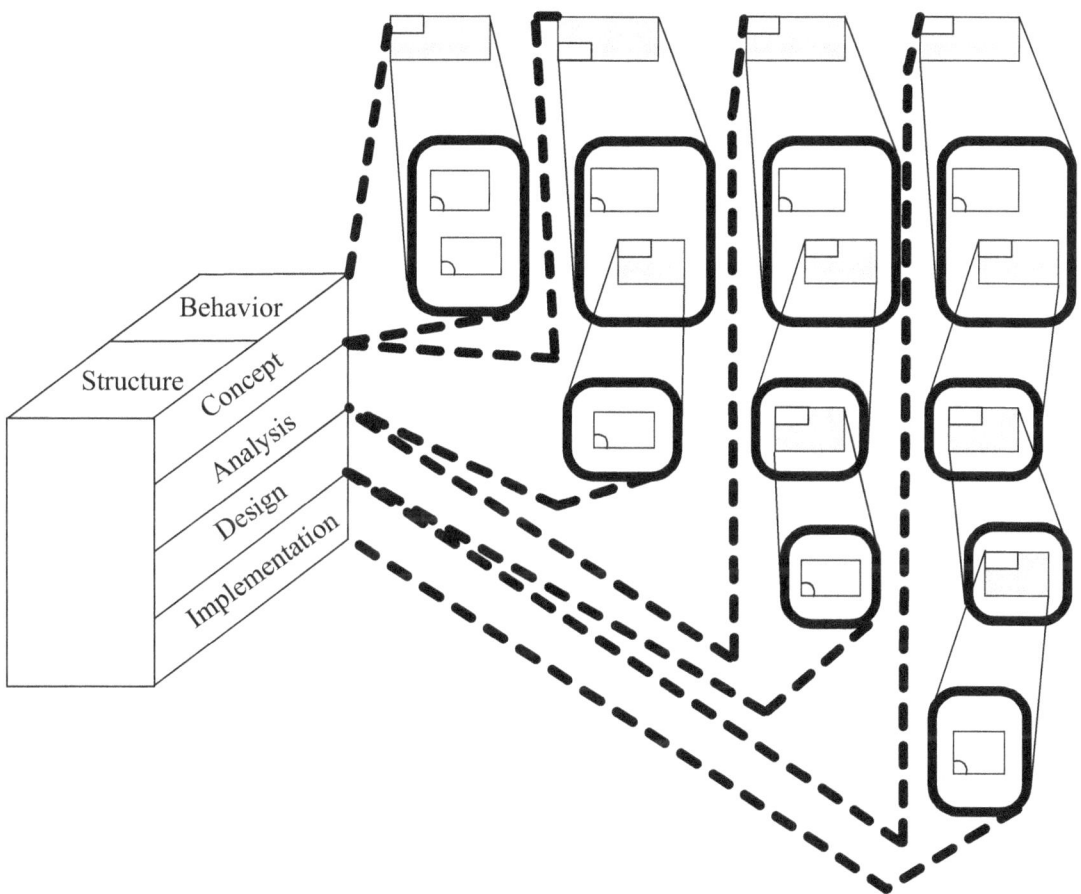

Figure 5-8   Multi-Level Architecture Relating to the Systemic View

In Figure 5-8, we observe that *Concept View* contains the *Concept's Systems Structure* and *Concept's Systems Behavior*; *Analysis View* contains the *Analysis' Systems Structure* and *Analysis' Systems Behavior*; *Design View* contains the *Design's Systems Structure* and *Design's Systems Behavior*; *Implementation View* contains the *Implementation's Systems Structure* and *Implementation's Systems Behavior*.

According to the approach of SBC architecture description language (SBC-

78

ADL), systems structure consists of AHD, FD, COD and CCD; systems behavior consists of SBCD and IFD. Adding these SBC-ADL systemic views to Figure 5-8, we then get the SBC multi-level view as shown in Figure 5-9.

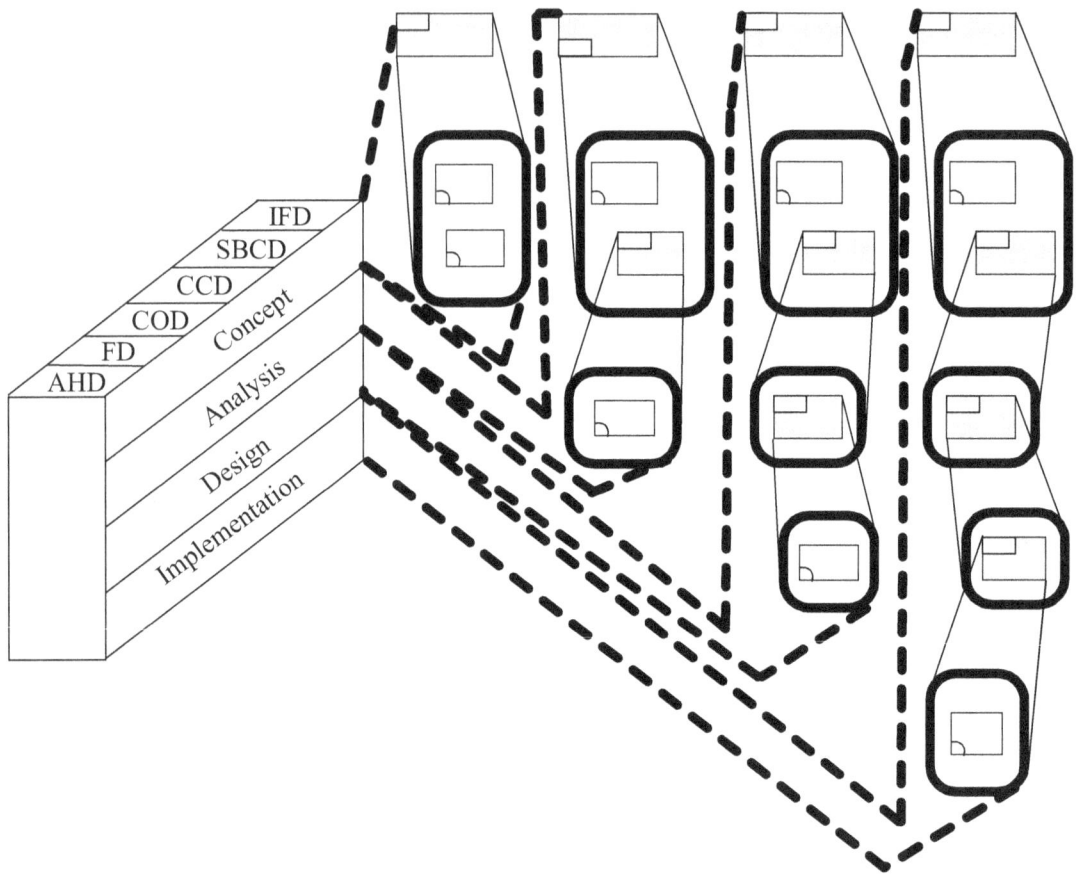

Figure 5-9   SBC Multi-Level View

5-3 Systemic View

In the SBC architecture, systemic view contains the structure and behavior views as shown in Figure 5-10.

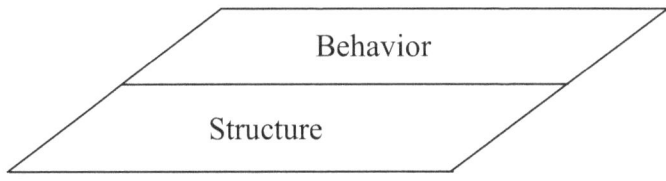

Figure 5-10   Systemic View

According to the approach of SBC architecture description language (SBC-

ADL), the structure view consists of AHD, FD, COD and CCD; the behavior view consists of SBCD and IFD. Also, FD consists of business layer, application layer, data layer and technology layer. Adding these ideas to Figure 5-10, we then get the systemic view of general systems theory 2.0 (general architectural theory), as shown in Figure 5-11.

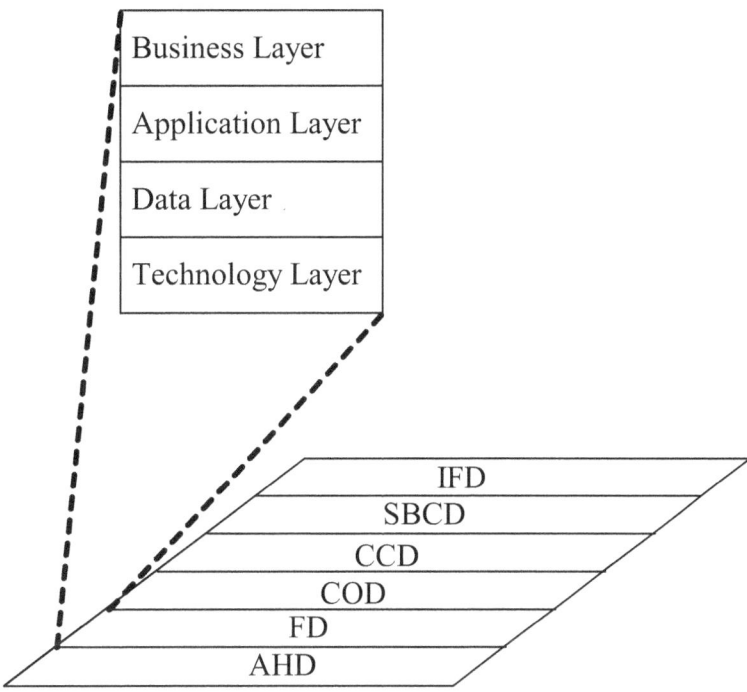

Figure 5-11   Systemic View of General of Systems Theory 2.0

PART II: SYSTEMIC VIEW

Chapter 6: Architecture Hierarchy Diagram

General systems theory 2.0 (general architectural theory) uses an architecture hierarchy diagram (AHD) to define the multi-level decomposition and composition of a system. AHD is the first fundamental diagram to achieve structure-behavior coalescence.

6-1 Decomposition and Composition

The following is an example of systems decomposition and composition. The *Computer* system consists of *Monitor*, *Keyboard*, *Mouse* and *Case*, as shown in Figure 6-1. The *Monitor*, *Keyboard*, *Mouse* and *Case* are subsystems that comprise the *Computer* system.

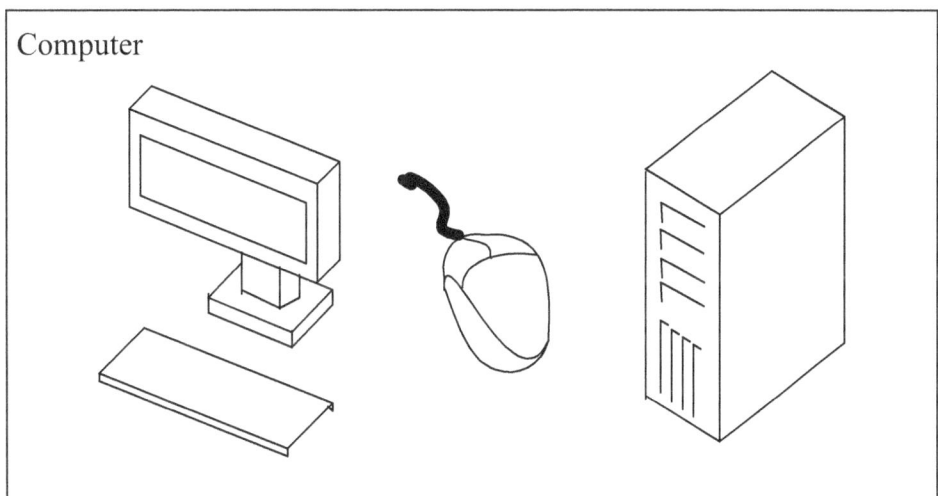

Figure 6-1   Decomposition and Composition of the *Computer* System

Another example indicates that the *Tree* system is composed of *Root* and *Stem*, as shown in Figure 6-2. In this example, we would say that the *Root* and *Stem* are subsystems, respectively, while the *Tree* system consists of its subsystems.

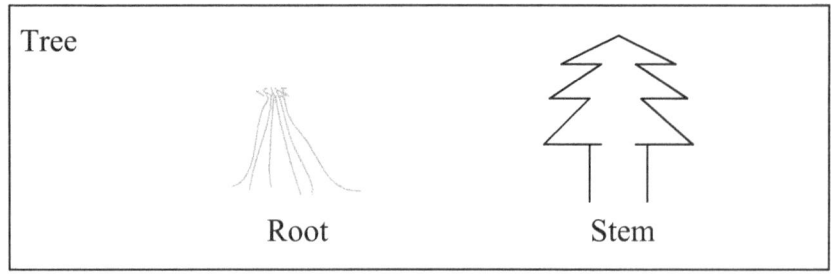

Figure 6-2   Decomposition and Composition of the *Tree* System

The last example demonstrates that the *SBC_Book* system is composed of *Chapter_1*, *Part_1* and *Part_2*, as shown in Figure 6-3. In this example, we would say that the *Chapter_1*, *Part_1* and *Part_2* are subsystems, respectively while the *SBC_Book* system consists of its subsystems.

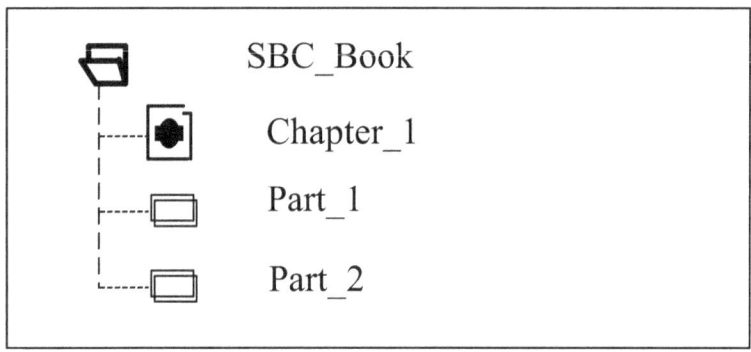

Figure 6-3   Decomposition and Composition of the *SBC_Book* System

Architecture hierarchy diagram (AHD) is used to define the decomposition and composition of a system. As an example, Figure 6-4 shows an AHD of the *Computer* system. We clearly observe that the *Computer* system is composed of *Monitor*, *Keyboard*, *Mouse* and *Case*.

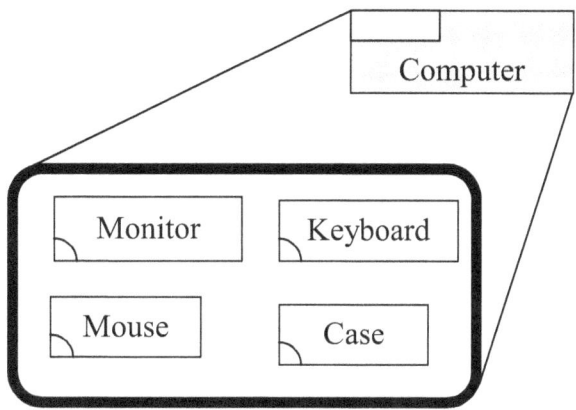

Figure 6-4   AHD of the *Computer* System

As a second example, Figure 6-5 shows an AHD of the *Tree* system. We clearly observe that the *Tree* system is composed of *Root* and *Stem*.

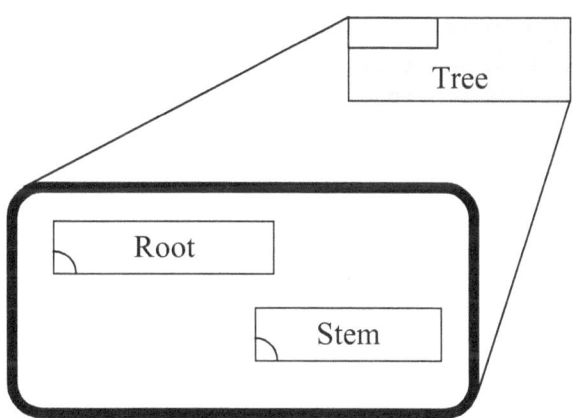

Figure 6-5   AHD of the *Tree* System

As a third example, Figure 6-6 shows an AHD of the *SBC_Book* system. We clearly observe that the *SBC_Book* is composed of *Chapter_1*, *Part_1* and *Part_2*.

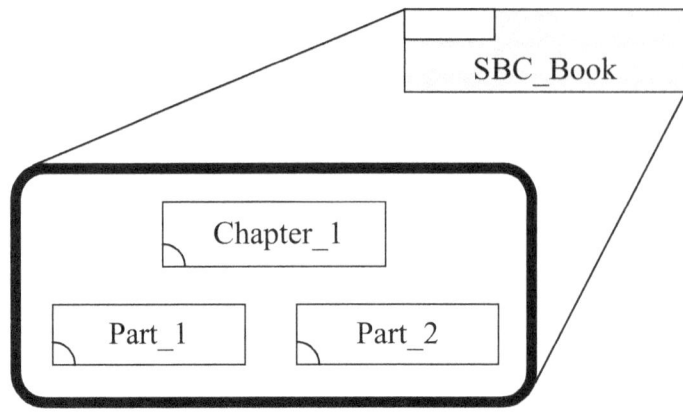

Figure 6-6   AHD of the *SBC_Book* system

6-2 Multi-Level Decomposition and Composition

The subsystem may also contain subsystems as we further decompose it. For example, *Case* is a subsystem of the *Computer*, and we further decompose it into *Motherboard*, *Hard_Disk*, *Power_Supply* and *DVD_Disk*, as shown in Figure 6-7.

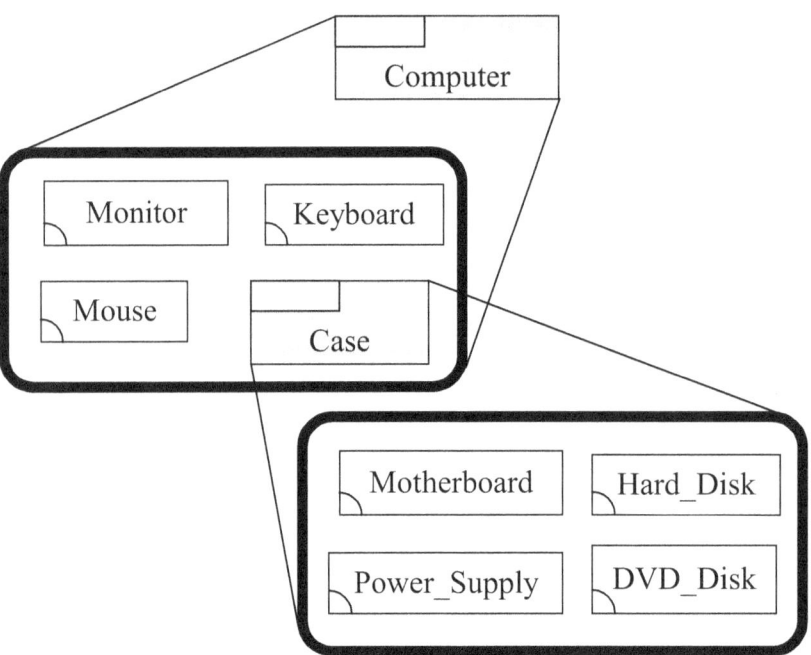

Figure 6-7   Multi-Level Decomposition/Composition of the *Computer* System

As a second example, *Stem* is a subsystem of the *Tree*, and we further decompose it into *Trunk* and *Leaf*, as shown in Figure 6-8.

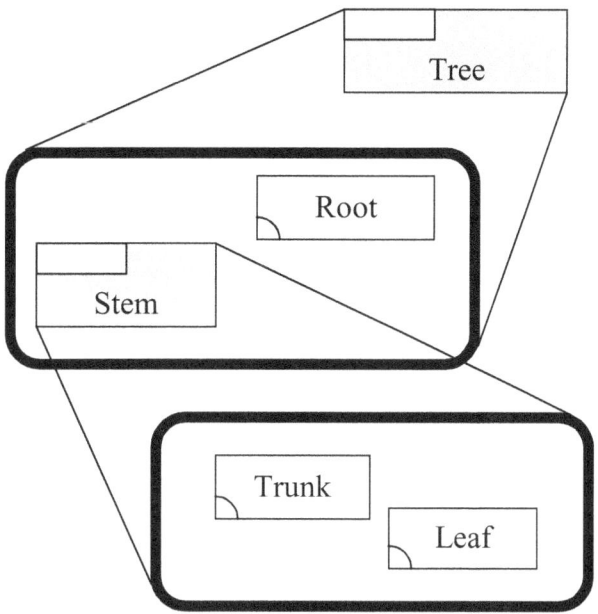

Figure 6-8   Multi-Level Decomposition/Composition of the *Tree* System

As a third example, *Part_1* is a subsystem of the *SBC_Book*, and we further decompose it into *Chapter_2* and *Chapter_3*; *Part_2* is also a subsystem of the *SBC_Book*, and we further decompose it into *Chapter_4* and *Chapter_5*, as shown in Figure 6-9.

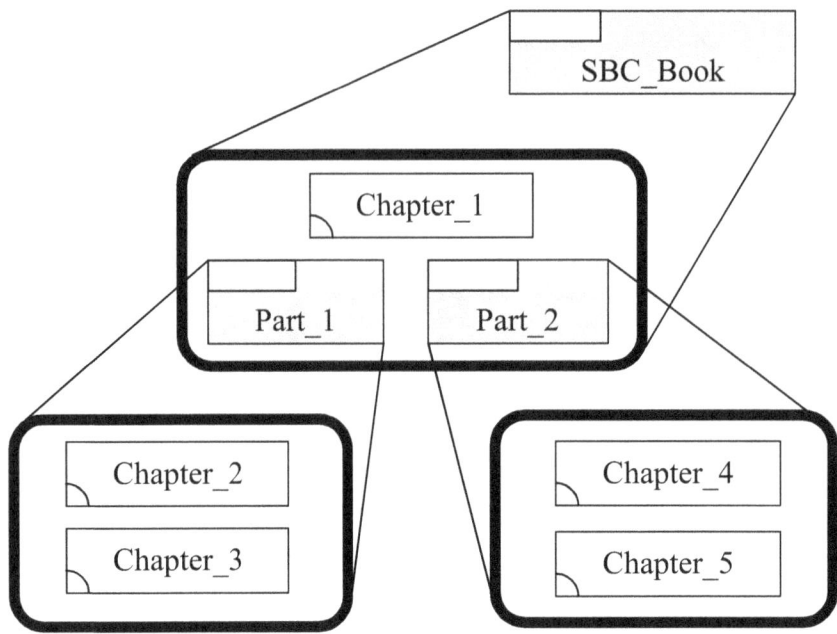

Figure 6-9   Multi-Level Decomposition/Composition of the *SBC_Book* System

Generally speaking, multi-level decomposition and composition of a system is applied often in defining a system. To make a complex system look simple, the mechanism of multi-level composition and decomposition should be utilized.

6-3 Aggregated and Non-Aggregated Systems

Any system (at any level) involved with multi-level decomposition and composition of a system is either aggregated or non-aggregated. The definition of aggregated and non-aggregated systems is shown in Figure 6-10.

Definition of Aggregated Systems:

A system (within an AHD) is aggregated if it is composed of any sub-system.

Definition of Non-aggregated Systems

A system (within an AHD) is non-aggregated if it is NOT composed of any sub-system.

Figure 6-10   Definition of Aggregated and Non-aggregated Systems

Non-aggregated systems are sometimes referred to as components, parts, entities, objects and building blocks [Chao09, Chao14a, Chao14b].

In the multi-level systems decomposition and composition, any system is either aggregated or non-aggregated, but not both. For example, in Figure 6-4, *Case* is a non-aggregated system, not an aggregated system. As an interesting contrast, in Figure 6-7, *Case* is an aggregated system, not a non-aggregated system.

As a second example, in Figure 6-5, *Stem* is a non-aggregated system, not an aggregated system. As an interesting contrast, in Figure 6-8, *Stem* is an aggregated system, not a non-aggregated system.

As a third example, in Figure 6-6, *Part_1* and *Part_2* are non-aggregated systems, not aggregated systems. As an interesting contrast, in Figure 6-9, *Part_1* and *Part_2* are aggregated systems, not non-aggregated systems.

Chapter 7: Framework Diagram

General systems theory 2.0 (general architectural theory) uses a framework diagram (FD) to define the multi-layer (also referred to as multi-tier) decomposition and composition of a system. FD is the second fundamental diagram to achieve structure-behavior coalescence.

7-1 Multi-Layer Decomposition and Composition

Decomposition and composition of a system can also be defined in a multi-layer manner. We draw a framework diagram (FD) for the multi-layer decomposition and composition of a system.

As an example, Figure 7-1 shows a FD of the *Computer* system. In the figure, *Technology_SubLayer_2* contains *Monitor*, *Keyboard* and *Mouse*; *Technology_SubLayer_1* contains *Motherboard*, *Hard_Disk*, *Power_Supply* and *DVD_Disk*.

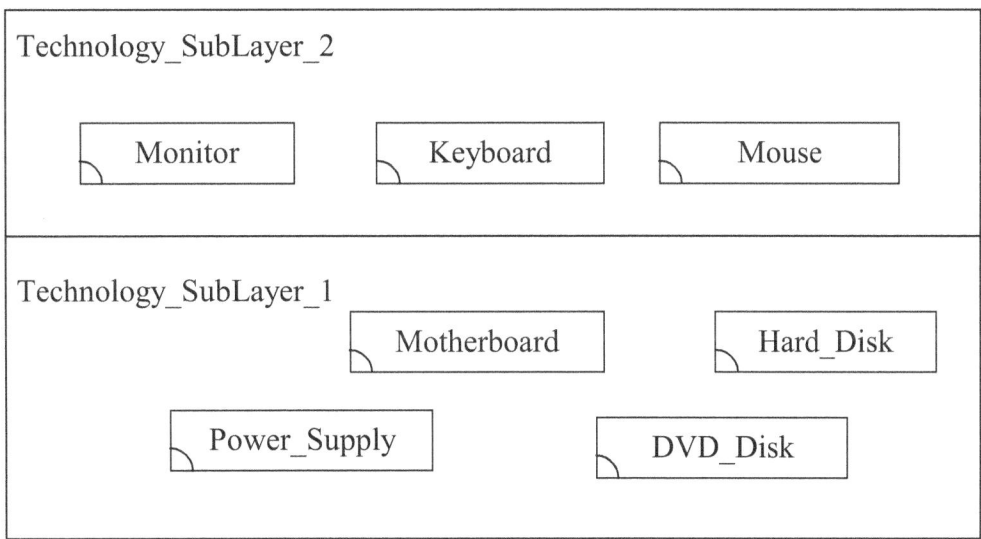

Figure 7-1   FD of the *Computer* System

As a second example, Figure 7-2 shows a FD of the *Tree* system. In the figure, *Technology_SubLayer_2* contains *Root*; *Technology_SubLayer_1* contains *Trunk* and *Leaf*.

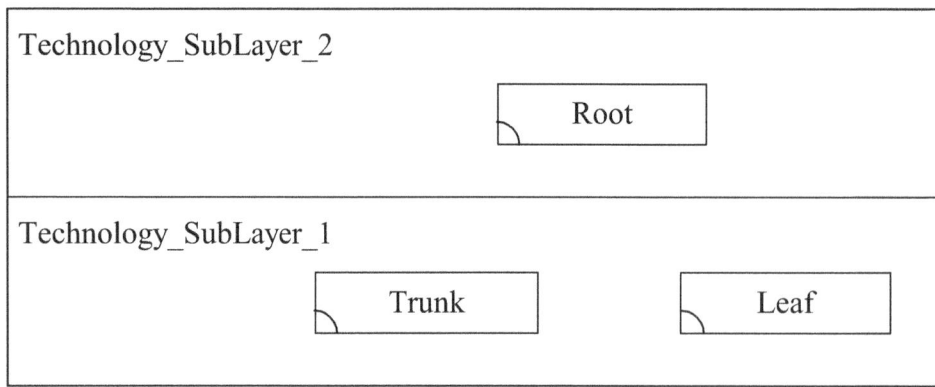

Figure 7-2   FD of the *Tree* System

As a third example, Figure 7-3 shows a FD of the *SBC_Book* system. In the figure, *Technology_SubLayer_2* contains *Chapter_1*; *Technology_SubLayer_1* contains *Chapter_2*, *Chapter_3*, *Chapter_4* and *Chapter_5*.

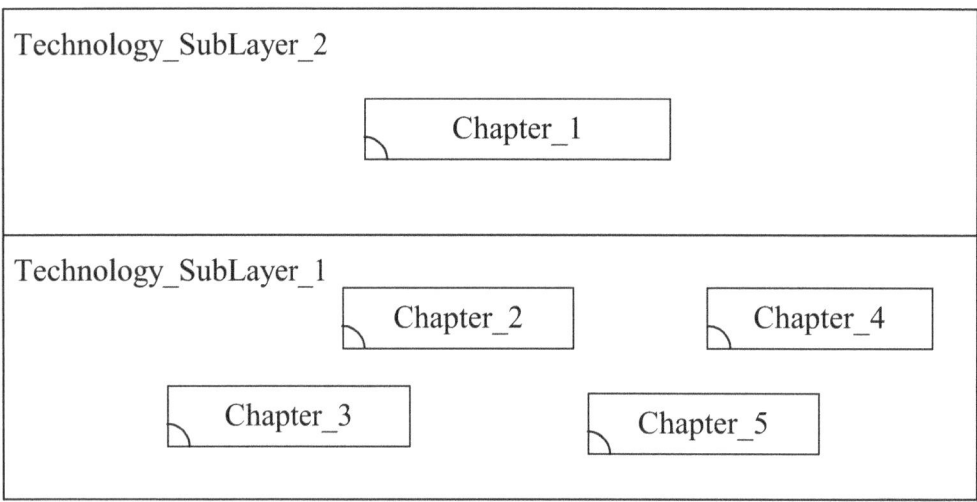

Figure 7-3   FD of the *SBC_Book* System

7-2 Only Non-Aggregated Systems Appearing in Framework Diagrams

Both aggregated and non-aggregated systems are displayed in the multi-level AHD decomposition and composition of a system. As an interesting contrast, only non-aggregated systems shall appear in the multi-layer FD decomposition and composition of a system.

For example, Figure 6-7 in the previous chapter shows an AHD of the *Computer* system in which both aggregated systems such as *Computer*, *Case* and non-aggregated systems such as *Monitor*, *Keyboard*, *Mouse*, *Motherboard*, *Hard_Disk*, *Power_Supply*, *DVD_Disk* are displayed. As an interesting contrast, Figure 7-1 in the previous section shows a FD of the *Computer* system in which only non-aggregated systems such as *Monitor*, *Keyboard*, *Mouse*, *Motherboard*, *Hard_Disk*, *Power_Supply* and *DVD_Disk* are displayed.

For a second example, Figure 6-8 in the previous chapter shows an AHD of the *Tree* system in which both aggregated systems such as *Tree*, *Stem* and non-aggregated systems such as *Root*, *Trunk*, *Leaf* are displayed. As an interesting contrast, Figure 7-2 in the previous section shows a FD of the *Tree* system in which only non-aggregated systems such as *Root*, *Trunk* and *Leaf* are displayed.

For a third example, Figure 6-9 in the previous chapter shows an AHD of the *SBC_Book* system in which both aggregated systems such as *SBC_Book*, *Part_1*, *Part_2* and non-aggregated systems such as *Chapter_1*, *Chapter_2*, *Chapter_3*, *Chapter_4*, *Chapter_5* are displayed. As an interesting contrast, Figure 7-3 in the previous section shows a FD of the *SBC_Book* system in which only non-aggregated systems such as *Chapter_1*, *Chapter_2*, *Chapter_3*, *Chapter_4* and *Chapter_5* are displayed.

Chapter 8: Component Operation Diagram

General systems theory 2.0 (general architectural theory) uses the component operation diagram (COD) to define all components' operations of a system. COD is the third fundamental diagram to achieve structure-behavior coalescence.

8-1 Operations of Each Component

An operation provided by each component represents a procedure or method or function of the component. If other components request this component to perform an operation, then shall use it to accomplish the operation request.

Each component in a system must possess at least one operation. A component should not exist in a system if it does not possess any operation. Figure 8-1 shows that component *SalePurchase_GUI* has four operations: *SaleInputClick*, *SalePrintClick*, *PurchaseInputClick* and *PurchasePrintClick*.

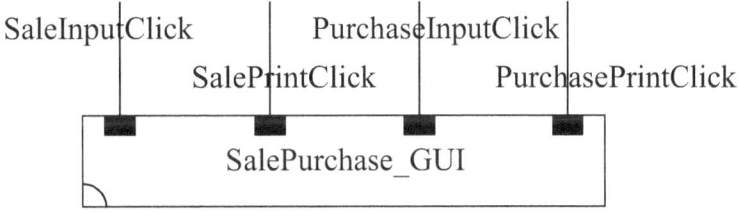

Figure 8-1   Four Operations of *SalePurchase_GUI*

An operation formula is utilized to fully define an operation. An operation formula includes a) operation name, b) input parameters and c) output parameters as shown in Figure 8-2.

$$\text{Operation\_Name (In } i_1, i_2, ..., i_m \text{ ; Out } o_1, o_2, ..., o_n)$$

Figure 8-2   Operation Formula

Operation name is the name of this operation. In a system, every operation name should be unique. Duplicate operation names shall not be allowed in any system.

An operation may have several input and output parameters. The input and output parameters, gathered from all operations, represent the input data and output data views of a system [Date03, Elma10]. As shown in Figure 8-3, component *SalePrint_GUI* possesses the *ShowModal* operation which has no input/output parameter; component *SalePrint_GUI* also possesses the *SalePrintButtonClick* operation which has the *sDate* and *sNo* input parameters (with the arrow direction pointing to the component) and the *s_report* output parameter (with the arrow direction opposite to the component).

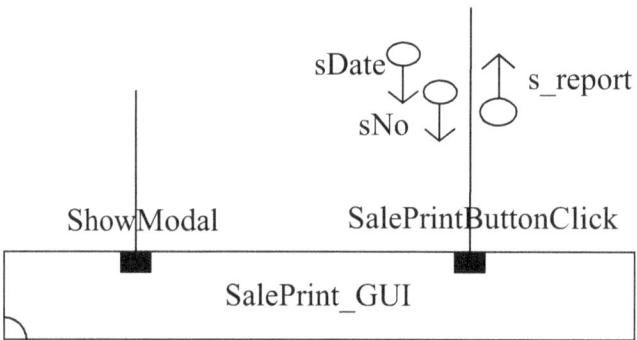

Figure 8-3   Input/Output Parameters of *SalePrintButtonClick*

Data formats of input and output parameters are defined by data type specifications. There are two groups of data types: primitive and composite [Date03, Elma10]. Figure 8-4 shows primitive data type specifications of the *sDate* and *sNo* input parameters occurring in the *SalePrintButtonClick(In sDate, sNo; Out s_report)* operation formula.

Parameter	Data Type	Instances
sDate	Text	20100517, 20100612
sNo	Text	001, 002

Figure 8-4   Primitive Data Type Specifications

Figure 8-5 shows composite data type specifications of the *s_report* output parameter occurring in the *SalePrintButtonClick(In sDate, sNo; Out s_report)* operation formula.

Parameter	*s_report*
Data Type	TABLE of Sale Date : Text Sale No : Text Customer : Text ProductNo : Text Quantity : Integer UnitPrice : Real Total : Real End TABLE;
Instances	Sale Date : 20100517 Sale No : 001 Customer : Larry Fink | ProductNo	Quantity	UnitPrice | | A12345 | 400 | 100.00 | | A00001 | 300 | 200.00 | Total : 100,000.00

Figure 8-5   Composite Data Type Specifications

8-2 Drawing the Component Operation Diagram

We use a component operation diagram (COD) to define all components' operations of a system. Figure 8-6 shows a COD of the *Multi-Tier Personal Data System*. In the figure, component *MTPDS_GUI* has two operations: *Calculate_AgeClick* and *Calculate_OverweightClick*; component *Age_Logic* has one operation: *Calculate_Age*; component *Overweight_Logic* has one operation: *Calculate_Overweight*; component *Personal_Database* has two operations: *Sql_DateOfBirth_Select* and *Sql_SexHeightWeight_Select*.

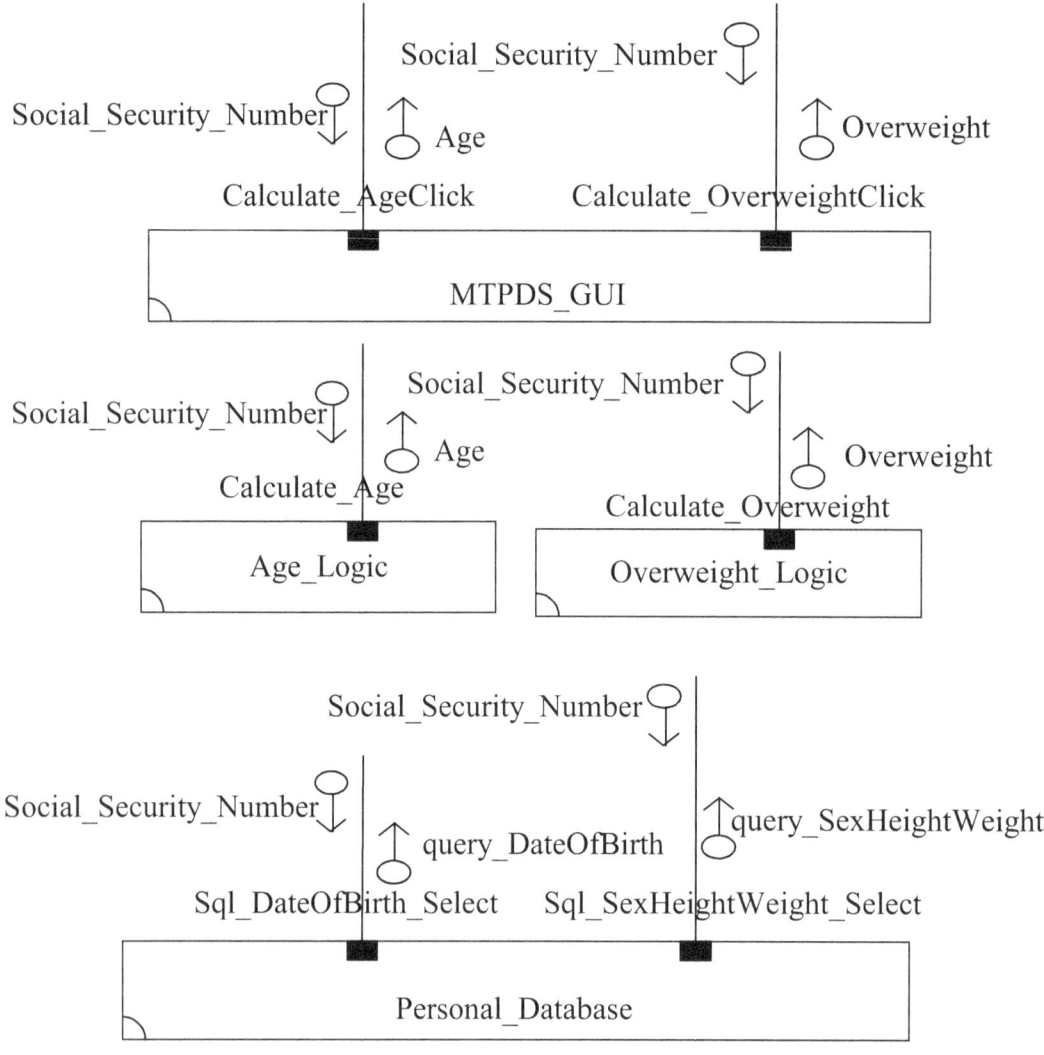

Figure 8-6   COD of the *Multi-Tier Personal Data System*

The operation formula of *Calculate_AgeClick* is *Calculate_AgeClick(In Social_Security_Number; Out Age)*. The operation formula of *Calculate_OverweightClick* is *Calculate_OverweightClick(In Social_Security_Number; Out Overweight)*. The operation formula of *Calculate_Age* is *Calculate_Age(In Social_Security_Number; Out Age)*. The operation formula of *Calculate_Overweight* is *Calculate_Overweight(In Social_Security_Number; Out Overweight)*. The operation formula of *Sql_DateOfBirth_Select* is *Sql_DateOfBirth_Select(In Social_Security_Number; Out query_DateOfBirth)*. The operation formula of *Sql_SexHeightWeight_Select* is *Sql_SexHeightWeight_Select(In Social_Security_Number; Out query_SexHeightWeight)*.

Figure 8-7 shows primitive data type specifications of the *Social_Security_Number* input parameter and the *Age, Overweight* output parameters.

Parameter	Data Type	Instances
Social_Security_Number	Text	424-99-9153, 512-24-3722
Age	Integer	28, 77
Overweight	Boolean	Yes, No

Figure 8-7   Primitive Data Type Specification

Figure 8-8 shows composite data type specifications of the *query_DateOfBirth* output parameter occurring in the *Sql_DateOfBirth_Select(In Social_Security_Number; Out query_DateOfBirth)* operation formula.

100

Parameter	*query_DateOfBirth*

Data Type	TABLE of
	   Social_Security_Number : Text
	   Age : Integer
	End TABLE ;

Instances	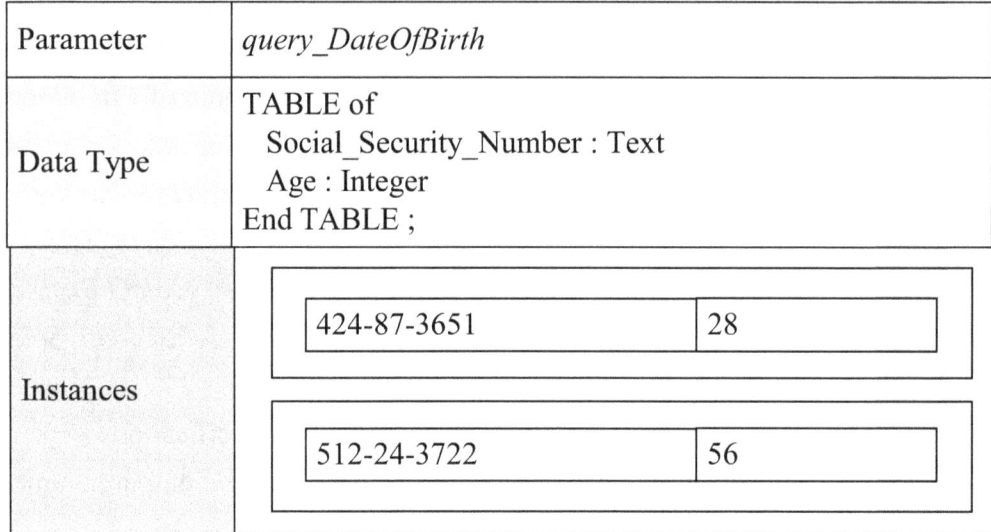

Figure 8-8   Composite Data Type Specifications

Figure 8-9 shows composite data type specifications of the *query_SexHeightWeight* output parameter occurring in the *Sql_SexHeightWeight_Select(In Social_Security_Number; Out query_SexHeightWeight)* operation formula.

Parameter	*query_SexHeightWeight*

Data Type	TABLE of
	   Social_Security_Number : Text
	   Sex : Text
	   Height : Number
	   Weight : Number
	End TABLE ;

Instances	424-87-3651 Female 162 76 512-24-3722 Male 180 80

Figure 8-9   Composite Data Type Specifications

Chapter 9: Component Connection Diagram

General systems theory 2.0 (general architectural theory) uses a component connection diagram (CCD) to define how all components and actors are connected within a system. CCD is the fourth fundamental diagram to achieve structure-behavior coalescence.

9-1 Essence of a Connection

A connection implies an operation request. When an operation is used by another subsystem then a connection appears. Accordingly, a connection is defined as the linkage that is constructed when an operation is used by another subsystem. Figure 9-1 shows that Subsystem_A uses the *Salary_Calculation* operation provided by the *Component_B* component.

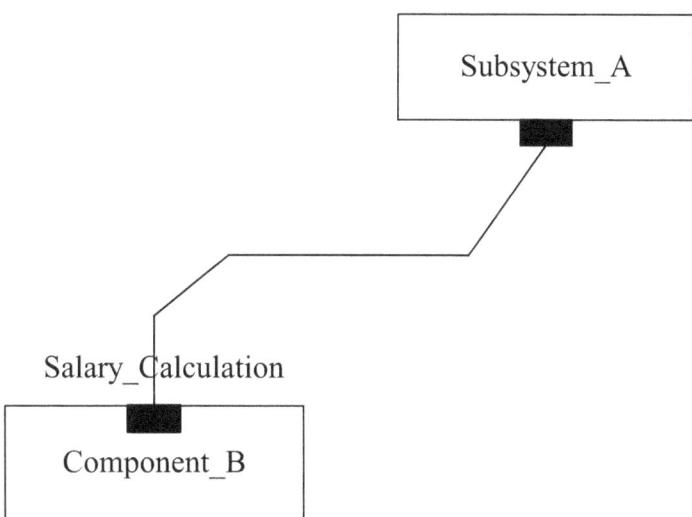

Figure 9-1   A Connection Appears When an Operation is Used

The above figure describes, sufficiently, the essence of a connection. However, we seldom use this kind of drawing. Instead, a simplified drawing of the above figure is often used as shown in Figure 9-2.

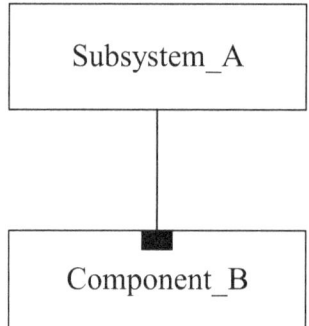

Figure 9-2   Simplified Drawing of a Connection

Since an operation is always provided by a component, there is no doubt that the *Component_B* operation provider is a component. On the contrary, the *Subsystem_A* operation user can be either a component (e.g., *Component_A*) or an actor (e.g., *Actor_A*) as shown in Figure 9-3. An actor belongs to the external environment of a system.

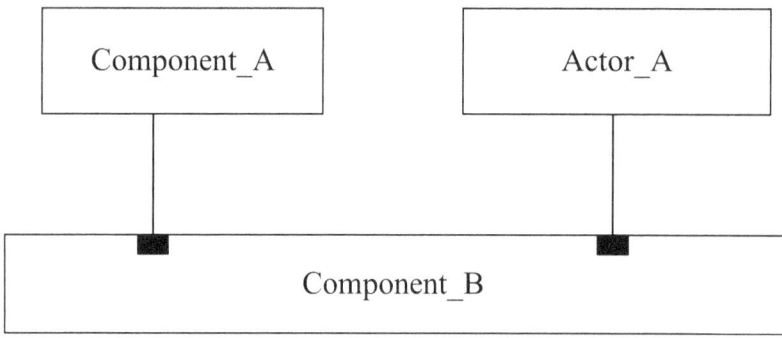

Figure 9-3   Operation User is Either a Component Or an Actor

Within a connection the subsystem (either a component or an actor) using the operation is always entitled the *Client* and the component which provides the operation is always entitled the *Server* as Figure 9-4 shows.

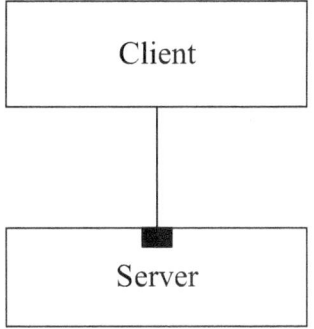

Figure 9-4   Roles of Client and Server Within a Connection

9-2 Drawing the Component Connection Diagram

A component connection diagram (CCD) is utilized to describe how all components and actors (in the external environment) are connected within a system. Figure 9-5 exhibits a CCD of the *Multi-Tier Personal Data System.*

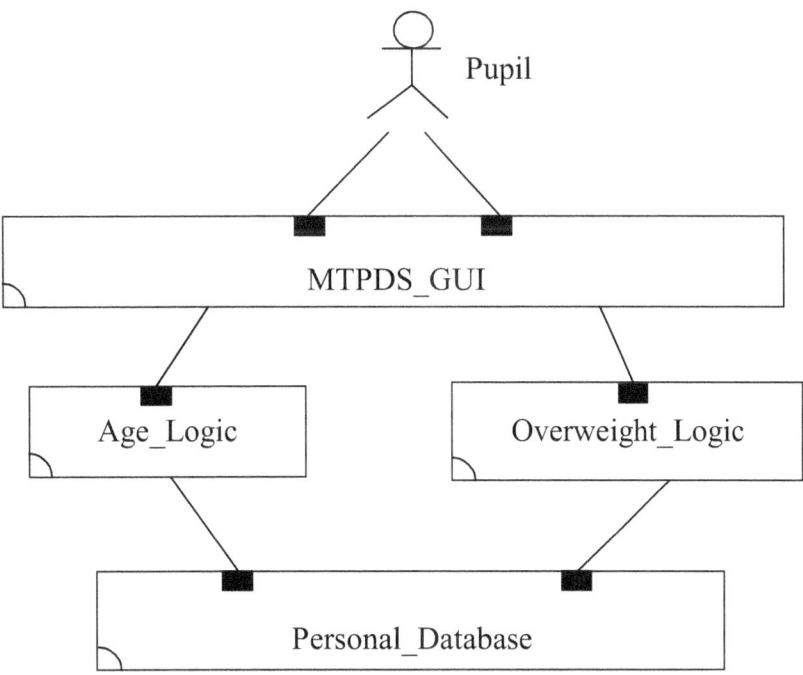

Figure 9-5   CCD of the *Multi-Tier Personal Data System*

In Figure 9-5, actor *Pupil* has two connections with the *MTPDS_GUI* component; component *MTPDS_GUI* has one connection with each of the *Age_Logic* and *Overweight_Logic* components; component *Age_Logic* has a connection with the *Personal_Database* component; component *Overweight_Logic* has a connection with the *Personal_Database* component.

After finishing the CCD, the formation pattern of *Multi-Tier Personal Data System* will be constructed; thus the systems structure of *Multi-Tier Personal Data System* becomes more transparent.

Chapter 10: Structure-Behavior Coalescence Diagram

General systems theory 2.0 (general architectural theory) uses a structure-behavior coalescence diagram (SBCD) to define the systems structure and systems behavior coexisting in a system. SBCD is the fifth fundamental diagram to achieve structure-behavior coalescence.

10-1 Purpose of Structure-Behavior Coalescence Diagram

The major aim of SBC architecture description language is to achieve the integration of systems structure and systems behavior within a system. SBCD enables us to observe the systems structure and systems behavior coexisting in a system. This is the purpose of utilizing SBCD when defining a system.

Figure 10-1 exhibits a SBCD of the *Multi-Tier Personal Data System*. In this example, interactions among the *Pupil* actor and the *MTPDS_GUI*, *Age_Logic*, *Overweight_Logic*, *Personal_Database* components shall draw forth the *AgeCalculation* and *OverweightCalculation* behaviors.

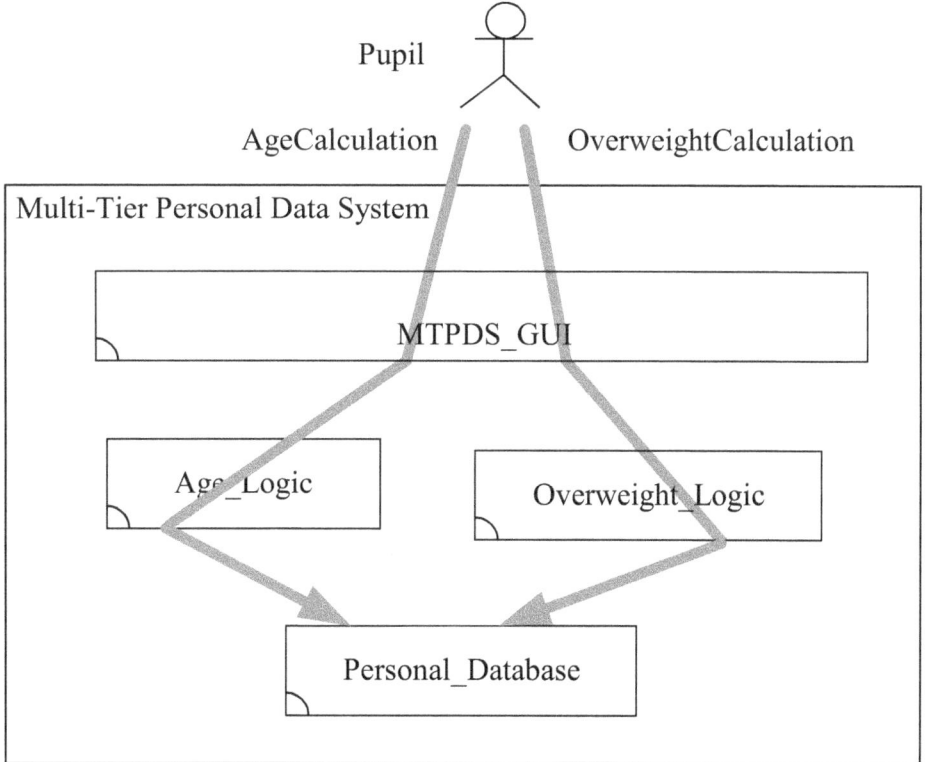

Figure 10-1   SBCD of the *Multi-Tier Personal Data System*

The overall behavior of a system is the aggregation of all its individual behaviors. All individual behaviors are mutually independent of each other. They tend to be executed concurrently [Hoar85, Miln89, Miln99]. For example, the overall behavior of *Multi-Tier Personal Data System* includes the *AgeCalculation* and *OverweightCalculation* behaviors. In other words, the *AgeCalculation* and *OverweightCalculation* behaviors are combined to produce the overall behavior of *Multi-Tier Personal Data System*.

The major purpose of using the SBC architecture description language is to achieve the integration of systems structure and systems behavior within a system. In Figure 10-1, we are able to define the systems structure and systems behavior coexisting in a SBCD. That is, in the *Multi-Tier Personal Data System*'s SBCD, we not only see its systems structure but also see (at the same time) its systems behavior.

10-2 Drawing the Structure-Behavior Coalescence Diagram

Let us now explain the usage of SBCD by constructing a SBCD step by step. The goal of having a SBCD is enabling us to see both the structure and behavior, simultaneously. In order to achieve this goal, a SBCD is drawn by first constructing all of the components, then describing the external environment's actors, and finally describing the interactions among these components and the external environment's actors.

For example, *Multi-Tier Personal Data System* has two behaviors: *AgeCalculation* and *OverweightCalculation*. After constructing the *Multi-Tier Personal Data System* with all its components, the external environment's actors and the *AgeCalculation* behavior, we obtain the graphical representation as shown in Figure 10-2. In this Figure, the *AgeCalculation* behavior indicates that actor *Pupil* interacts with the *MTPDS_GUI* component first, then component *MTPDS_GUI* interacts with the *Age_Logic* component later, then component *Age_Logic* interacts with the *Personal_Database* component finally.

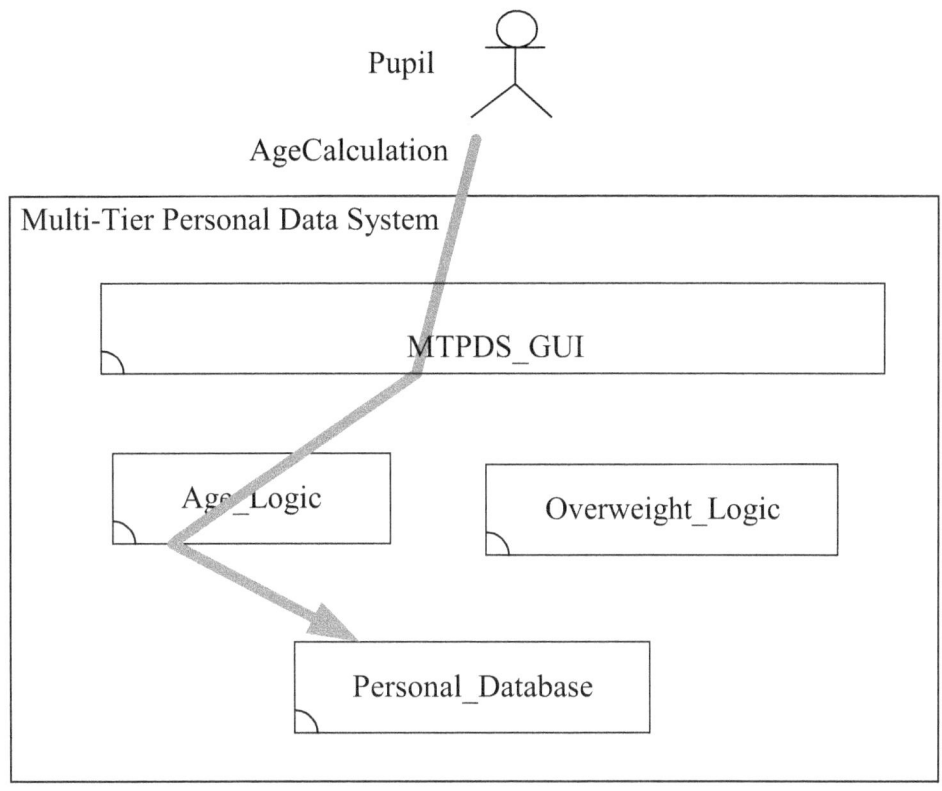

Figure 10-2   All Components, Actors, and the *AgeCalculation* Behavior

Adding the *OverweightCalculation* behavior to Figure 10-2, we then obtain the graphical representation shown in Figure 10-3. In this Figure, the *OverweightCalculation* behavior indicates that actor *Pupil* interacts with the *MTPDS_GUI* component first, then component *MTPDS_GUI* interacts with the *Overweight_Logic* component later, then component *Overweight_Logic* interacts with the *Personal_Database* component finally.

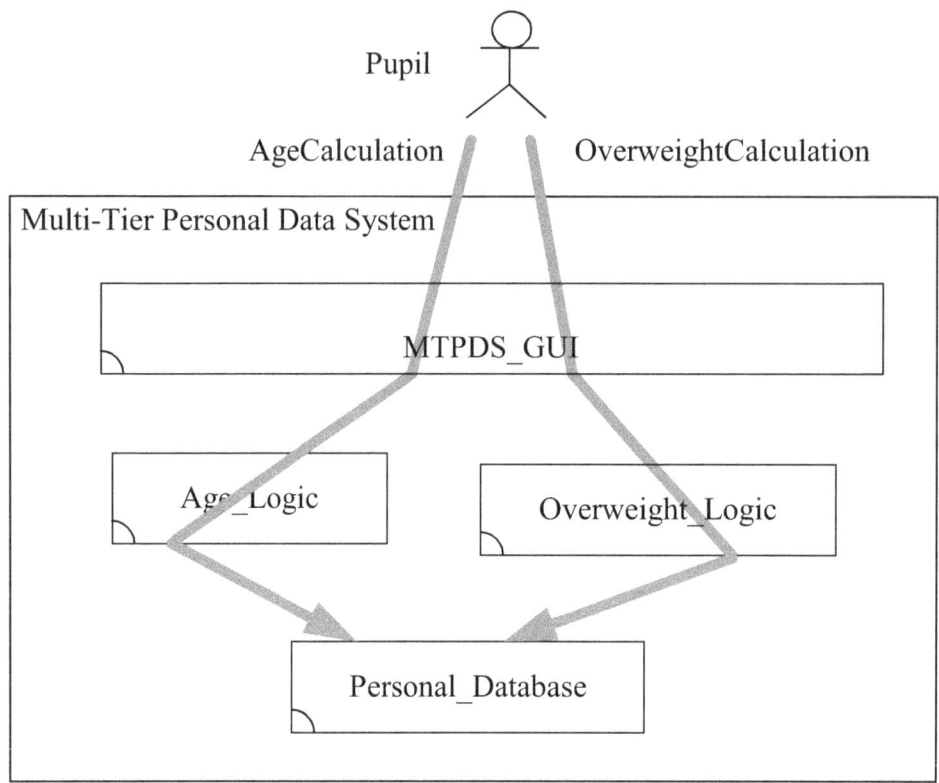

Figure 10-3   Adding the *OverweightCalculation* Behavior to Figure 10-2

After finishing Figure 10-3, we actually have accomplished all the works needed to draw the entire SBCD of the *Multi-Tier Personal Data System*. As a matter of fact, Figure 10-3 shows exactly the *Multi-Tier Personal Data System's* SBCD.

Chapter 11: Interaction Flow Diagram

General systems theory 2.0 (general architectural theory) uses an interaction flow diagram (IFD) to define each individual behavior of the overall behavior of a system. IFD is the sixth fundamental diagram to achieve structure-behavior coalescence.

11-1 Individual Behavior Represented by Interaction Flow Diagram

The overall behavior of a system consists of many individual behaviors. Each individual behavior represents an execution path. An IFD is used to define such an individual behavior.

Figure 11-1 demonstrates that the *Robot* system has two behaviors; thus, it has two IFDs.

System	IFD
Robot	Writing
	Walking

Figure 11-1   *Robot* System has Two IFDs

Figure 11-2 demonstrates that the *Purchasing Management* System has five behaviors; thus, it has five IFDs.

110

System	IFD
Purchasing Management	Purchase_Requisition
	Quotation
	Purchase_Order
	Purchase
	Collect_Supplier_Data

Figure 11-2   *Purchasing Management* System has Five IFDs

11-2 Drawing the Interaction Flow Diagram

Let us now explain the usage of interaction flow diagram (IFD) by drawing an IFD step by step. Figure 11-3 demonstrates an IFD of the *AgeCalculation* behavior. The X-axis direction is from the left side to right side and the Y-axis direction is from the above to the below. Inside an IFD, there are four elements: a) external environment's actor, b) components, c) interactions and d) input/output parameters. Participants of the interaction, such as the external environment's actor and each component, are laid aside along the X-axis direction on the top of the diagram. The external environment's actor which initiates the sequential interactions is always placed on the most left side of the X-axis. Then, interactions among the external environment's actor and components successively in turn decorate along the Y-axis direction. The first interaction is placed on the top of the Y-axis position. The last interaction is placed on the bottom of the Y-axis position. Each interaction may carry several input and/or output parameters.

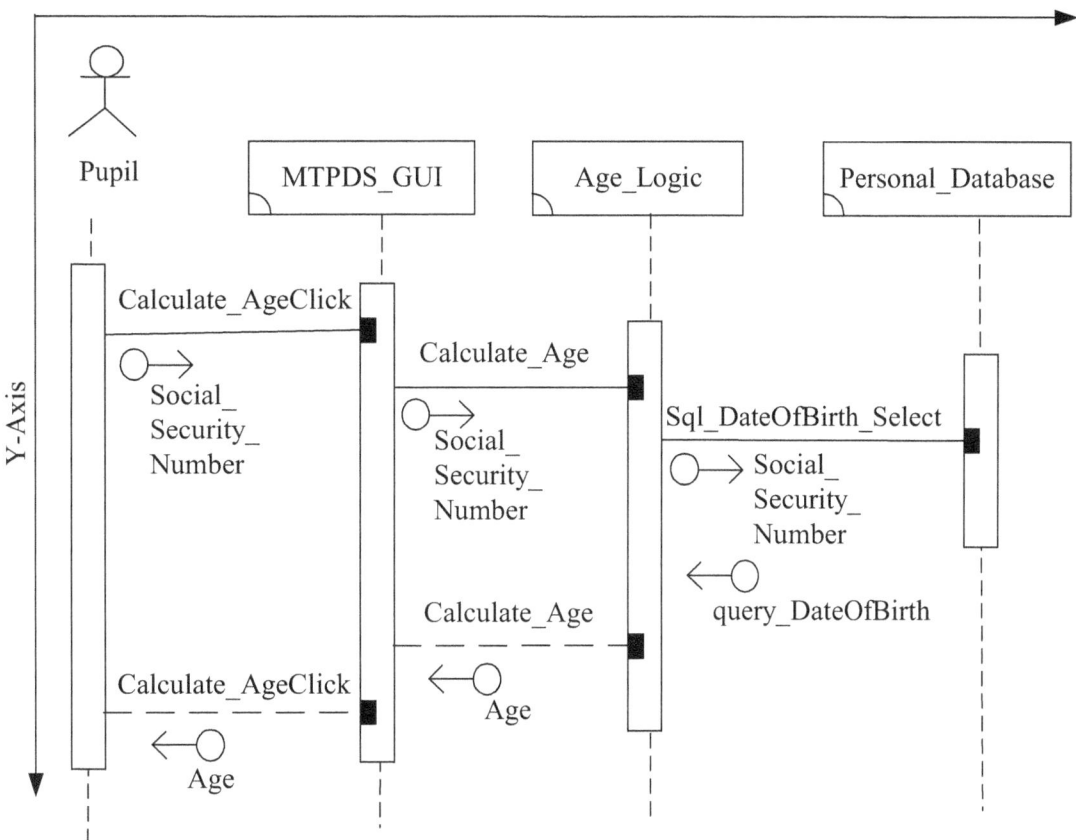

Figure 11-3   IFD of the *AgeCalculation* Behavior

In Figure 11-3, *Pupil* is an external environment's actor. *MTPDS_GUI*, *Age_Logic* and *Personal_Database* are components. *Calculate_AgeClick* is an operation, carrying the *Social_Security_Number* input parameter and *Age* output parameter, which is provided by the *MTPDS_GUI* component. *Calculate_Age* is an operation, carrying the *Social_Security_Number* input parameter and *Age* output parameter, which is provided by the *Age_Logic* component. *Sql_DateOfBirth_Select* is an operation, carrying the *Social_Security_Number* input parameter and *query_DateOfBirth* output parameter, which is provided by the *Personal_Database* component.

The execution path of Figure 11-3 is as follows. First, actor *Pupil* interacts with the *MTPDS_GUI* component through the *Calculate_AgeClick* operation call interaction, carrying the *Social_Security_Number* input parameter. Next, component *MTPDS_GUI* interacts with the *Age_Logic* component through the *Calculate_Age* operation call interaction, carrying the *Social_Security_Number* input parameter. Continuingly, component *Age_Logic* interacts with the *Personal_Database* component through the *Sql_DateOfBirth_Select* operation call interaction, carrying the *Social_Security_Number* input parameter and the *query_DateOfBirth* output parameter. Repeatedly, component *MTPDS_GUI* interacts with the *Age_Logic* component through the *Calculate_Age* operation return interaction, carrying the *Age* output parameter. Finally, actor *Pupil* interacts with the *MTPDS_GUI* component through the *Calculate_AgeClick* operation return interaction, carrying the *Age* output parameter.

For each interaction, the solid line stands for operation call while the dashed line stands for operation return. The operation call and operation return interactions, if using the same operation name, belong to the identical operation. Figure 11-4 exhibits two interactions (operation call interaction and operation return interaction) having the identical "*Request*" operation.

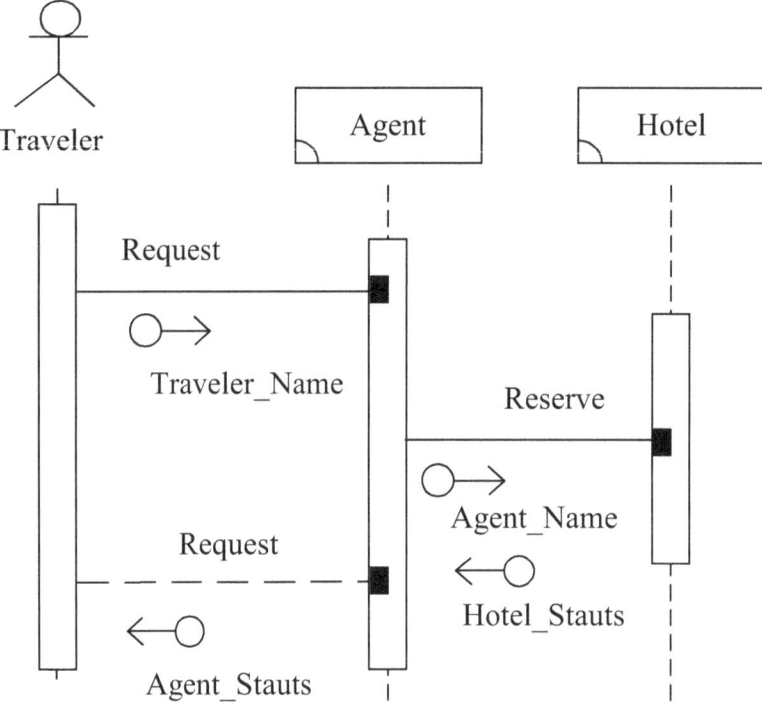

Figure 11-4　　Two Interactions Have the Identical Operation

The execution path of Figure 11-4 is as follows. First, external environment's actor *Traveler* interacts with the *Agent* component through the *Request* operation call interaction, carrying the *Traveler_Name* input parameter. Next, component *Agent* interacts with the *Hotel* component through the *Reserve* operation call interaction, carrying the *Agent_Name* input parameter and *Hotel_Stauts* output parameter. Finally, external environment's actor *Traveler* interacts with the *Agent* component through the *Request* operation return interaction, carrying the *Agent_Stauts* output parameter.

An interaction flow diagram may contain a conditional expression. Figure 11-5 shows such an example which has the following execution path. First, external environment's actor *Employee* interacts with the *Computer* component through the *Open* operation call interaction, carrying the *Task_No* input parameter. Next, if the *var_1 < 4 & var_2 > 7* condition is true then component *Computer* shall interact with the *Skype* component through the *Op_1* operation call interaction and component *Skype* shall interact with the *Earphone* component through the *Op_4* operation call interaction, carrying the *Skype_Earphone* output parameter; else if the *var_3 = 99* condition is true then component *Computer* shall interact with the *Skype* component through the *Op_2* operation call interaction and component *Skype* shall interact with the *Speaker* component through the *Op_5* operation call interaction, carrying the *Skype_Speaker* output parameter; else component *Computer* shall interact with the *Youtube* component through the *Op_3* operation call interaction and component *Youtube* shall interact with the *Speaker* component through the *Op_6* operation call interaction, carrying the *Youtube_Speaker* output parameter. Continuingly, if the *var_1 < 4 & var_2 > 7* condition is true then component *Computer* shall interact with the *Skype* component through the *Op_1* operation return interaction, carrying the *Status_1* output parameter; else if the *var_3 = 99* condition is true then component *Computer* shall interact with the *Skype* component through the *Op_2* operation return interaction, carrying the *Status_2* output parameter; else component *Computer* shall interact with the *Youtube* component through the *Op_3* operation return interaction, carrying the *Status_3* output parameter. Finally, external environment's actor *Employee* interacts with the *Computer* component through the *Open* operation return interaction, carrying the *Status* output parameter.

114

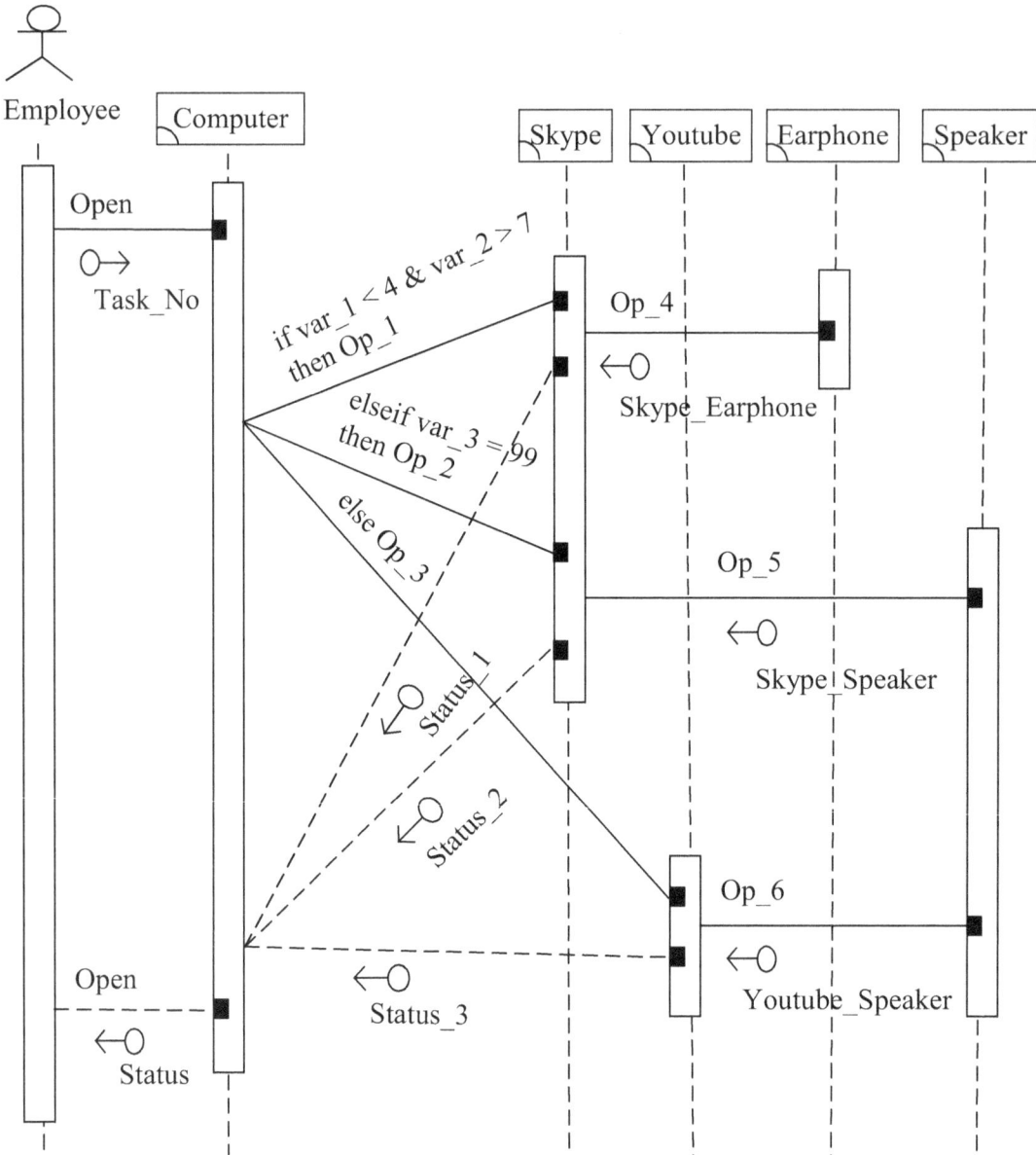

Figure 11-5   Conditional Expression

Several Boolean conditions are shown in Figure 11-5. They are "*var_1 < 4 & var_2 > 7*" and "*var_3 = 99*". Variables, such as *var_1*, *var_2* and *var_3*, appearing in the Boolean condition can be local or global variables [Prat00, Seth96].

PART III: MULTI-LEVEL VIEW

Chapter 12: Concept View

In the SBC view model, multi-level view contains the concept, analysis, design and implementation views. This chapter demonstrates how the general systems theory 2.0 (general architectural theory) defining the concept view.

12-1 Principle of Concept View

Concept view corresponds to an executive summary for an administrator who wants an overview or estimate of the scope of the system, what it would cost, and how it would relate to the general environment in which it will operate.

According to the SBC multi-level (hierarchical) view, systems architects shall construct the concept's systems architecture for the administrator to view. This concept's systems architecture is called the concept view as shown in Figure 12-1.

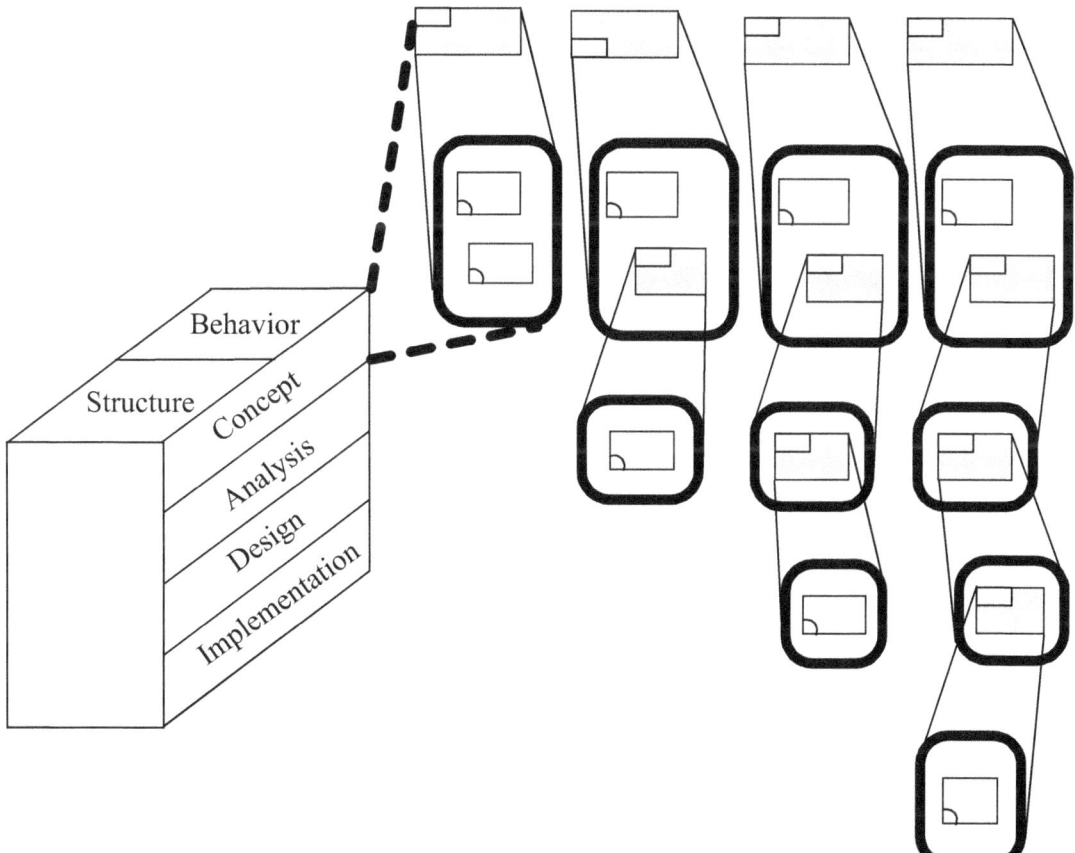

Figure 12-1   Concept View

The concept view consists of: a) concept's systems structure and b) concept's systems behavior.

12-2 Concept's Systems Structure

The entire SBC concept's systems structure includes: a) *Concept's AHD*, b) *Concept's FD*, c) *Concept's COD* and d) *Concept's CCD*, as shown in Figure 12-2.

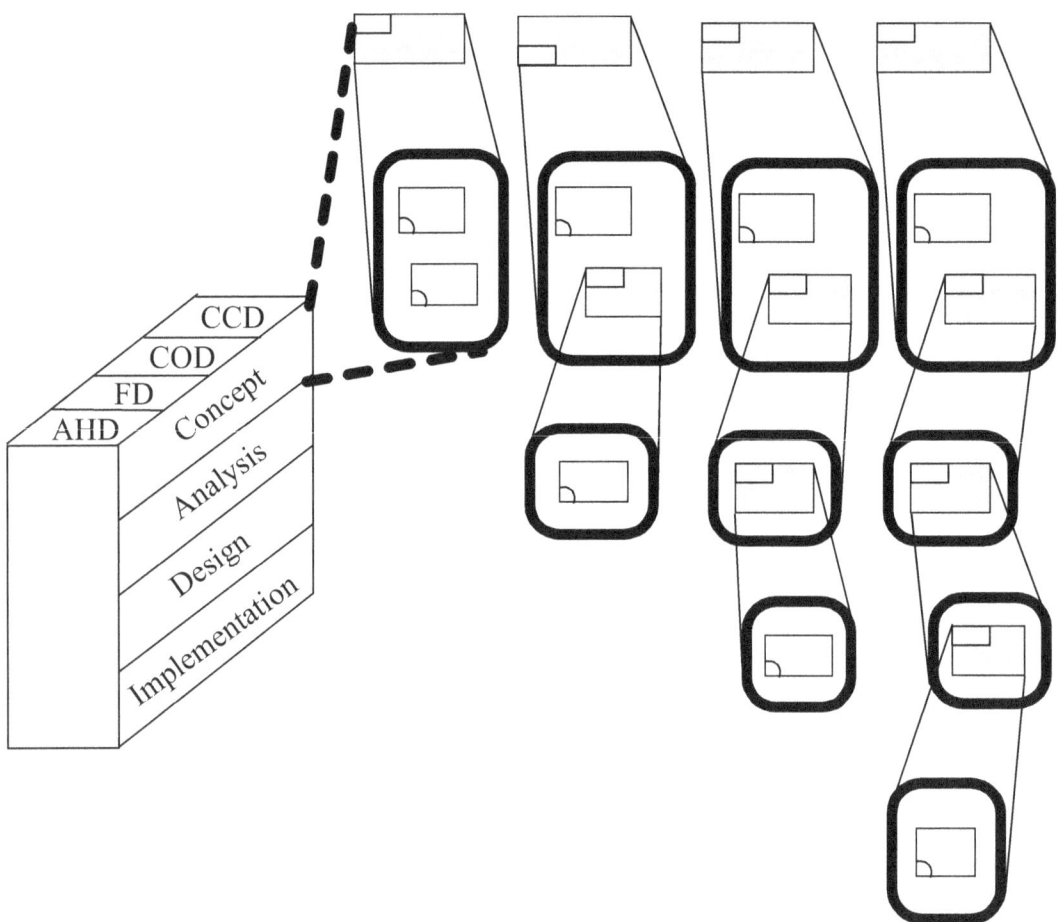

Figure 12-2   SBC Concept's Systems Structure

Another drawing can also be used to illustrate the SBC concept's systems structure as shown in Figure 12-3.

Systems Structure			
Architecture Hierarchy Diagram	Framework Diagram	Component Operation Diagram	Component Connection Diagram

Concept	Concept's AHD	Concept's FD	Concept's COD	Concept's CCD

Figure 12-3   SBC Concept's Systems Structure

12-2-1 Concept's AHD

Concept's AHD is the architecture hierarchy diagram we obtain after the concept phase is finished. Figure 12-4 shows the concept's AHD of the *QQQ* system. In the figure, *QQQ* is composed of *B1*, *B2*, *A1*, *D1*, *D2* and *T1*.

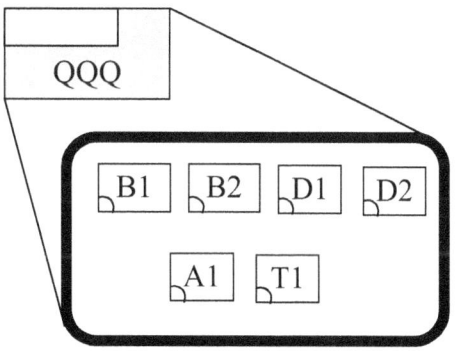

Figure 12-4   Concept's AHD of the *QQQ* system

In Figure 12-4, *QQQ* is an aggregated system while *B1*, *B2*, *A1*, *D1*, *D2* and *T1* are non-aggregated systems.

12-2-2 Concept's FD

Concept's FD is the framework diagram we obtain after the concept phase is finished. Figure 12-5 shows the concept's FD of the *QQQ* system. In the figure, *Business_Layer* contains the *B1* and *B2* components; *Application_Layer* contains the *A1* component; *Data_Layer* contains the *D1* and *D2* components; *Technology_Layer* contains the *T1* component.

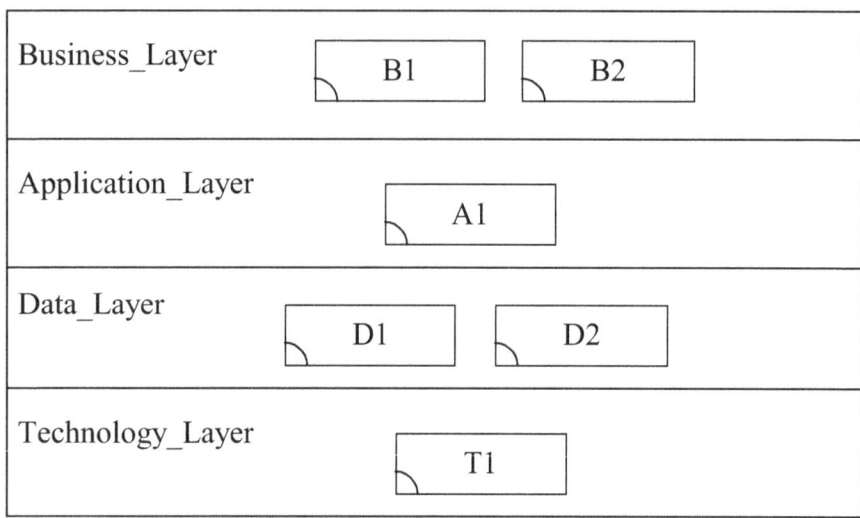

Figure 12-5   Concept's FD of the *QQQ* System

12-2-3 Concept's COD

Concept's COD is the component operation diagram we obtain after the concept phase is finished. Figure 12-6 shows the concept's COD of the *QQQ* system. In the figure, component *B1* has two operations: *op_01* and *op_02*; component *B2* has one operation: *op_03*; component *A1* has three operations: *op_04*, *op_05* and *op_06*; component *D1* has two operations: *op_07* and *op_08*; component *D2* has one operation: *op_9*; component *T1* has two operations: *op_10* and *op_11*.

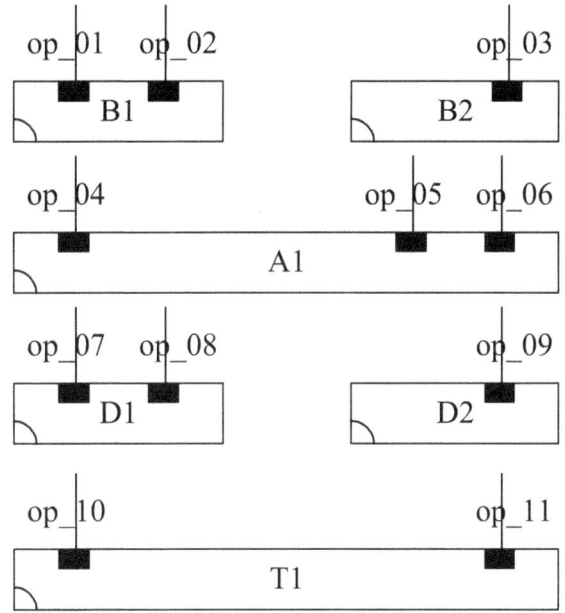

Figure 12-6   Concept's COD of the *QQQ* System

12-2-4 Concept's CCD

Concept's CCD is the component connection diagram we obtain after the concept phase is finished. Figure 12-7 shows the concept's CCD of the *QQQ* system.

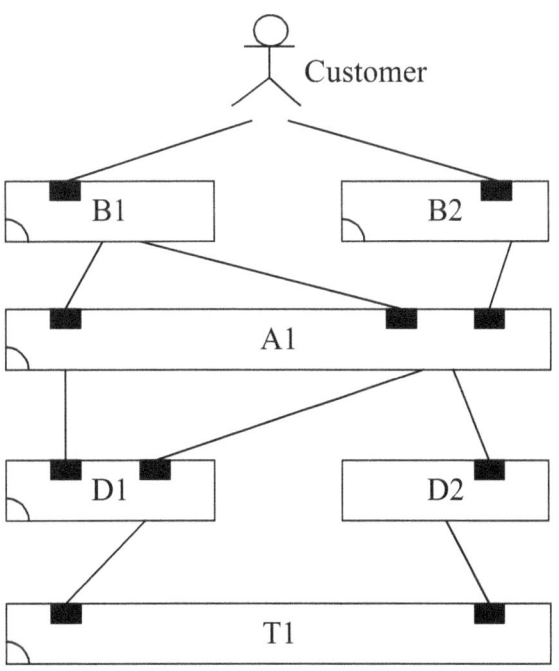

Figure 12-7   Concept's CCD of the *QQQ* System

In Figure 12-7, we see that actor *Customer* has a connection with each of the *B1* and *B2* components; component *B1* has two connections with the *A1* component; component *B2* has a connection with the *A1* component; component *A1* has two connections with the *D1* component; component *A1* has a connection with the *D2* components; component *D1* has a connection with the *T1* component; component *D2* has a connection with the *T1* component.

12-3 Concept's Systems Behavior

The entire SBC concept's systems behavior includes: a) *Concept's SBCD* and b) *Concept's IFD* as shown in Figure 12-8.

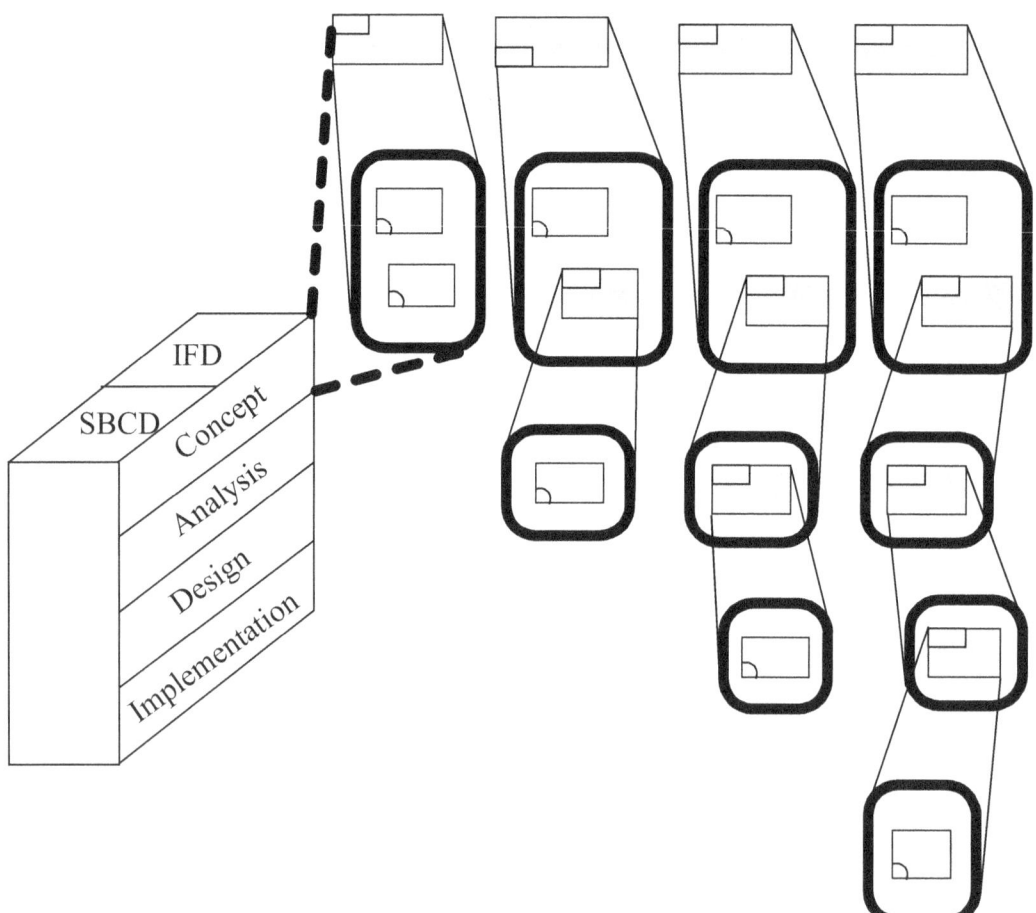

Figure 12-8   SBC Concept's Systems Behavior

Let us use another drawing to illustrate the SBC concept's systems behavior as shown in Figure 12-9.

	Systems Behavior	
	Structure-Behavior Coalescence Diagram	Interaction Flow Diagram
Concept	Concept's SBCD	Concept's IFD

Figure 12-9   SBC Concept's Systems Behavior

12-3-1 Concept's SBCD

Concept's SBCD is the structure-behavior coalescence diagram we obtain after the concept phase is finished. Figure 12-10 shows the concept's SBCD of the *QQQ* system. In this example, an actor interacting with six components shall describe the overall concept's systems behavior. Interactions among the *Customer* actor and the *B1*, *A1*, *D1* components draw forth the *qqq_1* behavior. Interactions among the *Customer* actor and the *B1*, *A1*, *D1*, *T1* components draw forth the *qqq_2* behavior. Interactions among the *Customer* actor and the *B2*, *A1*, *D2*, *T1* components draw forth the *qqq_3* behavior.

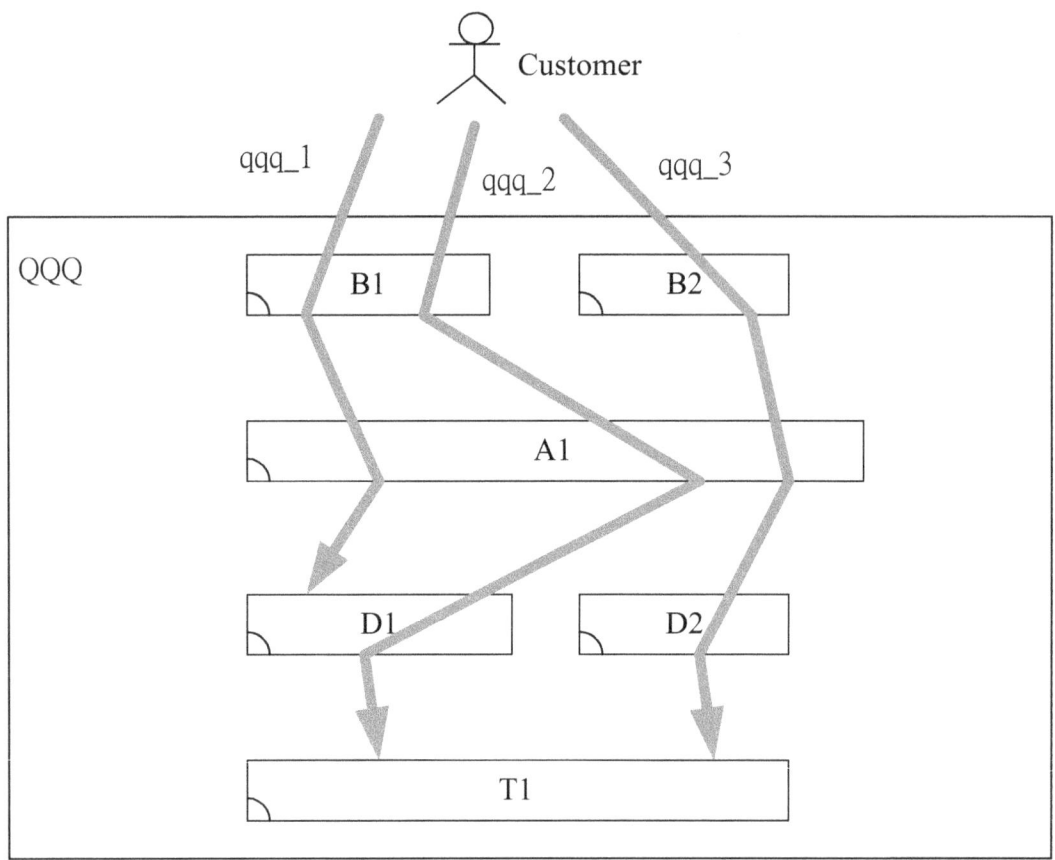

Figure 12-10   Concept's SBCD of the *QQQ* System

The overall behavior of a system is a collection of all its individual behaviors. All individual behaviors are mutually independent of each other. They tend to be executed concurrently [Hoar85, Miln89, Miln99]. For example, the overall concept's behavior of the *QQQ* system includes the *qqq_1*, *qqq_2* and *qqq_3* behaviors. In other words, the behaviors *qqq_1*, *qqq_2* and *qqq_3* are combined to produce the overall concept's behavior of the *QQQ* system.

The major purpose of using the architectural approach, instead of separating the structure model from the behavior model, is to achieve one single coalesced model. In Figure 12-10, systems architects are able to see that the systems structure and behavior coexist in the concept's SBCD. That is, in the concept's SBCD of the *QQQ* system, systems architects not only see its systems structure but also see (at the same time) its systems behavior.

12-3-2 Concept's IFD

Concept's IFDs are the interaction flow diagrams we obtain after the concept phase is finished. The overall concept's behavior of the *QQQ* system includes three behaviors: *qqq_1*, *qqq_2* and *qqq_3*. Each of them is described by an individual IFD. Figure 12-11 shows the concept's IFD of the *qqq_1* behavior. First, actor *Customer* interacts with the *B1* component through the *op_01* operation call interaction. Next, component *B1* interacts with the *A1* component through the *op_04* operation call interaction. Finally, component *A1* interacts with the *D1* component through the *op_07* operation call interaction.

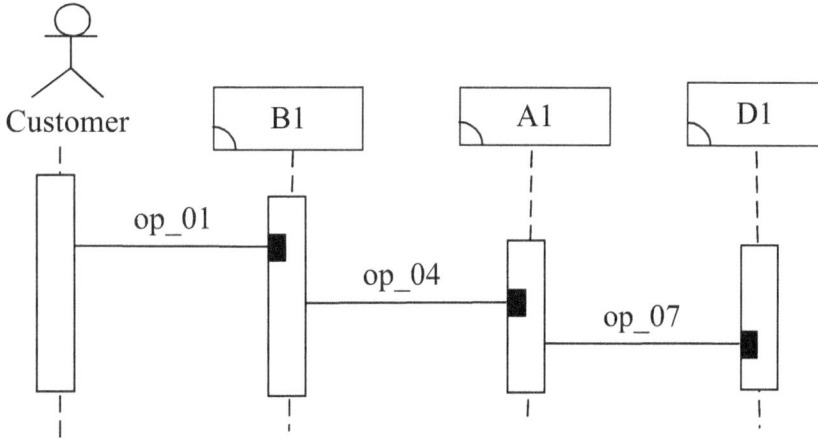

Figure 12-11   Concept's IFD of the *qqq_1 Behavior*

Figure 12-12 shows the concept's IFD of the *qqq_2* behavior. First, actor *Customer* interacts with the *B1* component through the *op_02* operation call interaction. Next, component *B1* interacts with the *A1* component through the *op_05* operation call interaction. Continuingly, component *A1* interacts with the *D1* component through the *op_08* operation call interaction. Finally, component *D1* interacts with the *T1* component through the *op_10* operation call interaction.

126

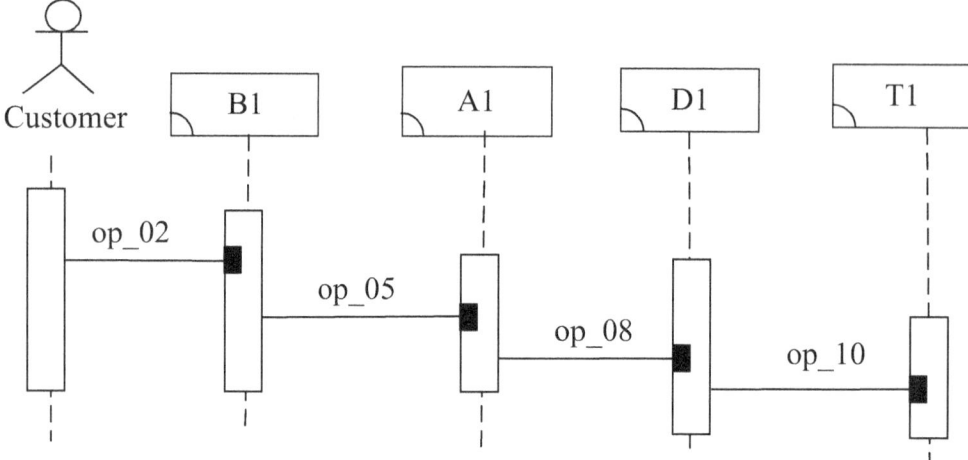

Figure 12-12   Concept's IFD of the *qqq_2* Behavior

Figure 12-13 shows the concept's IFD of the *qqq_3* behavior. First, actor *Customer* interacts with the *B2* component through the *op_03* operation call interaction. Next, component *B2* interacts with the *A1* component through the *op_06* operation call interaction. Continuingly, component *A1* interacts with the *D2* component through the *op_09* operation call interaction. Finally, component *D2* interacts with the *T1* component through the *op_11* operation call interaction.

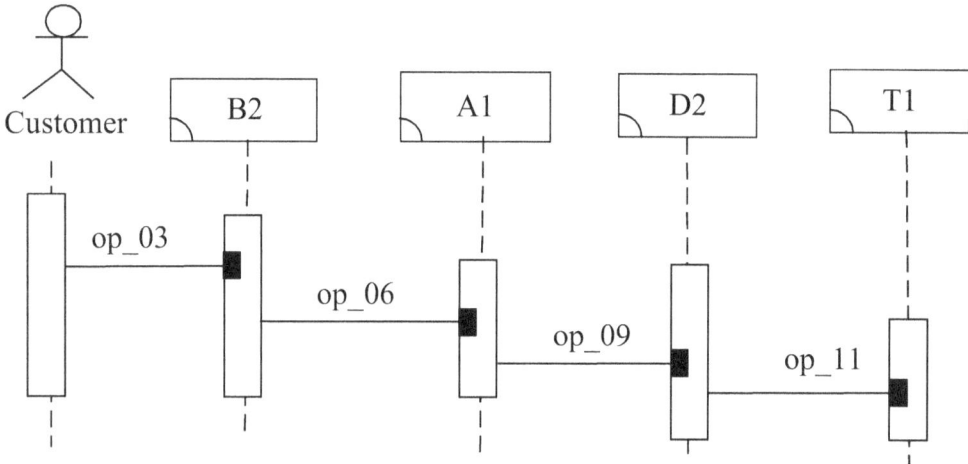

Figure 12-13   Concept's IFD of the *qqq_3* Behavior

Chapter 13: Analysis View

In the SBC view model, multi-level view contains the concept, analysis, design and implementation views. This chapter demonstrates how the general systems theory 2.0 (general architectural theory) defining the analysis view.

13-1 Principle of Analysis View

Analysis view corresponds to a summary for an analyzer who works on the analysis of a system. The analysis is mainly to find out what the system is. When working on the analysis, we only ask what this system is about, but may not provide sufficient focus on how the system is actually designed.

Analysis view is one level down structural decomposition (with observation congruence verification) of the concept view [Chao15a, Chao15b, Chao15c, Chao15d, Chao15e]. That is, we shall not create the analysis view from the scratch. Instead, we will construct the analysis view by decomposing the concept view.

According to the SBC multi-level (hierarchical) view, systems architects shall construct the analysis' systems architecture for the analyzer to view. This analysis' systems architecture is called the analysis view as shown in Figure 13-1.

128

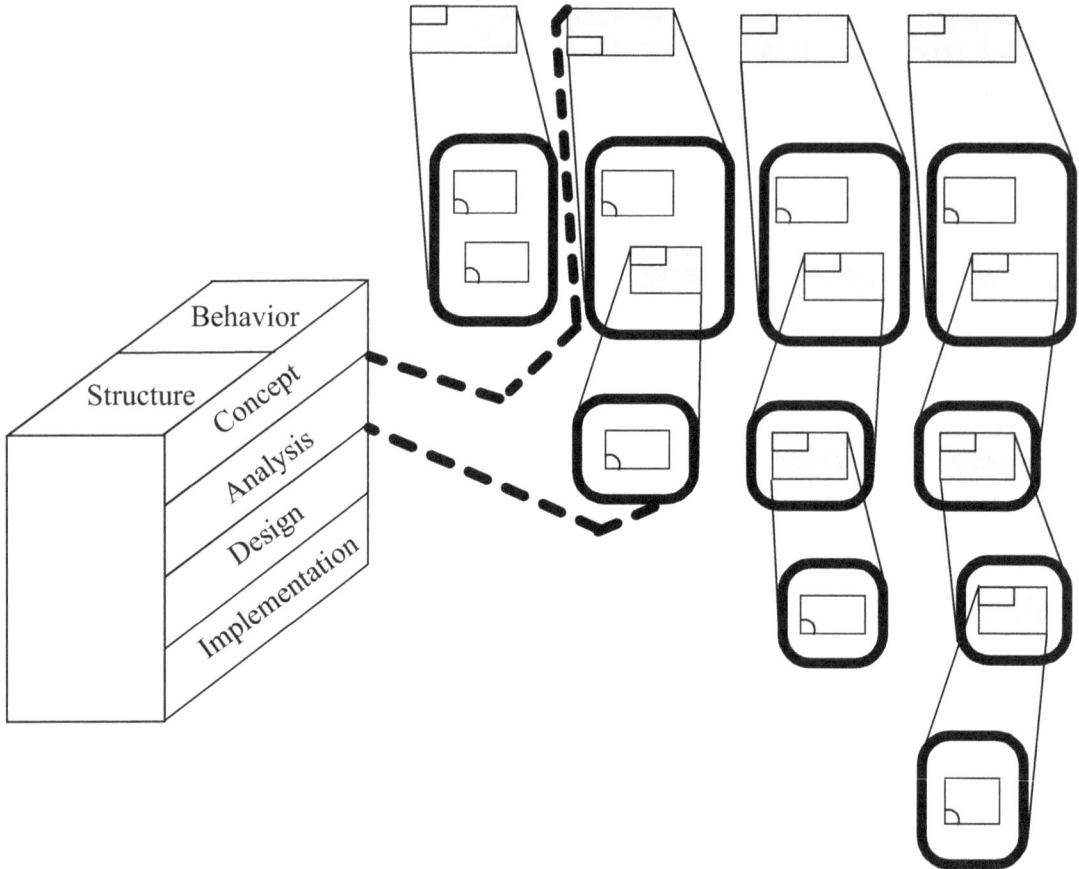

Figure 13-1   Analysis View

The analysis view consists of: a) analysis' systems structure and b) analysis' systems behavior.

13-2 Analysis' Systems Structure

The entire SBC analysis' systems structure includes: a) *Analysis' AHD*, b) *Analysis' FD*, c) *Analysis' COD* and d) *Analysis' CCD*, as shown in Figure 13-2.

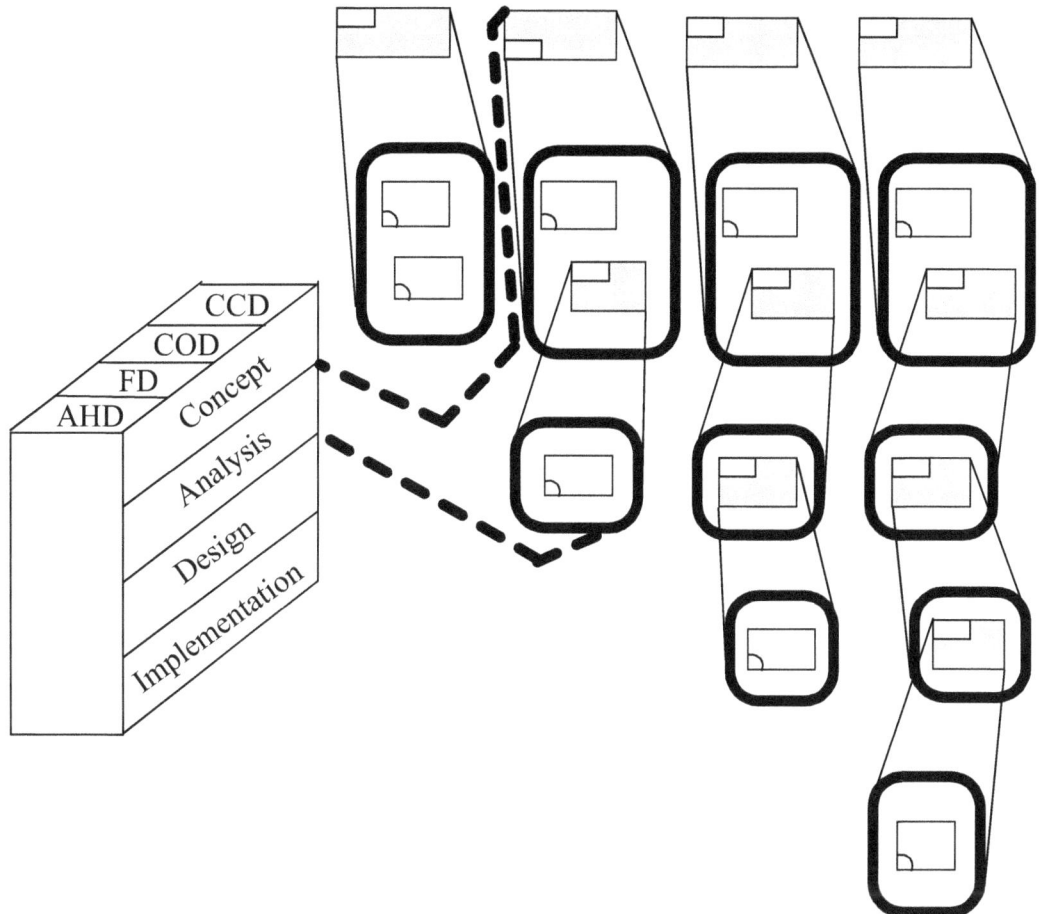

Figure 13-2   SBC Analysis' Systems Structure

Another drawing can also be used to illustrate the SBC analysis' systems structure as shown in Figure 13-3.

Systems Structure			
Architecture Hierarchy Diagram	Framework Diagram	Component Operation Diagram	Component Connection Diagram
Analysis	Analysis' AHD	Analysis' FD	Analysis' COD	Analysis' CCD

Figure 13-3   SBC Analysis' Systems Structure

13-2-1 Analysis' AHD

Analysis' AHD is the architecture hierarchy diagram we obtain after the analysis phase is finished. Figure 13-4 shows the analysis' AHD of the *QQQ* system. In the figure, *QQQ* is composed of *B1*, *B2*, *A1*, *D1*, *D2* and *T1*; *A1* is composed of *A11*; *T1* is composed of *T11* and *T12*.

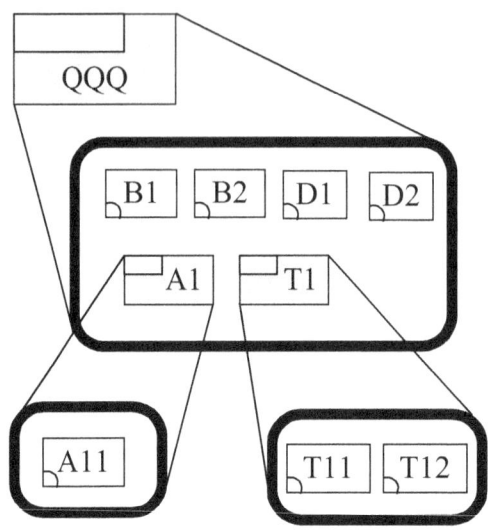

Figure 13-4   Analysis' AHD of the *QQQ* System

In Figure 13-4, *QQQ*, *A1* and *T1* are aggregated systems while *B1*, *B2*, *A11*, *D1*, *D2*, *T11* and *T12* are non-aggregated systems.

We validate our claim that the analysis view is one level down structural decomposition of the concept view by comparing Figure 12-4 with Figure 13-4. In Figure 12-4, *A1* and *T1* are non-aggregated systems. As an interesting contrast, in Figure 13-4, *A1* becomes an aggregated system and is composed of *A11*; *T1* also becomes an aggregated system and is composed of *T11* and *T12*.

13-2-2 Analysis' FD

Analysis' FD is the framework diagram we obtain after the analysis phase is finished. Figure 13-5 shows the analysis' FD of the *QQQ* system. In the figure, *Business_Layer* contains the *B1* and *B2* components; *Application_Layer* contains the *A11* component; *Data_Layer* contains the *D1* and *D2* components; *Technology_Layer* contains the *T11* and *T12* components.

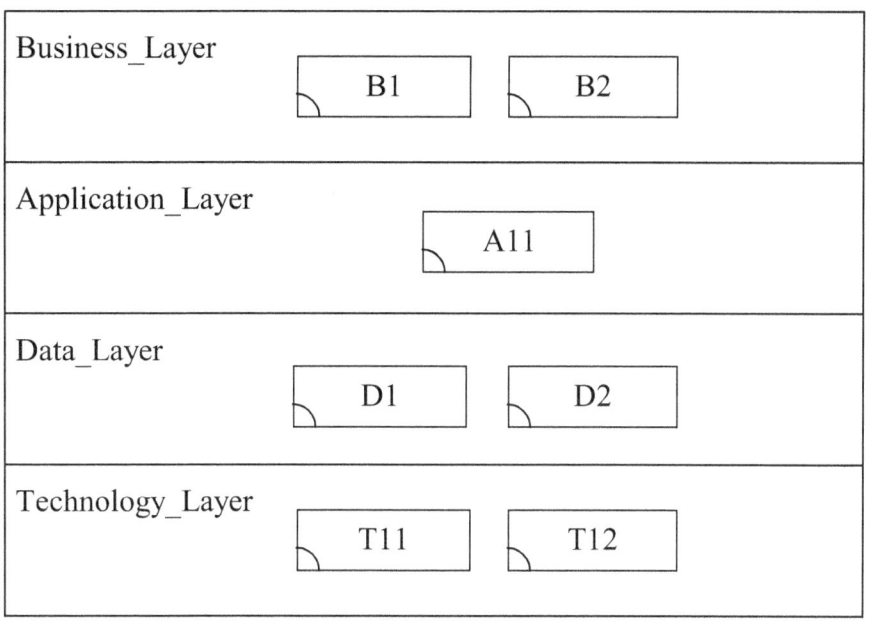

Figure 13-5   Analysis' FD of the *QQQ* System

13-2-3 Analysis' COD

Analysis' COD is the component operation diagram we obtain after the analysis phase is finished. Figure 13-6 shows the analysis' COD of the *QQQ* system. In the figure, component *B1* has two operations: *op_01* and *op_02*; component *B2* has one operation: *op_03*; component *A11* has three operations: *op_04*, *op_05 and op_06*; component *D1* has two operations: *op_07* and *op_08*; component *D2* has one operation: *op_9*; component *T11* has one operation: *op_10*; component *T12* has one operation: *op_11*.

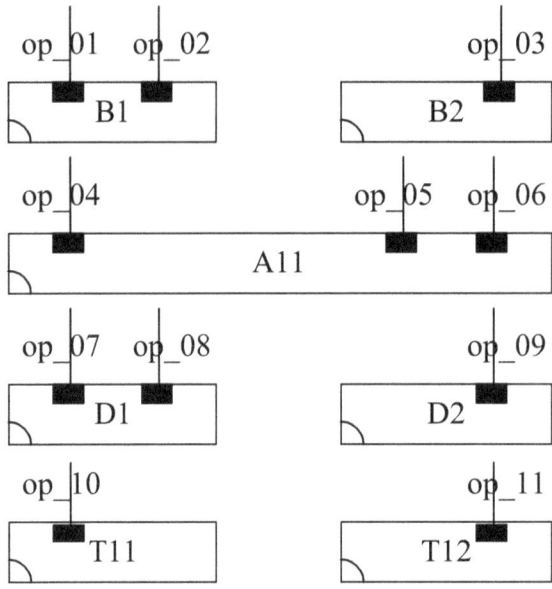

Figure 13-6   Analysis' COD of the *QQQ* System

13-2-4 Analysis' CCD

Analysis' CCD is the component connection diagram we obtain after the analysis phase is finished. Figure 13-7 shows the analysis' CCD of the *QQQ* system.

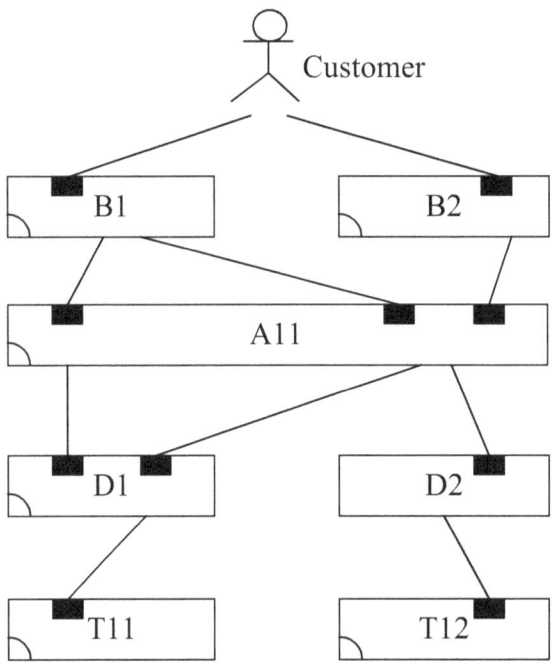

Figure 13-7   Analysis' CCD of the *QQQ* System

In Figure 13-7, we see that actor *Customer* has a connection with each one of the *B1* and *B2* components; component *B1* has two connections with the *A11* component; component *B2* has a connection with the *A11* component; component *A11* has two connections with the *D1* component; component *A11* has a connection with the *D2* component; component *D1* has a connection with the *T11* component; component *D2* has a connection with the *T12* component.

13-3 Analysis' Systems Behavior

The entire SBC analysis' systems behavior includes: a) *Analysis' SBCD* and *Analysis' IFD* as shown in Figure 13-8.

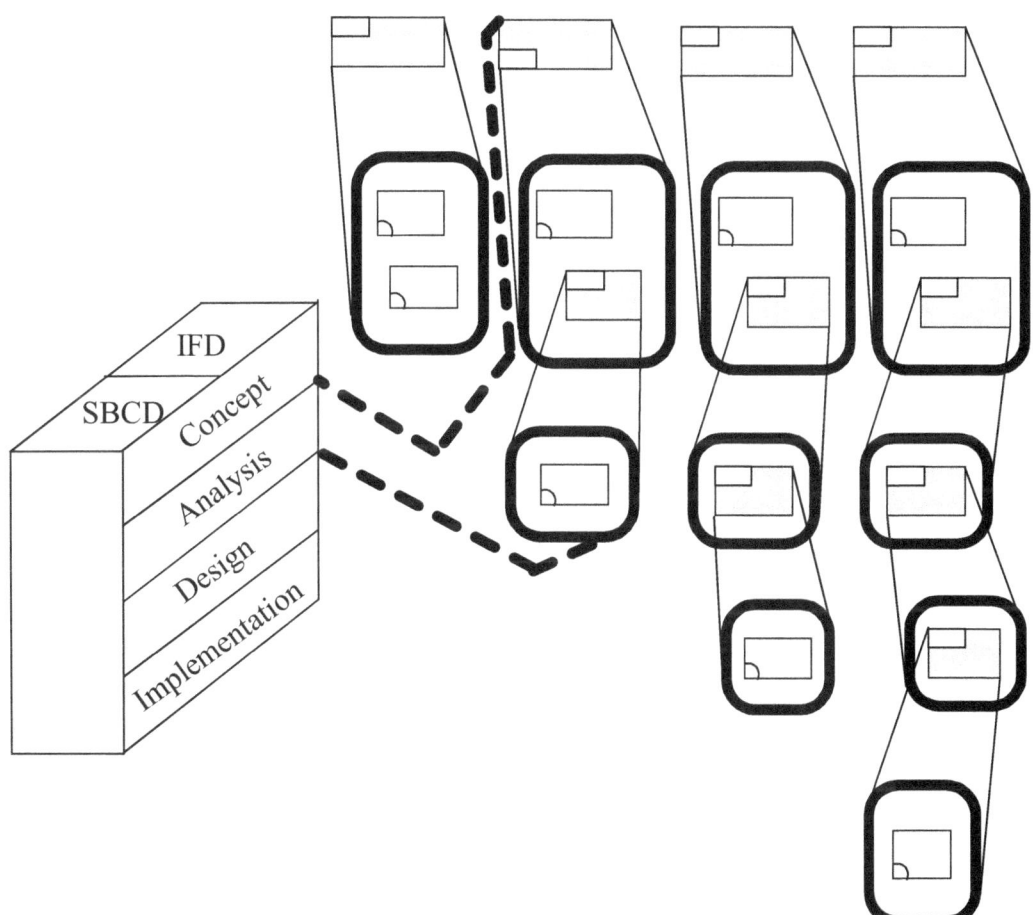

Figure 13-8   SBC Analysis' Systems Behavior

Let us use another drawing to illustrate the SBC analysis' systems behavior as shown in Figure 13-9.

Systems Behavior		
	Structure-Behavior Coalescence Diagram	Interaction Flow Diagram
Analysis	Analysis' SBCD	Analysis' IFD

Figure 13-9   SBC Analysis' Systems Behavior

13-3-1 Analysis' SBCD

Analysis' SBCD is the structure-behavior coalescence diagram we obtain after the analysis phase is finished. Figure 13-10 shows the analysis' SBCD of the *QQQ* system. In this example, an actor interacting with seven components shall describe the overall analysis' systems behavior. Interactions among the *Customer* actor and the *B1*, *A11*, *D1* components draw forth the *qqq_1* behavior. Interactions among the *Customer* actor and the *B1*, *A11*, *D1*, *T11* components draw forth the *qqq_2* behavior. Interactions among the *Customer* actor and the *B2*, *A11*, *D2*, *T12* components draw forth the *qqq_3* behavior.

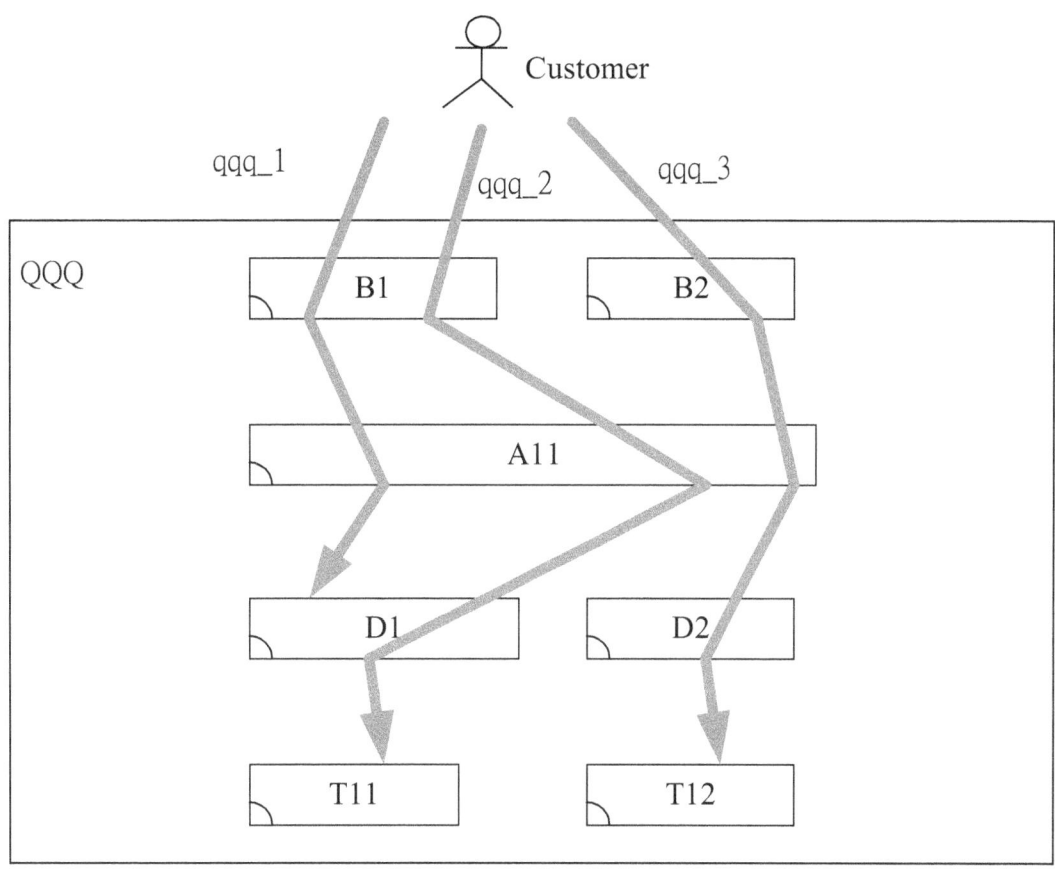

Figure 13-10   Analysis' SBCD of the *QQQ* System

The overall behavior of a system is a collection of all its individual behaviors. All individual behaviors are mutually independent of each other. They tend to be executed concurrently. For example, the overall analysis' behavior of the *QQQ* system includes the *qqq_1*, *qqq_2* and *qqq_3* behaviors. In other words, the *qqq_1*, *qqq_2* and *qqq_3* behaviors are combined to produce the overall analysis' behavior of the *QQQ* system.

The major purpose of using the architectural approach, instead of separating the structure model from the behavior model, is to achieve one single coalesced model. In Figure 13-10, systems architects are able to see that the systems structure and systems behavior coexist in the analysis' SBCD. That is, in the analysis' SBCD of the *QQQ* system, systems architects not only see its systems structure but also see (at the same time) its systems behavior.

13-3-2 Analysis' IFD

Analysis' IFDs are the interaction flow diagrams we obtain after the analysis phase is finished. The overall analysis' behavior of the *QQQ* system includes three behaviors: *qqq_1*, *qqq_2* and *qqq_3*. Each of them is described by an individual IFD. Figure 13-11 shows the analysis' IFD of the *qqq_1* behavior. First, actor *Customer* interacts with the *B1* component through the *op_01* operation call interaction. Next, component *B1* interacts with the *A11* component through the *op_04* operation call interaction. Finally, component *A11* interacts with the *D1* component through the *op_07* operation call interaction.

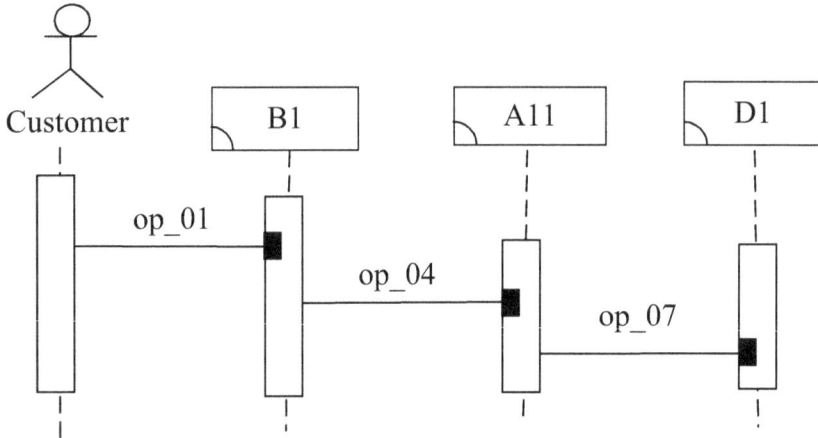

Figure 13-11   Analysis' IFD of the *qqq_1* Behavior

Figure 13-12 shows the analysis' IFD of the *qqq_2* behavior. First, actor *Customer* interacts with the *B1* component through the *op_02* operation call interaction. Next, component *B1* interacts with the *A11* component through the *op_05* operation call interaction. Continuingly, component *A11* interacts with the *D1* component through the *op_08* operation call interaction. Finally, component *D1* interacts with the *T11* component through the *op_10* operation call interaction.

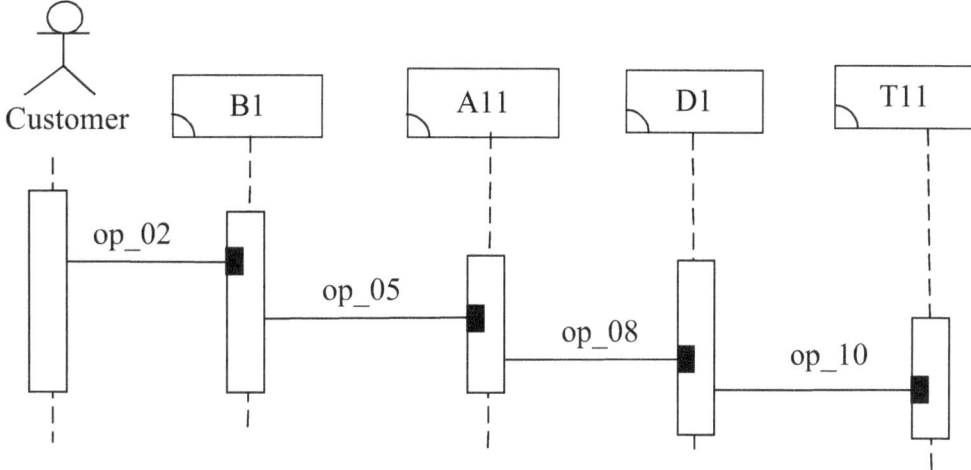

Figure 13-12   Analysis' IFD of the *qqq_2* Behavior

Figure 13-13 shows the analysis' IFD of the *qqq_3* behavior. First, actor *Customer* interacts with the *B2* component through the *op_03* operation call interaction. Next, component *B2* interacts with the *A11* component through the *op_06* operation call interaction. Continuingly, component *A11* interacts with the *D2* component through the *op_09* operation call interaction. Finally, component *D2* interacts with the *T12* component through the *op_11* operation call interaction.

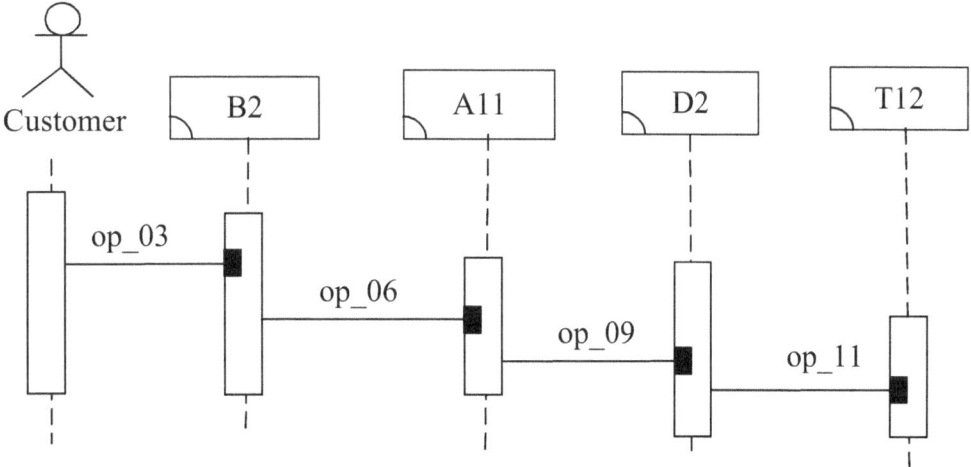

Figure 13-13   Analysis' IFD of the *qqq_3* Behavior

Chapter 14: Design View

In the SBC view model, multi-level view contains the concept, analysis, design and implementation views. This chapter demonstrates how the general systems theory 2.0 (general architectural theory) defining the design view.

14-1 Principle of Design View

Design view describes what a designer has accomplished for his work. Design view is one level down structural decomposition (with observation congruence verification) of the analysis view [Chao15a, Chao15b, Chao15c, Chao15d, Chao15e]. That is, we shall not create the design view from the scratch. Instead, we will construct the design view by decomposing the analysis view.

According to SBC multi-level (hierarchical) view, systems architects construct the design's systems architecture for the designer to view. This design's systems architecture is called the design view as shown in Figure14-1.

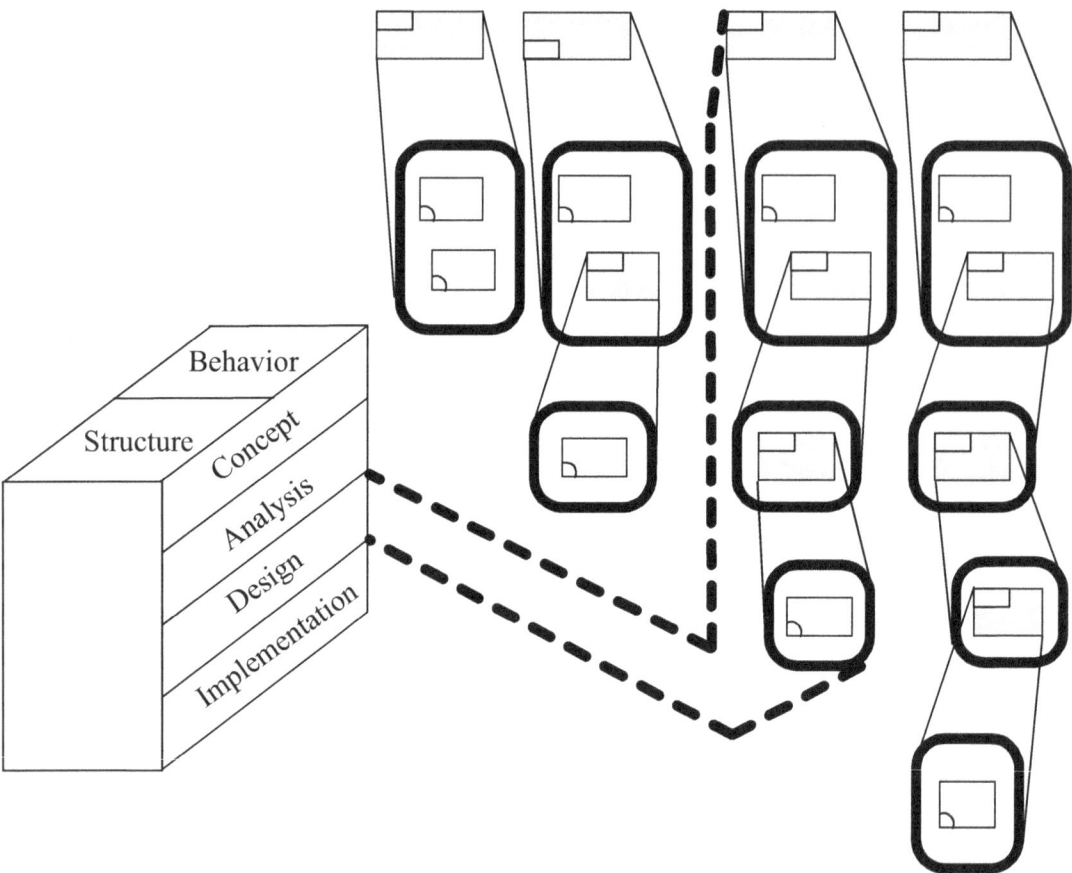

Figure 14-1   Design View

The design view consists of: a) design's systems structure and b) design's systems behavior.

14-2 Design's Systems Structure

The entire SBC design's systems structure includes: a) *Design's AHD*, b) *Design's FD*, c) *Design's COD* and d) *Design's CCD*, as shown in Figure 14-2.

141

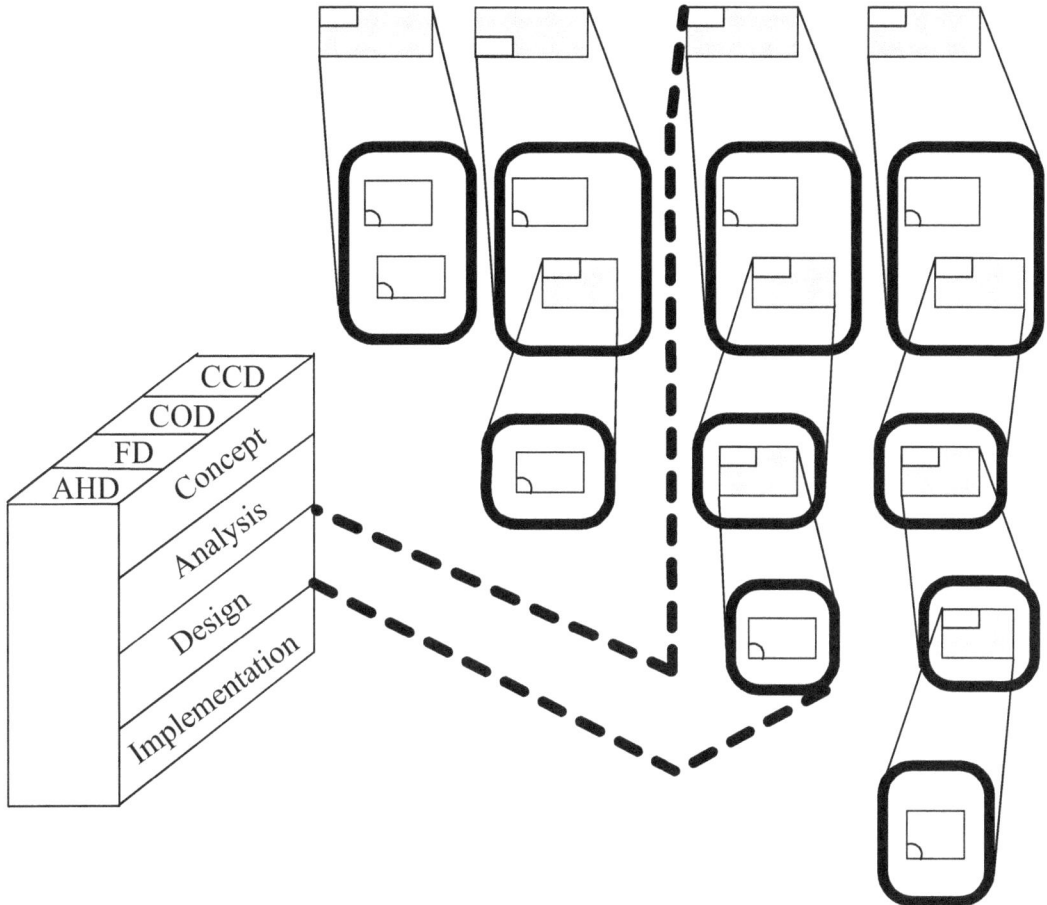

Figure 14-2   SBC Design's Systems Structure

Another drawing can also be used to illustrate the SBC design's systems structure as shown in Figure 14-3.

Systems Structure			
Architecture Hierarchy Diagram	Framework Diagram	Component Operation Diagram	Component Connection Diagram
Design	Design's AHD	Design's FD	Design's COD	Design's CCD

Figure 14-3   SBC Design's Systems Structure

14-2-1 Design's AHD

Design's AHD is the architecture hierarchy diagram we obtain after the designing phase is finished. Figure 14-4 shows the design's AHD of the *QQQ* system. In the figure, *QQQ* is composed of *B1*, *B2*, *A1*, *D1*, *D2* and *T1*; *A1* is composed of *A11*; *T1* is composed of *T11* and *T12*; *A11* is composed of *A111*.

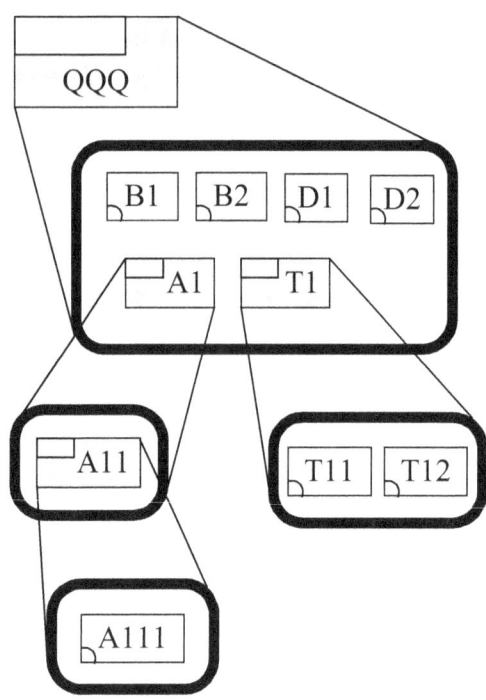

Figure 14-4   Design's AHD of the *QQQ* System

In Figure 14-4, *QQQ*, *A1*, *T1* and *A11* are aggregated systems while *B1*, *B2*, *A111*, *D1*, *D2*, *T11* and *T12* are non-aggregated systems.

We validate our claim that the design view is one level down structural decomposition of the analysis view by comparing Figure 13-4 with Figure 14-4. In Figure 13-4, *A11* is a non-aggregated system. As an interesting contrast, in Figure 14-4, *A11* becomes an aggregated system and is composed of *A111*.

14-2-2 Design's FD

Design's FD is the framework diagram we obtain after the designing phase is finished. Figure 14-5 shows the design's FD of the *QQQ* system. In the figure, *Business_Layer* contains the *B1* and *B2* components; *Application_Layer* contains the *A111* component; *Data_Layer* contains the *D1* and *D2* components; *Technology_Layer* contains the *T11* and *T12* components.

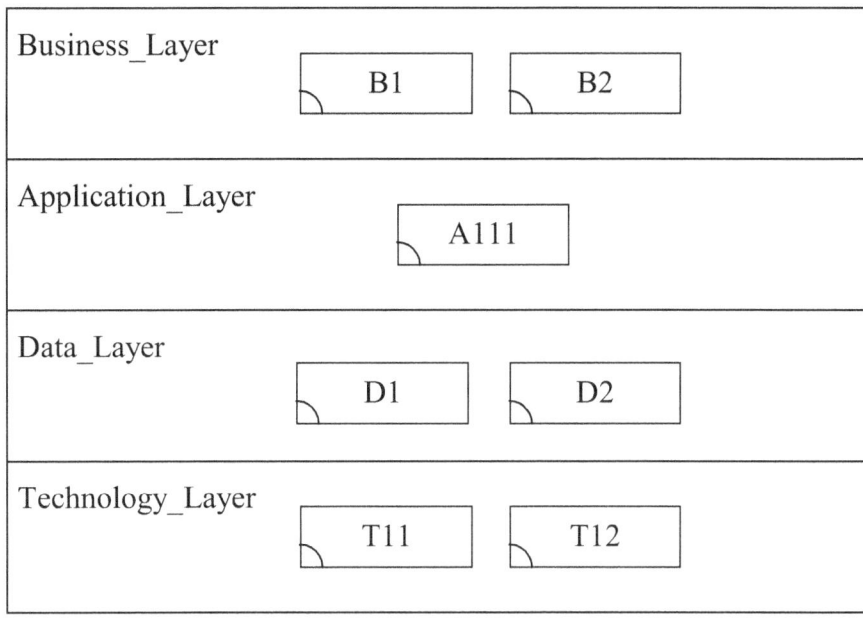

Figure 14-5   Design's FD of the *QQQ* System

14-2-3 Design's COD

Design's COD is the component operation diagram we obtain after the designing phase is finished. Figure 14-6 shows the design's COD of the *QQQ* system. In the figure, component *B1* has two operations: *op_01* and *op_02*; component *B2* has one operation: *op_03*; component *A111* has three operations: *op_04, op_05 and op_06*; component *D1* has two operations: *op_07* and *op_08*; component *D2* has one operation: *op_9*; component *T11* has one operation: *op_10*; component *T12* has one operation: *op_11*.

144

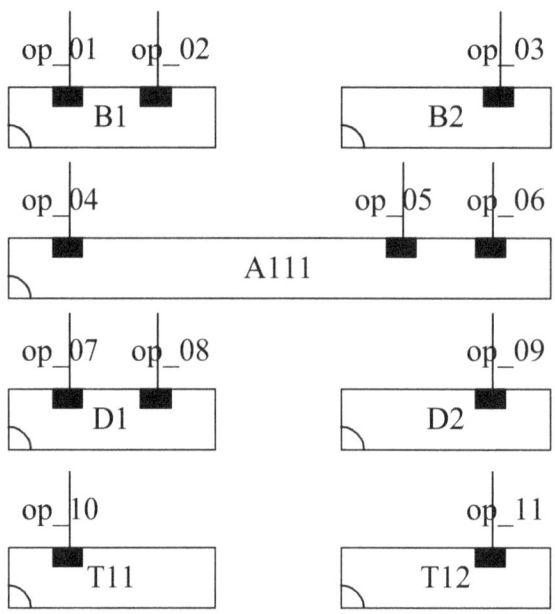

Figure 14-6　Design's COD of the *QQQ* System

14-2-4 Design's CCD

Design's CCD is the component connection diagram we obtain after the designing phase is finished. Figure 14-7 shows the design's CCD of the *QQQ* system.

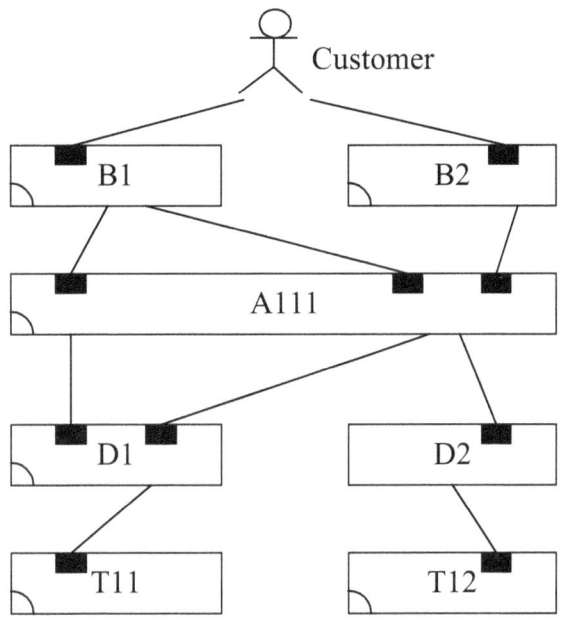

Figure 14-7   Design's CCD of the *QQQ* System

In Figure 14-7, we see that actor *Customer* has a connection with each one of the *B1* and *B2* components; component *B1* has two connections with the *A111* component; component *B2* has a connection with the *A111* component; component *A111* has two connections with the *D1* component; component *A111* has a connection with the *D2* component; component *D1* has a connection with the *T11* component; component *D2* has a connection with the *T12* component.

14-3 Design's Systems Behavior

The entire SBC design's systems behavior includes: a) *Design's SBCD* and b) *Design's IFD* as shown in Figure 14-8.

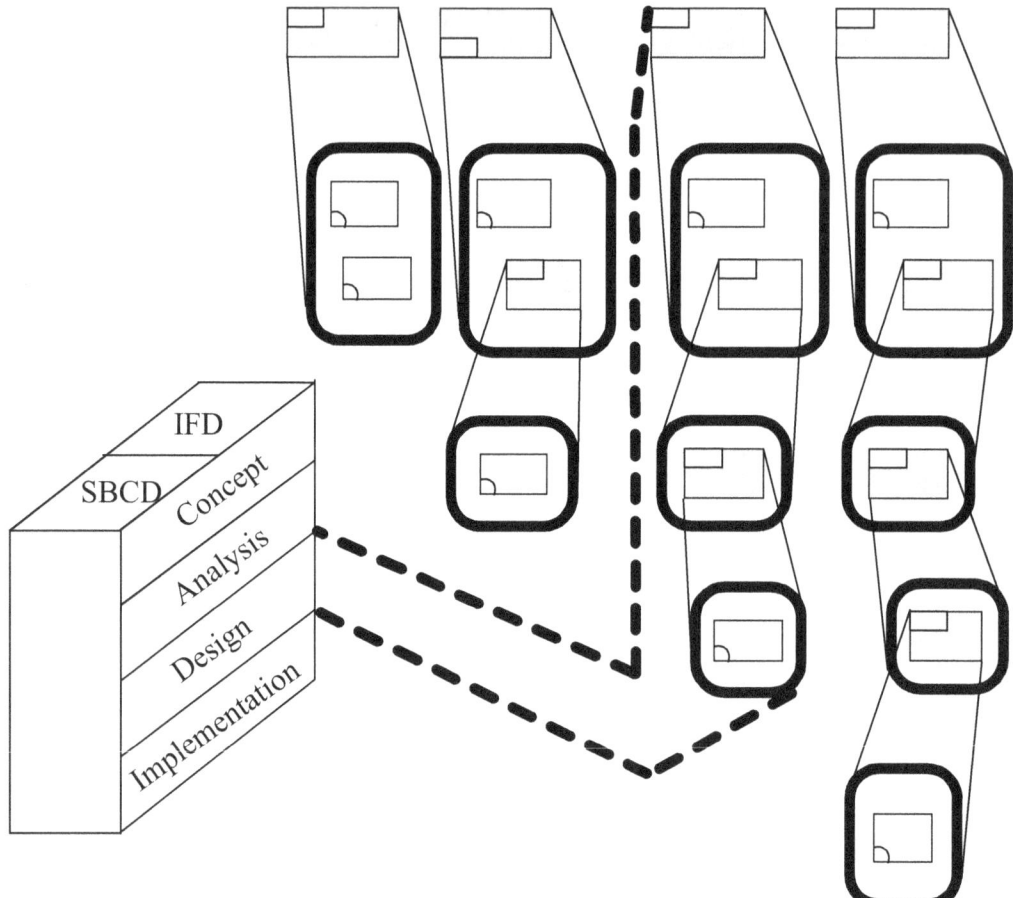

Figure 14-8   SBC Design's Systems Behavior

Let us use another drawing to illustrate the SBC design's systems behavior as shown in Figure 14-9.

	Systems Behavior	
	Structure-Behavior Coalescence Diagram	Interaction Flow Diagram
Design	Design's SBCD	Design's IFD

Figure 14-9   SBC Design's Systems Behavior

14-3-1 Design's SBCD

Design's SBCD is the structure-behavior coalescence diagram we obtain after the designing phase is finished. Figure 14-10 shows the design's SBCD of the *QQQ* system. In this example, an actor interacting with seven components shall describe the overall design's systems behavior. Interactions among the *Customer* actor and the *B1*, *A111*, *D1* components draw forth the *qqq_1* behavior. Interactions among the *Customer* actor and the *B1*, *A111*, *D1*, *T11* components draw forth the *qqq_2* behavior. Interactions among the *Customer* actor and the *B2*, *A111*, *D2*, *T12* components draw forth the *qqq_3* behavior.

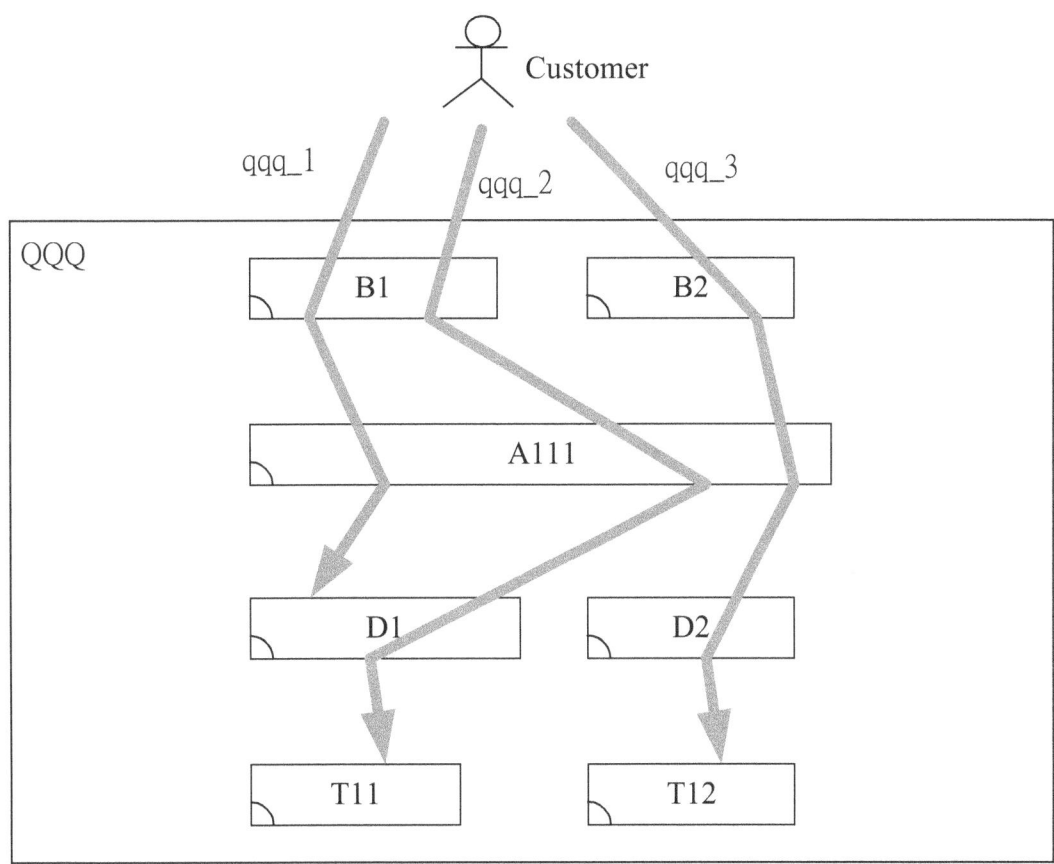

Figure 14-10   Design's SBCD of the *QQQ* System

The overall behavior of a system is a collection of all its individual behaviors. All individual behaviors are mutually independent of each other. They tend to be executed concurrently. For example, the overall design's behavior of the *QQQ* system includes the *qqq_1*, *qqq_2* and *qqq_3* behaviors. In other words, the *qqq_1*, *qqq_2* and *qqq_3* behaviors are combined to produce the overall design's behavior of the

QQQ system.

The major purpose of using the architectural approach, instead of separating the structure model from the behavior model, is to achieve one single coalesced model. In Figure 14-10, systems architects are able to see that the systems structure and systems behavior coexist in the design's SBCD. That is, in the design's SBCD of the *QQQ* system, systems architects not only see its systems structure but also see (at the same time) its systems behavior.

14-3-2 Design's IFD

Design's IFDs are the interaction flow diagrams we obtain after the designing phase is finished. The overall design's behavior of the *QQQ* system includes three behaviors: *qqq_1*, *qqq_2* and *qqq_3*. Each of them is described by an individual IFD. Figure 14-11 shows the design's IFD of the *qqq_1* behavior. First, actor *Customer* interacts with the *B1* component through the *op_01* operation call interaction. Next, component *B1* interacts with the *A111* component through the *op_04* operation call interaction. Finally, component *A111* interacts with the *D1* component through the *op_07* operation call interaction.

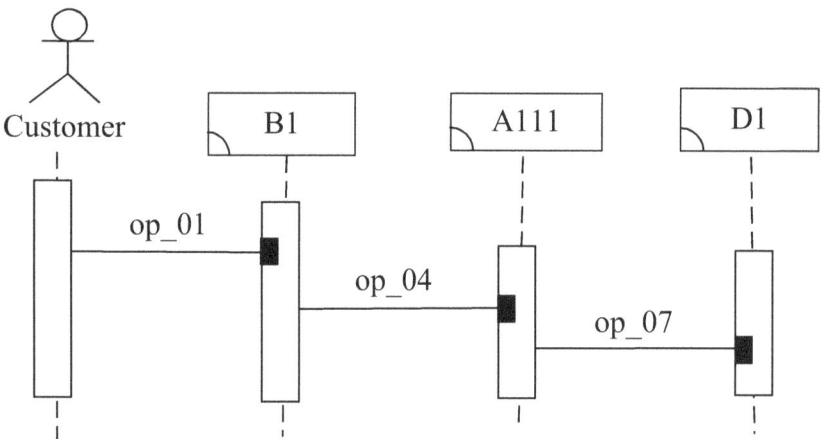

Figure 14-11   Design's IFD of the *qqq_1* Behavior

Figure 14-12 shows the design's IFD of the *qqq_2* behavior. First, actor *Customer* interacts with the *B1* component through the *op_02* operation call interaction. Next, component *B1* interacts with the *A111* component through the *op_05* operation call interaction. Continuingly, component *A111* interacts with the *D1* component through the *op_08* operation call interaction. Finally, component *D1* interacts with the *T11* component through the *op_10* operation call interaction.

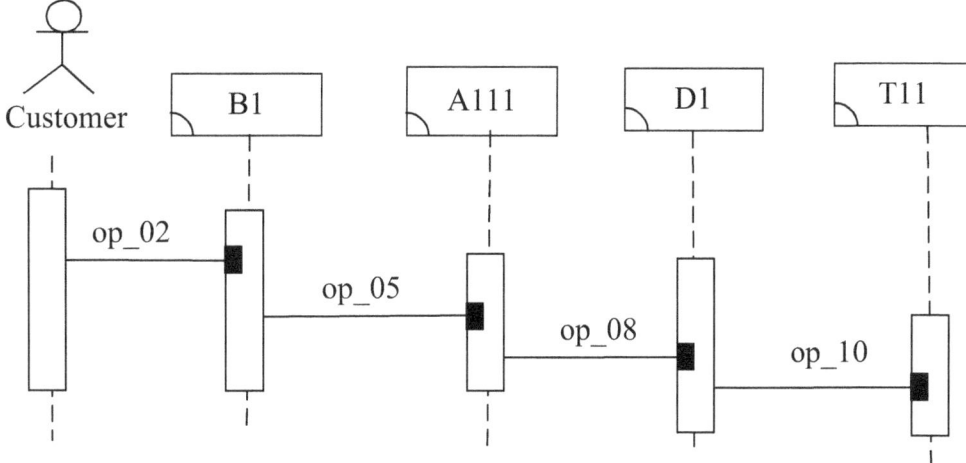

Figure 14-12   Design's IFD of the *qqq_2* Behavior

Figure 14-13 shows the design's IFD of the *qqq_3* behavior. First, actor *Customer* interacts with the *B2* component through the *op_03* operation call interaction. Next, component *B2* interacts with the *A111* component through the *op_06* operation call interaction. Continuingly, component *A111* interacts with the *D2* component through the *op_09* operation call interaction. Finally, component *D2* interacts with the *T12* component through the *op_11* operation call interaction.

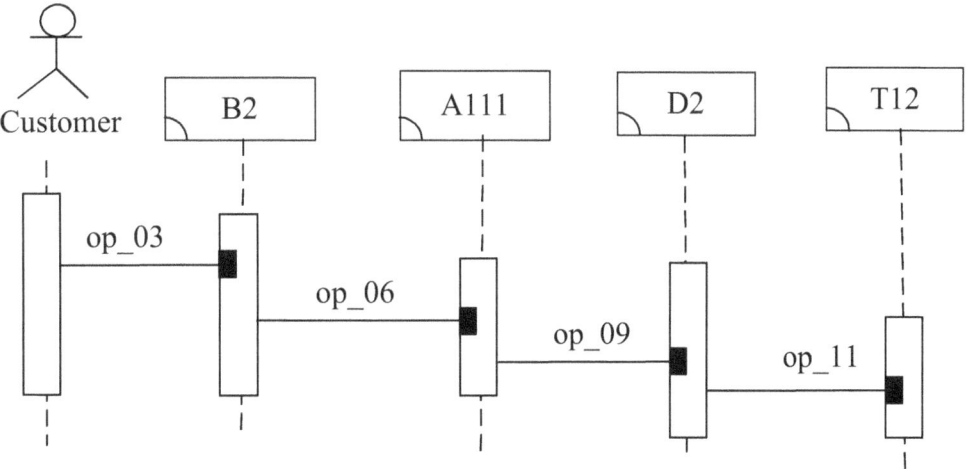

Figure 14-13   Design's IFD of the *qqq_3* Behavior

Chapter 15: Implementation View

In the SBC view model, multi-level view contains the concept, analysis, design and implementation views. This chapter demonstrates how the general systems theory 2.0 (general architectural theory) defining the implementation view.

15-1 Principle of Implementation View

Implementation view shows what an implementer has done for his work. Implementation view is one level down structural decomposition (with observation congruence verification) of the design view [Chao15a, Chao15b, Chao15c, Chao15d, Chao15e]. That is, we shall not create the implementation view from the scratch. Instead, we will construct the implementation view by decomposing the design view.

According to SBC multi-level (hierarchical) view, systems architects construct the implementation's systems architecture for the implementer to view. This implementation's systems architecture is called the implementation view as shown in Figure 15-1.

152

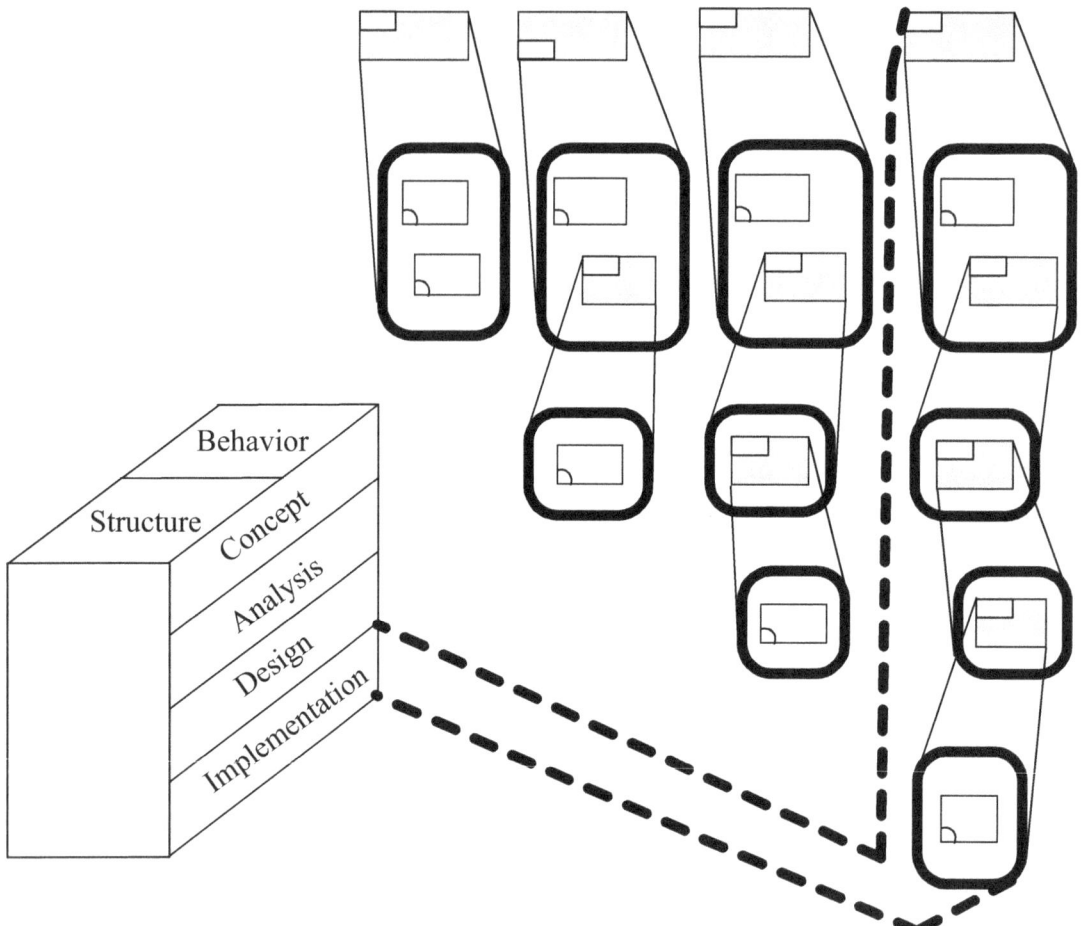

Figure 15-1   Implementation View

The implementation view consists of: a) implementation's systems structure and b) implementation's systems behavior.

15-2 Implementation's Systems Structure

The entire SBC implementation's systems structure includes: a) *Implementation's AHD*, b) *Implementation's FD*, c) *Implementation's COD* and d) *Implementation's CCD*, as shown in Figure 15-2.

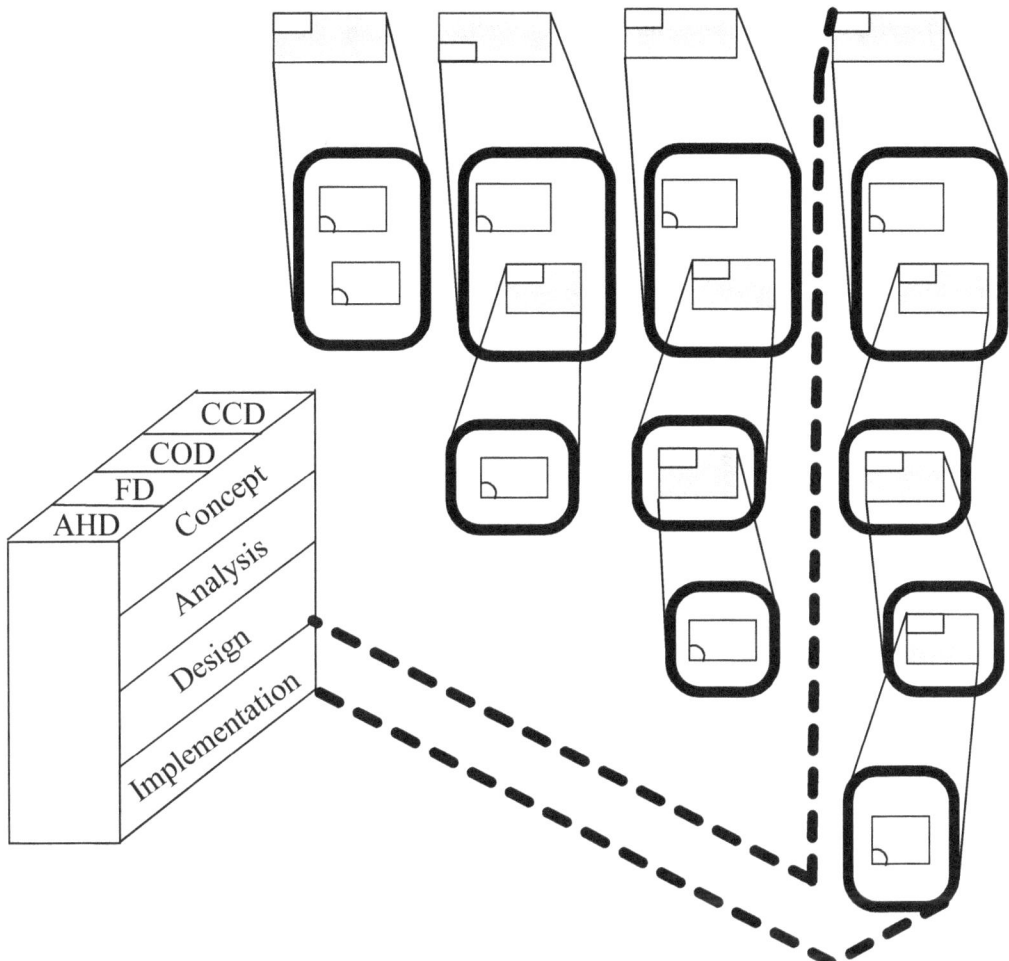

Figure 15-2   SBC Implementation's Systems Structure

Another drawing can also be used to illustrate the SBC implementation's systems structure as shown in Figure 15-3.

	Systems Structure			
	Architecture Hierarchy Diagram	Framework Diagram	Component Operation Diagram	Component Connection Diagram
Implementation	Implementation's AHD	Implementation's FD	Implementation's COD	Implementation's CCD

Figure 15-3   SBC Implementation's Systems Structure

154

15-2-1 Implementation's AHD

Implementation's AHD is the architecture hierarchy diagram we obtain after the implementation phase is finished. Figure 15-4 shows the implementation's AHD of the *QQQ* system. In the figure, *QQQ* is composed of *B1*, *B2*, *A1*, *D1*, *D2* and *T1*; *A1* is composed of *A11*; *T1* is composed of *T11* and *T12*; *A11* is composed of *A111*; *A111* is composed of *A1111* and *A1112*.

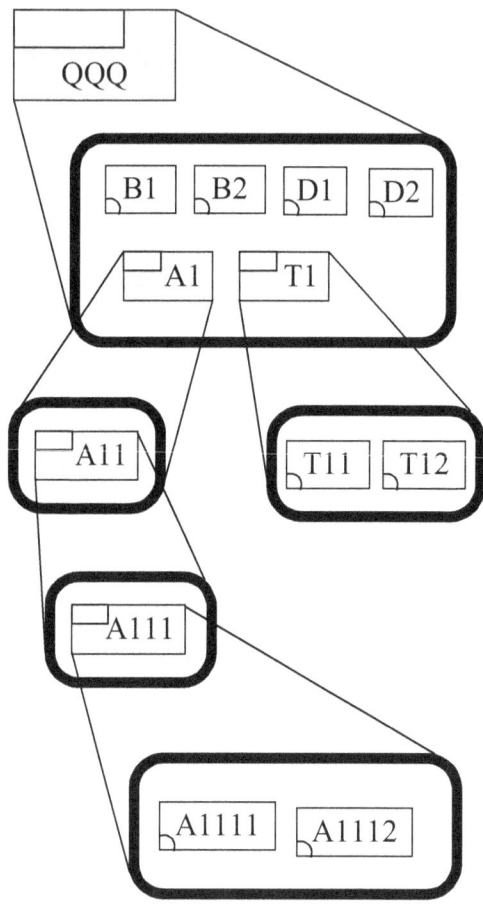

Figure 15-4   Implementation's AHD of the *QQQ* System

In Figure 15-4, *QQQ*, *A1*, *T1*, *A11* and *A111* are aggregated systems while *B1*, *B2*, *A1111*, *A1112*, *D1*, *D2*, *T11* and *T12* are non-aggregated systems.

We validate our claim that the implementation view is one level down structural decomposition of the design view by comparing Figure 14-4 with Figure 15-4. In Figure 14-4, *A111* is a non-aggregated system. As an interesting contrast, in Figure 15-4, *A111* becomes an aggregated system and is composed of *A1111* and *A1112*.

15-2-2 Implementation's FD

Implementation's FD is the framework diagram we obtain after the implementation phase is finished. Figure 15-5 shows the implementation's FD of the *QQQ* system. In the figure, *Business_Layer* contains the *B1* and *B2* components; *Application_Layer* contains the *A1111* and *A1112* components; *Data_Layer* contains the *D1* and *D2* components; *Technology_Layer* contains the *T11* and *T12* components.

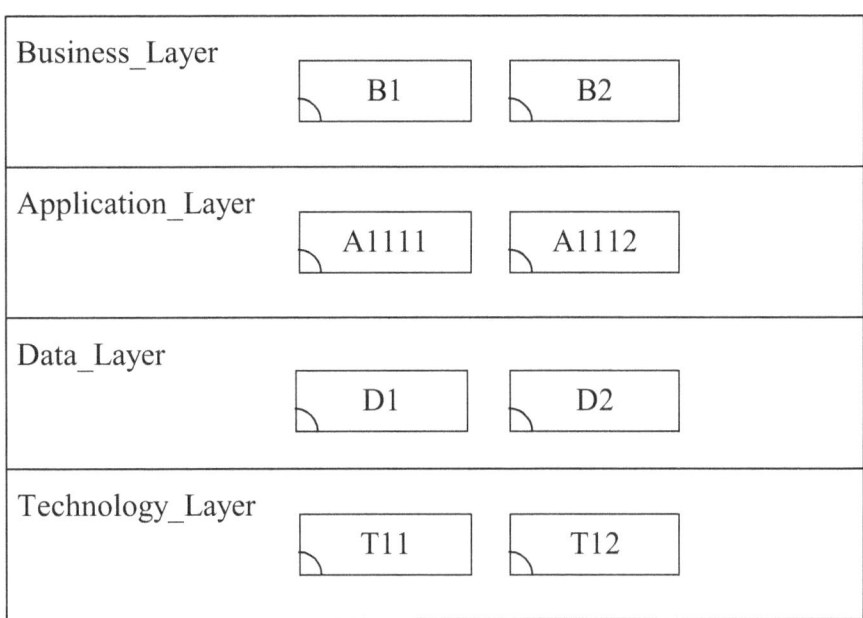

Figure 15-5　Implementation's FD of the *QQQ* System

15-2-3 Implementation's COD

Implementation's COD is the component operation diagram we obtain after the implementation phase is finished. Figure 15-6 shows the implementation's COD of the *QQQ* system. In the figure, component *B1* has two operations: *op_01* and *op_02*; component *B2* has one operation: *op_03*; component *A1111* has one operation: *op_04*; component *A1112* has two operations: *op_05* and *op_06*; component *D1* has two operations: *op_07* and *op_08*; component *D2* has one operation: *op_9*; component *T11* has one operation: *op_10*; component *T12* has one operation: *op_11*.

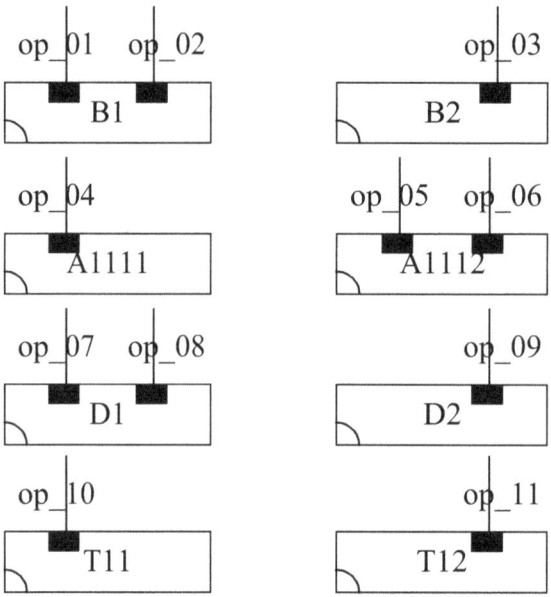

Figure 15-6   Implementation's COD of the *QQQ* System

15-2-4 Implementation's CCD

Implementation's CCD is the component connection diagram we obtain after the implementation phase is finished. Figure 15-7 shows the implementation's CCD of the *QQQ* system.

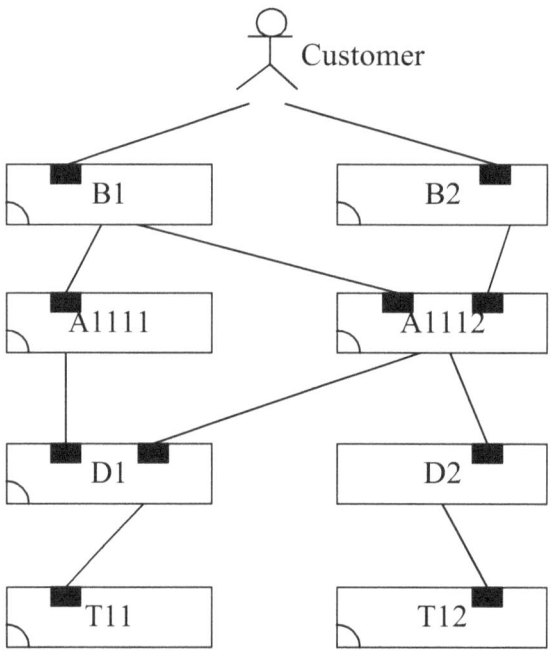

Figure 15-7   Implementation's CCD of the *QQQ* System

In Figure 15-7, we see that actor *Customer* has a connection with each one of the *B1* and *B2* components; component *B1* has a connection with each one of the *A1111* and *A1112* components; component *B2* has a connection with the *A1112* component; component *A1111* has a connection with the *D1* component; component *A1112* has a connection with each one of the *D1* and *D2* components; component *D1* has a connection with the *T11* component; component *D2* has a connection with the *T12* component.

15-3 Implementation's Systems Behavior

The entire SBC implementation's systems behavior includes: a) *Implementation's SBCD* and b) *Implementation's IFD* as shown in Figure 15-8.

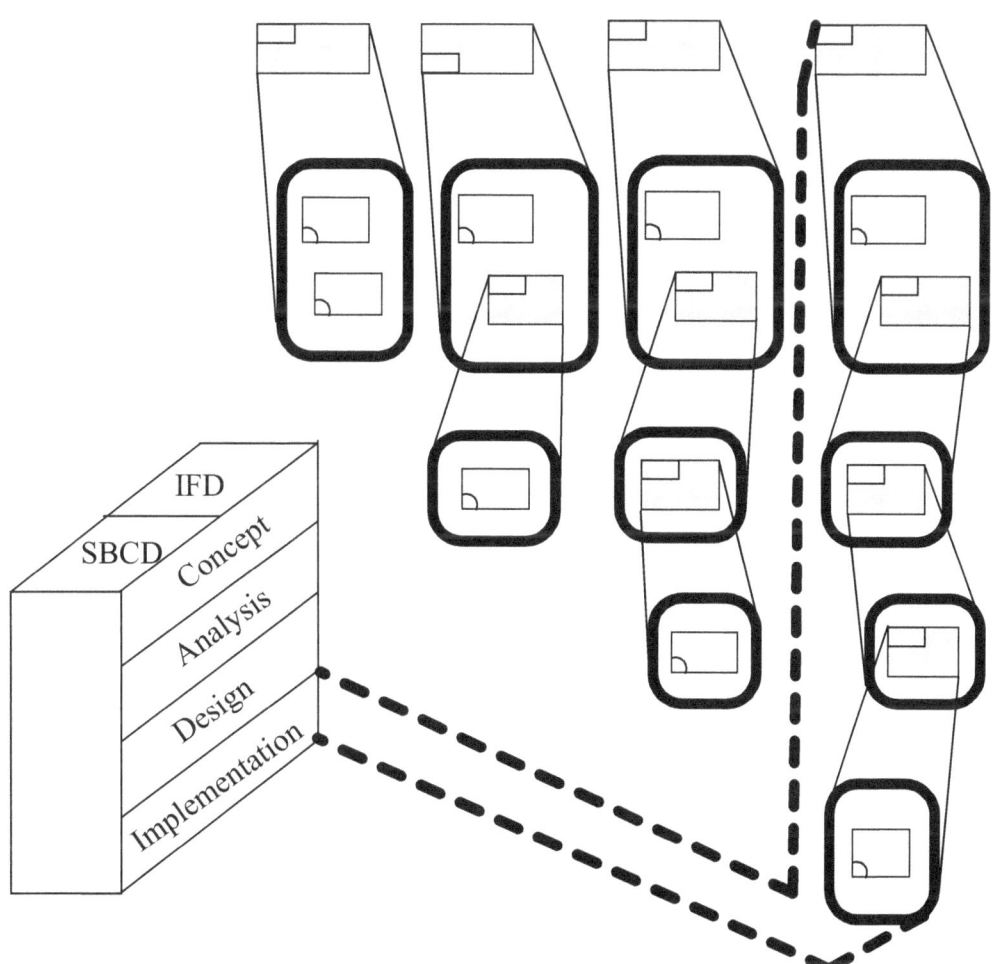

Figure 15-8   SBC Implementation's Systems Behavior

Let us use another drawing to illustrate the SBC implementation's systems behavior as shown in Figure 15-9.

Systems Behavior		
	Structure-Behavior Coalescence Diagram	Interaction Flow Diagram
Implementation	Implementation's SBCD	Implementation's IFD

Figure 15-9   SBC Implementation's Systems Behavior

15-3-1 Implementation's SBCD

Implementation's SBCD is the structure-behavior coalescence diagram we obtain after the implementation phase is finished. Figure 15-10 shows the implementation's SBCD of the *QQQ* system. In this example, an actor interacting with eight components shall describe the overall implementation's systems behavior. Interactions among the *Customer* actor and the *B1*, *A1111*, *D1* components draw forth the *qqq_1* behavior. Interactions among the *Customer* actor and the *B1*, *A1112*, *D1*, *T11* components draw forth the *qqq_2* behavior. Interactions among the *Customer* actor and the *B2*, *A1112*, *D2*, *T12* components draw forth the *qqq_3* behavior.

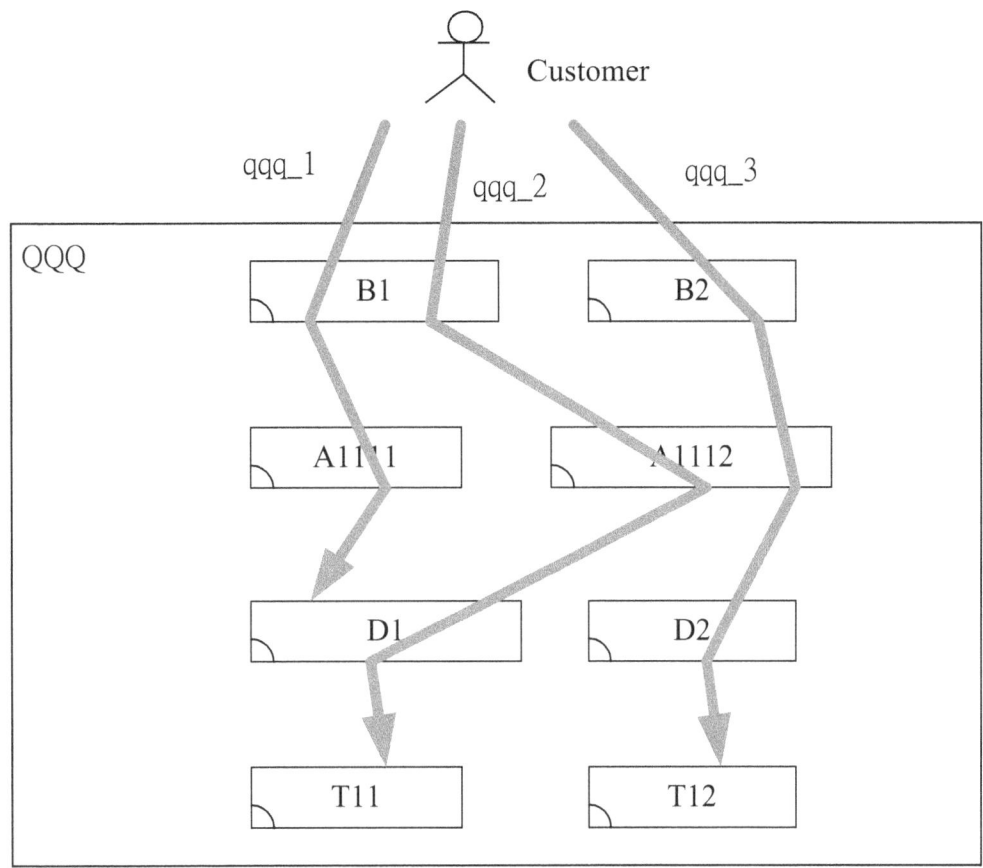

Figure 15-10 Implementation's SBCD of the *QQQ* System

The overall behavior of a system is a collection of all its individual behaviors. All individual behaviors are mutually independent of each other. They tend to be executed concurrently. For example, the overall implementation's behavior of the *QQQ* system includes the *qqq_1*, *qqq_2* and *qqq_3* behaviors. In other words, the *qqq_1*, *qqq_2* and *qqq_3* behaviors are combined to produce the overall implementation's behavior of the *QQQ* system.

The major purpose of using the architectural approach, instead of separating the structure model from the behavior model, is to achieve one single coalesced model. In Figure 15-10, we are able to see that the systems structure and systems behavior coexist in the implementation's SBCD. That is, in the implementation's SBCD of the *QQQ* system, a systems architect not only sees its systems structure but also sees (at the same time) its systems behavior.

15-3-2 Implementation's IFD

Implementation's IFDs are the interaction flow diagrams we obtain after the implementation phase is finished. The overall implementation's behavior of the *QQQ* system includes three behaviors: *qqq_1*, *qqq_2* and *qqq_3*. Each of them is described by an individual IFD. Figure 15-11 shows the implementation's IFD of the *qqq_1* behavior. First, actor *Customer* interacts with the *B1* component through the *op_01* operation call interaction. Next, component *B1* interacts with the *A1111* component through the *op_04* operation call interaction. Finally, component *A1111* interacts with the *D1* component through the *op_07* operation call interaction.

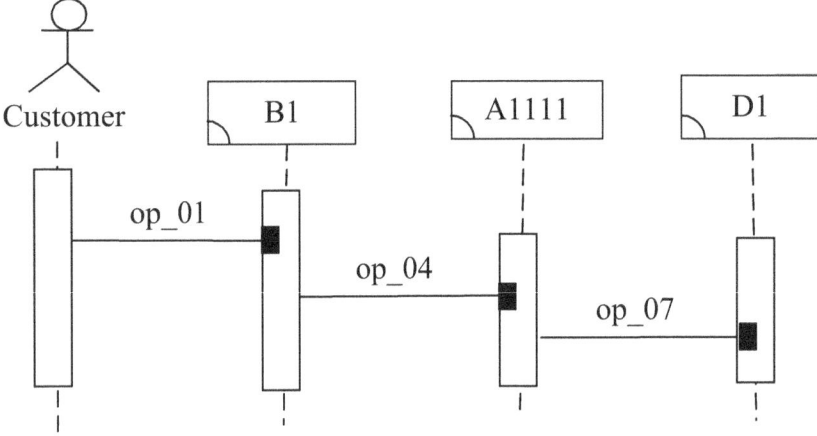

Figure 15-11   Implementation's IFD of the *qqq_1* Behavior

Figure 15-12 shows the implementation's IFD of the *qqq_2* behavior. First, actor *Customer* interacts with the *B1* component through the *op_02* operation call interaction. Next, component *B1* interacts with the *A1112* component through the *op_05* operation call interaction. Continuingly, component *A1112* interacts with the *D1* component through the *op_08* operation call interaction. Finally, component *D1* interacts with the *T11* component through the *op_10* operation call interaction.

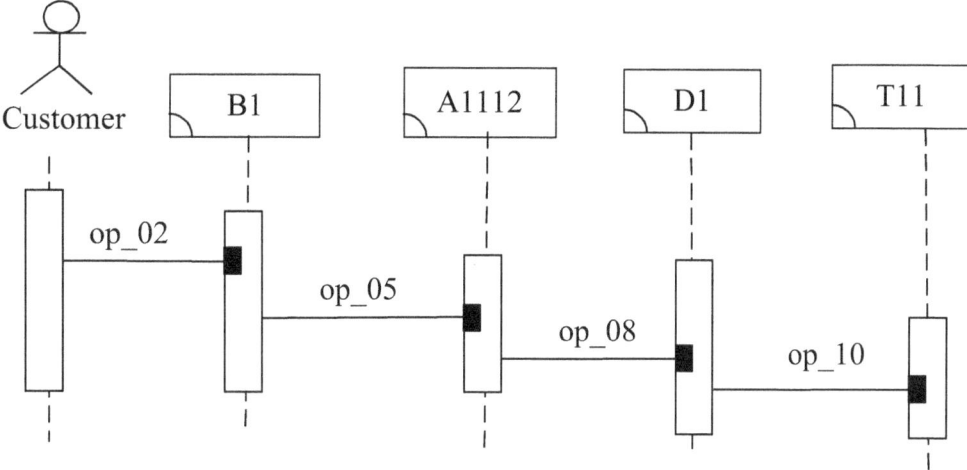

Figure 15-12   Implementation's IFD of the *qqq_2* Behavior

Figure 15-13 shows the implementation's IFD of the *qqq_3* behavior. First, actor *Customer* interacts with the *B2* component through the *op_03* operation call interaction. Next, component *B2* interacts with the *A1112* component through the *op_06* operation call interaction. Continuingly, component *A1112* interacts with the *D2* component through the *op_09* operation call interaction. Finally, component *D2* interacts with the *T12* component through the *op_11* operation call interaction.

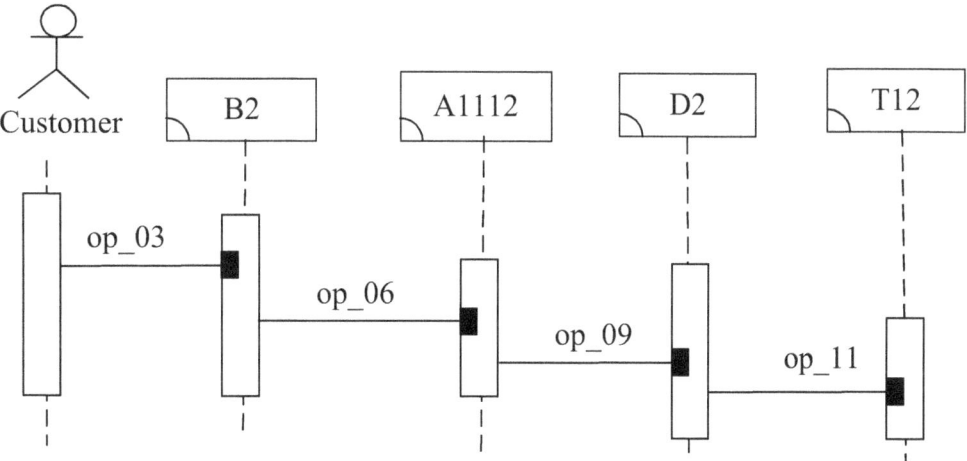

Figure 15-13   Implementation's IFD of the *qqq_3* Behavior

PART IV:
EVOLUTION&MOTIVATION VIEW

Chapter 16: Higher-Order Systems

Higher-order systems interact with the environment through the exchange of not only matter, energy, data, information, or message but also systems. In this chapter, after introducing higher-order functions, second-order logic and higher-order systems, we then will give some examples of higher-order systems.

16-1 Higher-Order Functions

In mathematics, a higher-order function is a function that does at least one of the following: a) takes one or more functions as an input, b) outputs a function [Bare84, Hend80, Sang03]. All other functions are first-order functions.

A first-order function takes a combination of two (or more) sets of data to a single set of data, as shown in Figure 16-1. In the figure, input *a*, *b* and output *c* are data.

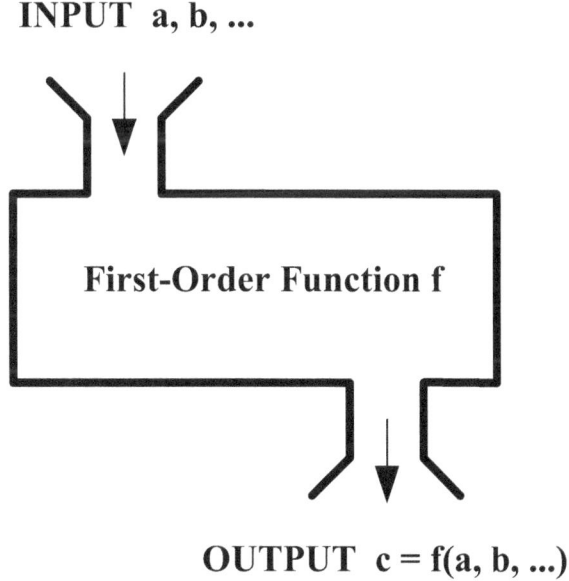

INPUT a, b, ...

First-Order Function f

OUTPUT c = f(a, b, ...)

Figure 16-1   First-Order Function

A higher-order function takes functions as input and return functions as output [Bare84, Hend80, Sang03], as shown in Figure 16-2. In the figure, input *f1*, *f2* and output *f3* are functions.

166

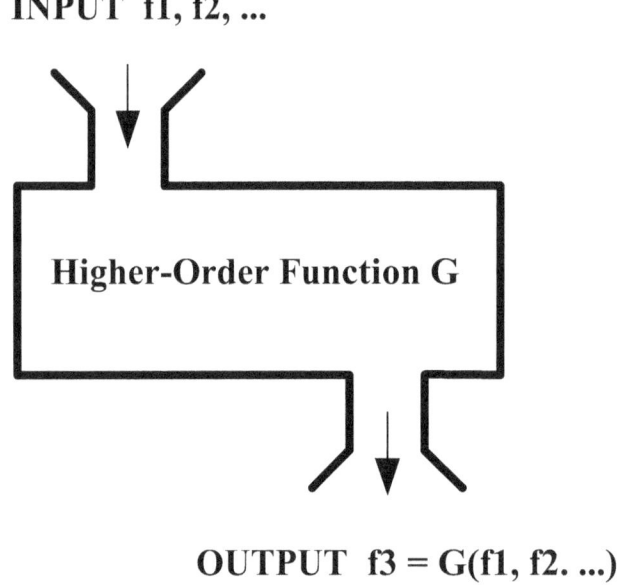

INPUT f1, f2, ...

Higher-Order Function G

OUTPUT f3 = G(f1, f2. ...)

Figure 16-2   Higher-Order Function

For a higher-order function is computable, it must be monotonic and continuous [Bare84, Cohe63, Mann74, Scot67].

16-2 Second-Order Logic

In logic, second-order logic is an extension of first-order logic, which itself is an extension of propositional logic [Mann74, Shap00].

First-order logic quantifies only variables that range over elements of the domain of discourse, as shown in Figure 16-3.

$$\forall x \; \exists y \; (x \longrightarrow P\,(y))$$

Figure 16-3   First-Order Logic

In contrast, second-order logic, in addition to first-order logic, also quantifies over relations [Mann74, Shap00], as shown in Figure 16-4.

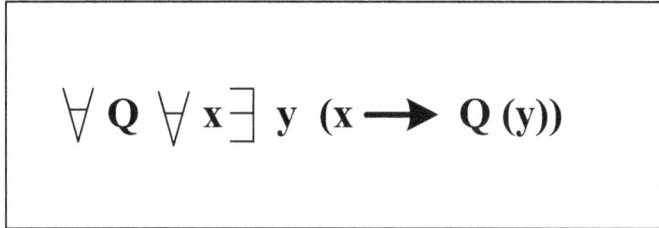

Figure 16-4   Second-Order Logic

For a second-order logic is computable, it must be monotonic and continuous [Bare84, Mann74, Scot67].

16-3 Higher-Order Systems

First-order systems interact with the environment through the exchange of matter, energy, data, information, or message, as shown in Figure 16-5.

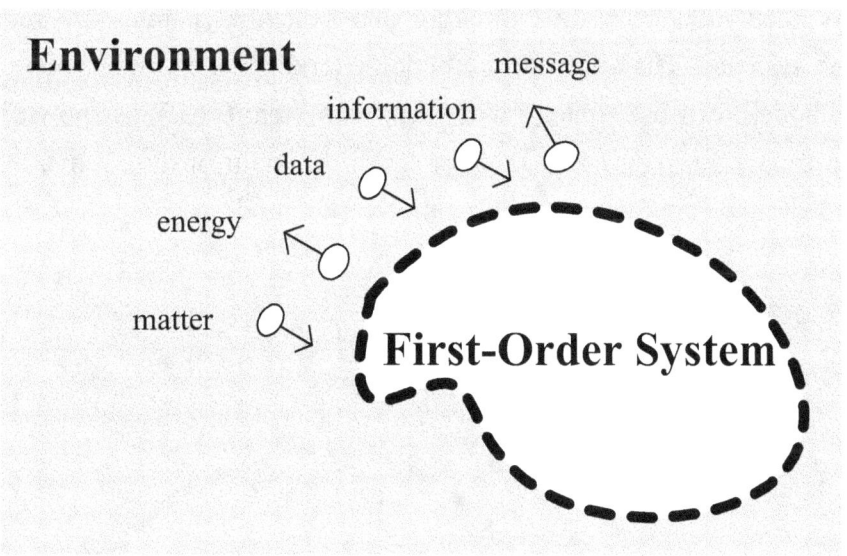

Figure 16-5   First-Order System

Higher-order systems, also known as second-order systems, interact with the environment through the exchange of not only matter, energy, data, information, or message but also systems, as shown in Figure 16-6.

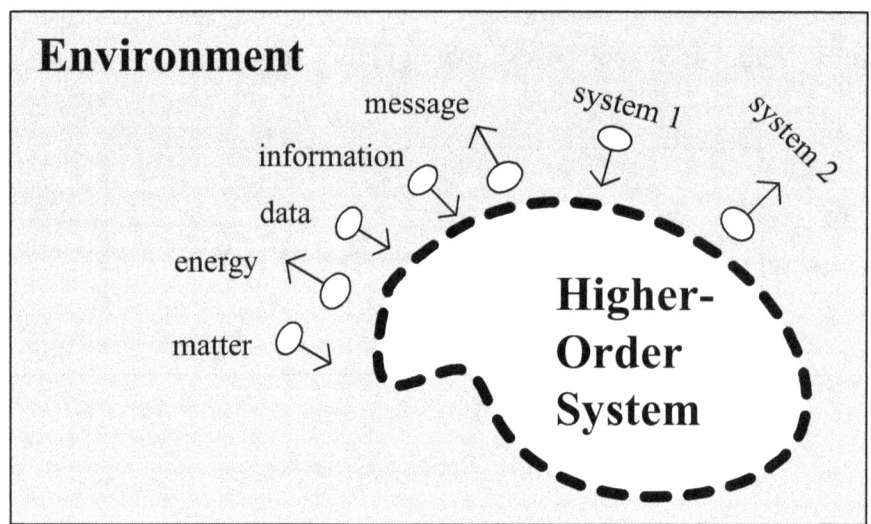

Figure 16-6   Higher-Order System

For a higher-order system to be computable, it must be monotonic and continuous [Bare84, Mann74, Scot67].

16-4 Examples of Higher-Order Systems

Motivation model, creative thinking and system dynamics are regarded as higher-order systems. The motivation model [Berk08] is a higher-order system. Motivation model, for each strategy will output a system (goal), as shown in Figure 16-7. In the figure, *strategy 1* and *strategy n* are the input of the motivation model; *system 1* and *system n* are the output of the motivation model.

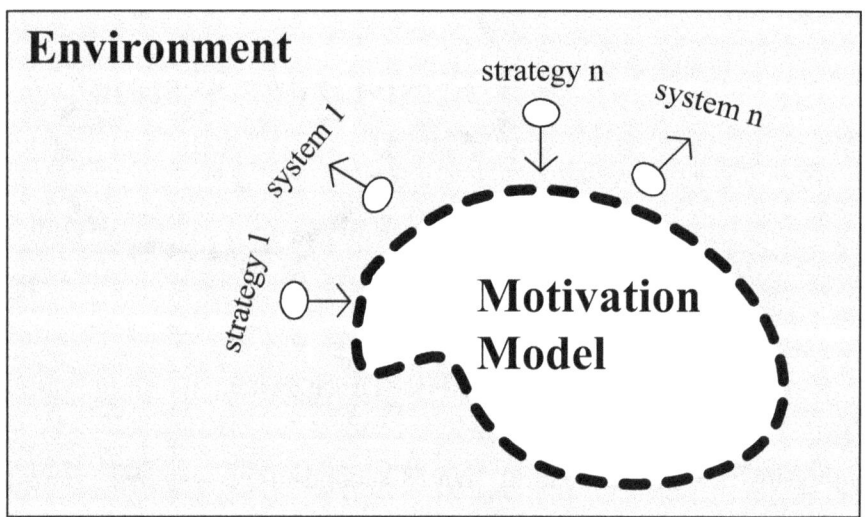

Figure 16-7   Motivation Model is a Higher-Order System

Creative thinking is a higher-order system, because it will be creating a large number of systems then chooses the best one, as shown in Figure 16-8. In the figure, *system 1*, *system 2* and *system n* are the output of the creative thinking.

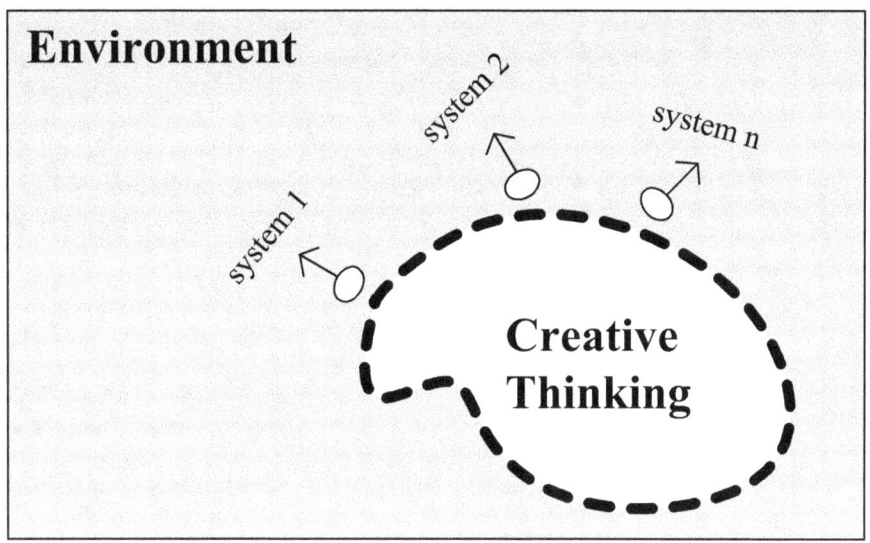

Figure 16-8   Creative Thinking is a Higher-Order System

System dynamics [Forr61, Ogat03, Palm09] is also a higher-order system, because it dynamically simulates the causal relationship among a large number of systems as shown in Figure 16-9. From these simulated systems, decision makers thus are able to strategically choose the most appropriate one.

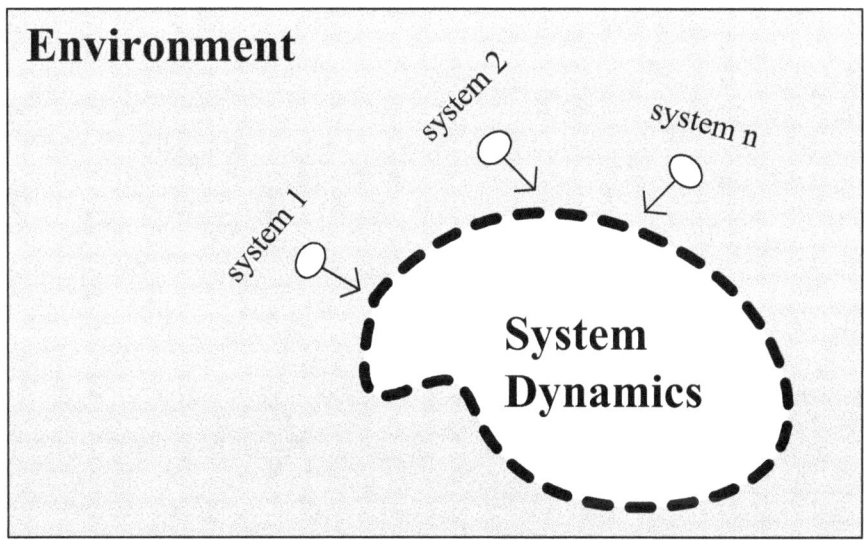

Figure 16-9   System Dynamics is a Higher-Order System

Chapter 17: Strategic Management

Strategic management is concerned primarily with responses to external issues such as in understanding the needs of customers and responding to competitive forces.

Strategic management provides overall direction to the enterprise. In short, it entails specifying the organizations goals, developing means designed to achieve these goals, and then allocating resources to implement the means.

This chapter first introduces what a strategy is. Then it discusses the motivation model is a higher-order system. Last, this chapter will work on the strategic means.

17-1 Strategy

Strategy is included in a high level plan to achieve one or more goals under conditions of uncertainty [Free13, Mcke12]. Strategy is important because the resources available to achieve these goals are usually limited.

In the business motivation model [Berk08], as shown in Figure 17-1, strategy is the human attempt to get to "desirable ends with available means".

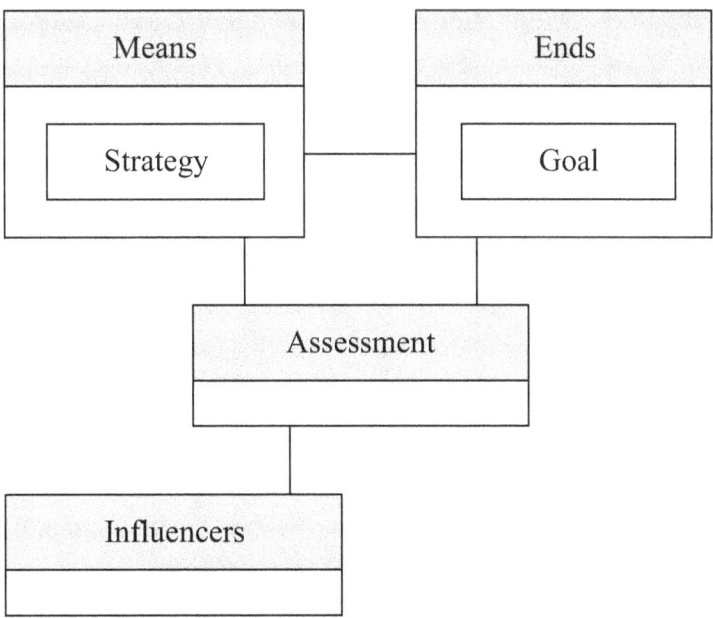

Figure 17-1   Business Motivation Model

In the above figure, available means are related with the strategy; desirable ends are related with the goal.

17-2 Motivation Model is a Higher-Order System

As discussed in the previous chapter, motivation model is a higher-order system. Motivation model will output a system (goal) for each strategy as shown in Figure 17-2.

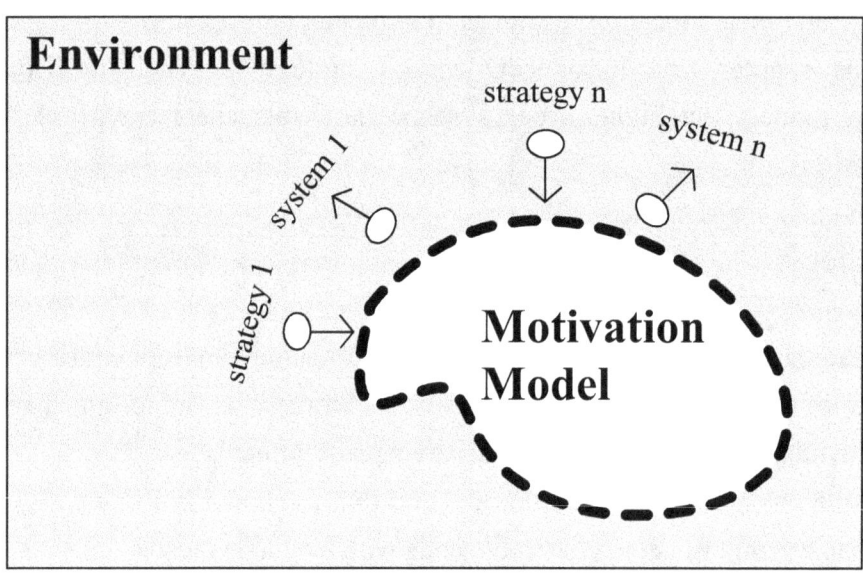

Figure 17-2   Motivation Model is a Higher-Order System

In general, each strategy is mapped to a system as shown in Figure 17-3. That is, the motivation model will output a system for each strategy.

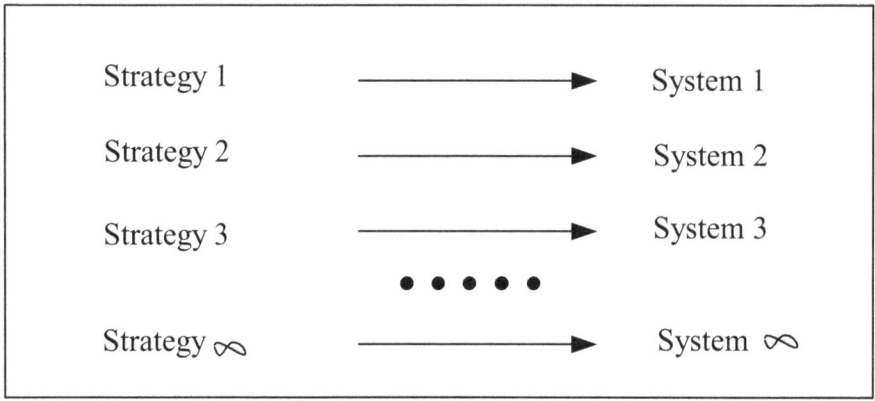

Figure 17-3   Each Strategy is Mapped to a System

17-3 Strategic Means

In the business motivation model, strategy is the human attempt to get to "desirable ends with available means". Strategic means analyzes the major initiatives taken by a company's top management on behalf of business owners, involving resources and performance in internal and external environments.

Strategic means include: (a) goal drivers, (b) goal assumptions, (c) goal constraints and (d) SWOT (strengths, weaknesses, opportunities, threat) analysis, etc. We use these strategic means to achieve the desirable ends.

Goal drivers are up from the policy considerations, the goal driver is kind of why we want to have those desirable ends. Goal assumptions are taking into account of those assumptions that have a positive impact on these desirable ends. Goal constraints are up from the policy considerations, the goal constraints are related to those restrictions which have a negative impact on those desirable ends. SWOT analysis is to analyze the internal strengths, weaknesses, opportunities and threats, and so for executing this strategy.

18-2 Change from the External Force

The change cause may come from the internal or external forces of the system. Repair workers doing the tire rotation, as shown in Figure 18-3, is an example of change from the external force.

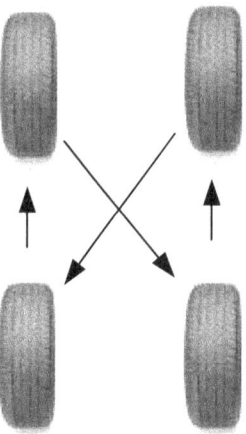

Figure 18-3   Tire Rotation

As a second example, workers doing road maintenance as shown in Figure 18-4, is another instance of change from the external force.

Figure 18-4   Road Maintenance

18-3 Result of Systems Evolution

A system evolves when it changes. Evolution of a system is shown in Figure 18-5.

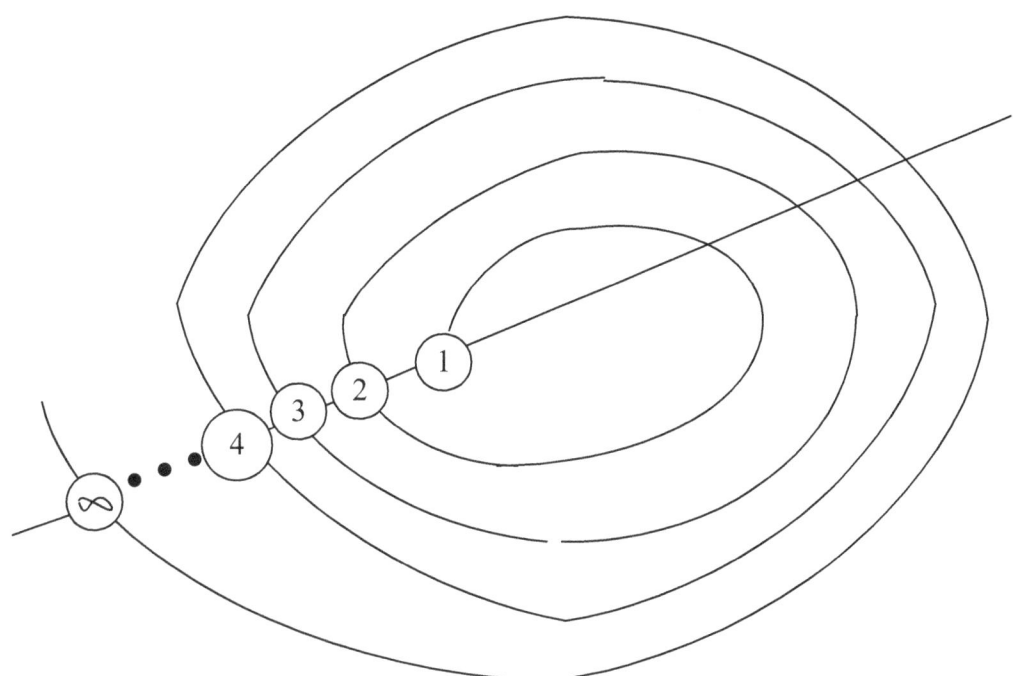

① : Systems Definition Version 1

② : Systems Definition Version 2

③ : Systems Definition Version 3

④ : Systems Definition Version 4

• • •

∞ : Systems Definition Version ∞

Figure 18-5   Evolution of a System

Each time when a system changes or evolves, we shall get a new version of its systems definition as the result of systems evolution. In the above figure, *version 1* stands for the original systems definition of a system and evolves into the *version 2*, *version 3*,..., and *version ∞* gradually.

According the motivation model, one strategy is mapped to one version for any systems definition, as shown in Figure 18-6.

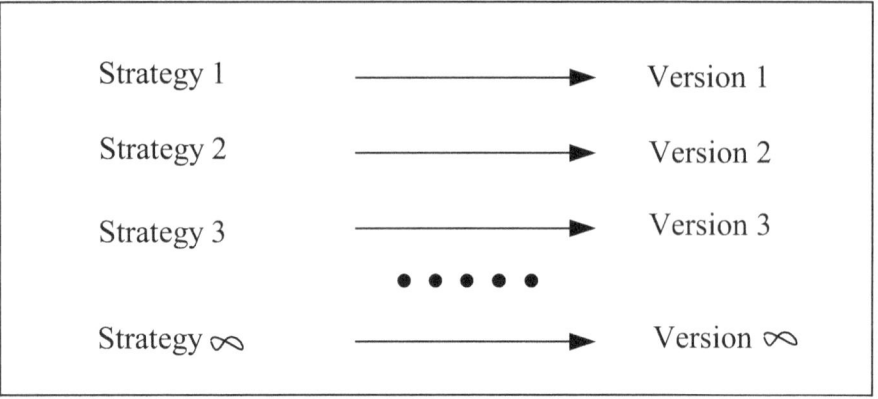

Figure 18-6   One Strategy is Mapped to One Version

Evolution of a system is represented, by general systems theory 2.0 (general architectural theory), as the SBC evolution&motivation view shown in Figure 18-7. In the figure, we see that, for any systems definition, one strategy is mapped to one version.

Dimension 1:
Evolution&Motivation View

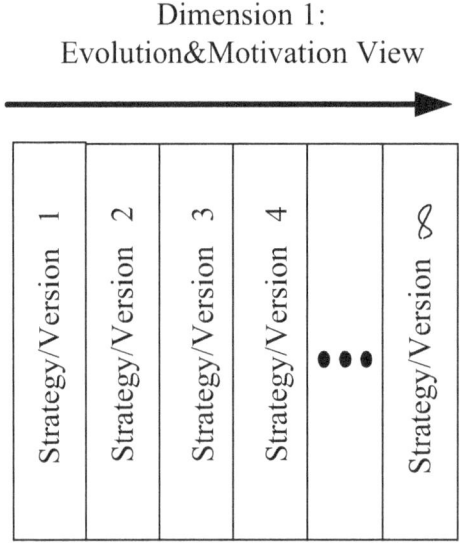

Figure 18-7   SBC Evolution&Motivation View

180

PART V: CASE STUDIES

Chapter 19: General Systems Theory 2.0 Defining the Robot

This chapter demonstrates how to achieve general systems theory 2.0 (general architectural theory) defining the *robot*, through the application of SBC architecture.

Based on the SBC architecture, we define the *robot* as shown in Figure 19-1.

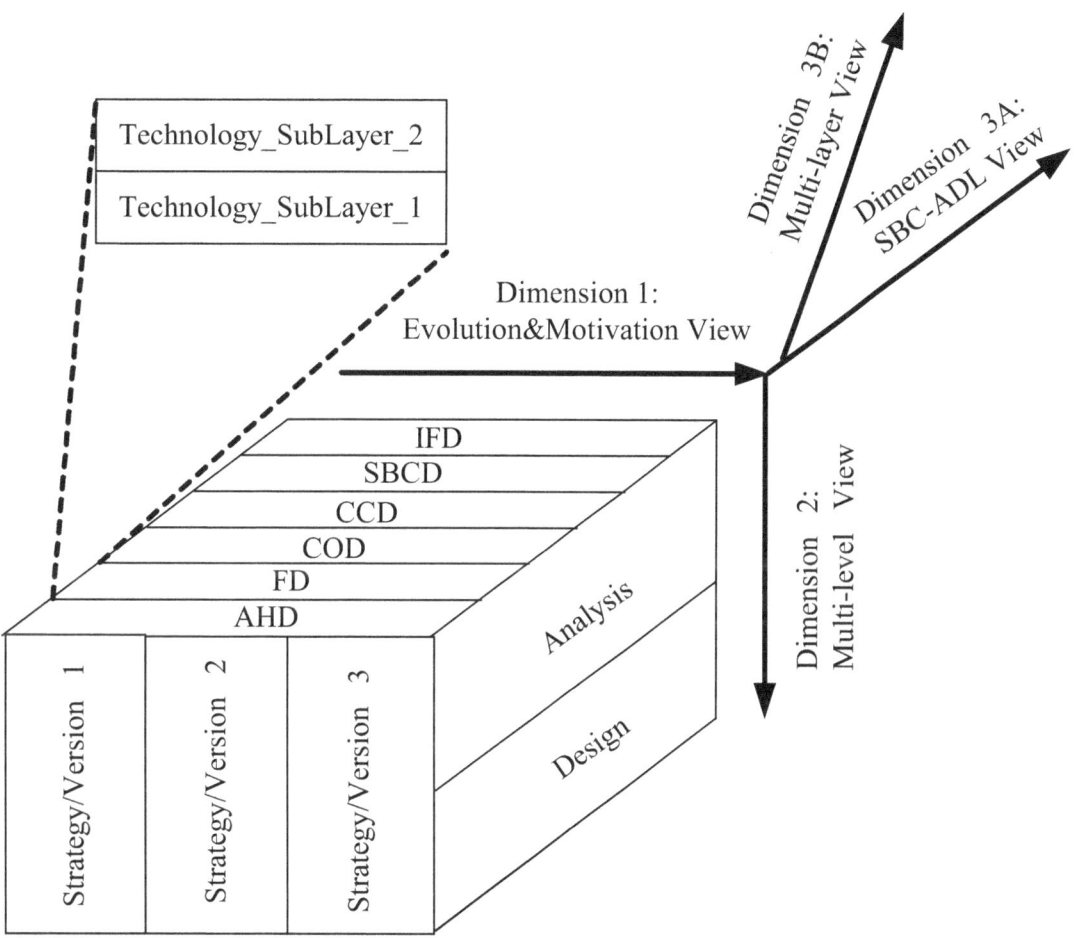

Figure 19-1   General Systems Theory 2.0 Defining the *Robot*

Using the SBC evolution&motivation view, we shall go through: a) robot systems definition strategy/version 1, b) robot systems definition strategy/version 2 and c) robot systems definition strategy/version 3, to accomplish general systems theory 2.0 (general architectural theory) systems definition of the *robot*.

19-1 Robot Systems Definition Strategy/Version 1

The *robot* motivation model shown in Figure 19-2, being a higher-order system, has the *robot* systems definition strategy 1 as its input and the *robot* systems definition version 1 as its output.

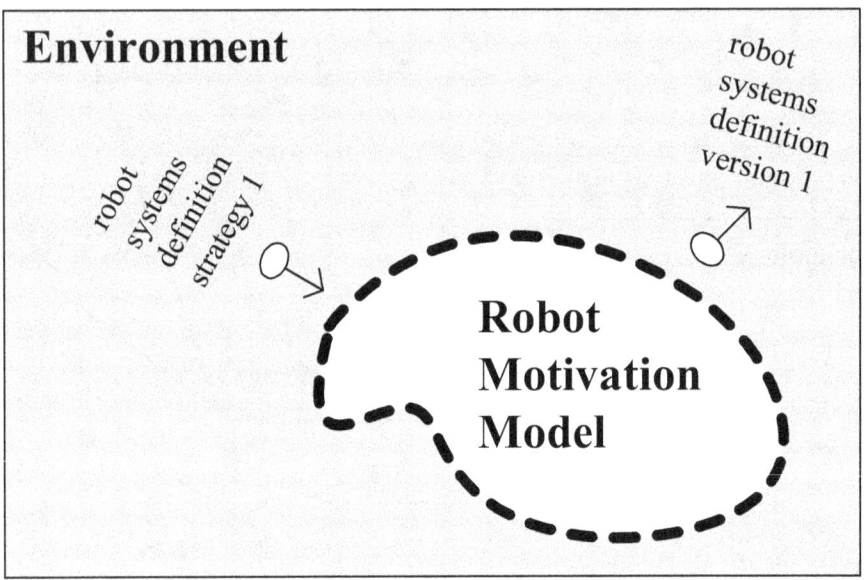

Figure 19-2   *Robot* Motivation Model is a Higher-Order System

The *robot* systems definition strategy 1 mapped to the *robot* systems definition version 1 can be represented as an ordered pair, as shown in Figure 19-3.

(*Robot* Systems Definition Strategy 1   ⟶   *Robot* Systems Definition Version 1)

Figure 19-3   An Ordered Pair

19-1-1 Robot Systems Definition Strategy 1

Here, we use (a) goal drivers, (b) goal assumptions, (c) goal constraints and (d) SWOT analysis, to illustrate the strategic means of the "*robot* systems definition strategy 1".

Goal drivers are up from the policy considerations, the goal driver is kind of why we want to have this *robot* systems definition version 1. The goal drivers of *robot* systems definition version 1 are: nowadays, the robot system can write and walk is very attractive to most customers; therefore, producing that kind of robot system can easily lead to the company's fortune.

Goal assumptions are taking into account of those assumptions that have a positive impact on the *robot* systems definition version 1. We assume that if the robot system's price is affordable, then every customer will have a great desire to buy one. This is the major goal assumption of this strategy.

Goal constraints are up from the policy considerations, the goal constraints are related to those restrictions which have a negative impact on the *robot* systems definition version 1. If the company is short of fund to invest in this new project, then this would become the goal constraint of this strategy.

SWOT analysis is to analyze the internal strengths, weaknesses, opportunities and threats, and so for executing this *robot* systems definition strategy 1. Being a leader of robot systems manufacturer, it should be trivial for the company to produce the robot system which is able to write and walk. This is the internal strength of this company. However, kind of bulky makes the company may not react fast enough to carry out this new project. This is the internal weakness of this company.

19-1-2 Robot Systems Definition Version 1

Using the SBC multi-level (hierarchical) view, an architect goes through: a) analysis view and b) design view for the *robot* systems definition version 1 as shown in Figure 19-4.

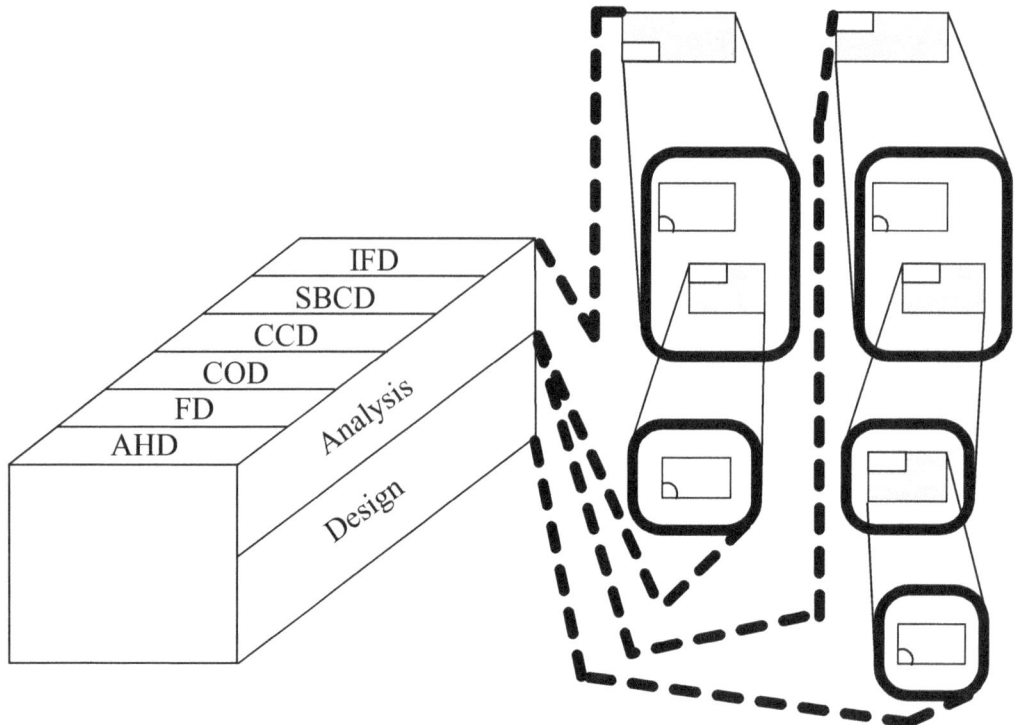

Figure 19-4   *Robot* Systems Definition Version 1

19-1-2-1 Analysis View of the Robot Systems Definition Version 1

The analysis view of the *robot* systems definition version 1 is shown in Figure 19-5.

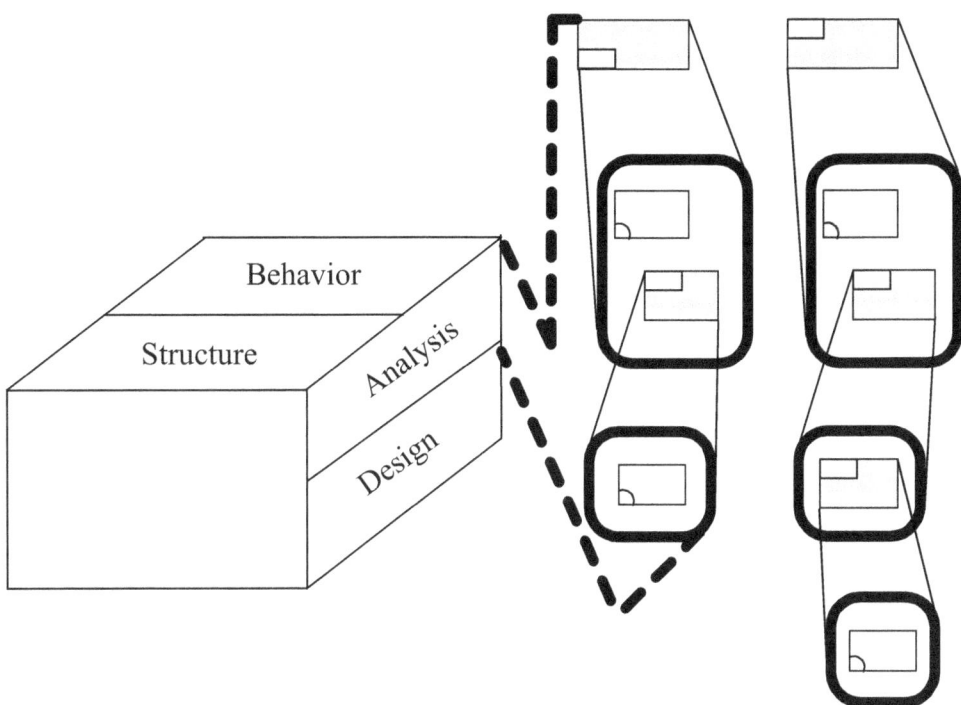

Figure 19-5   Analysis View of the *Robot* Systems Definition Version 1

The analysis view of the *robot* systems definition version 1 consists of: a) analysis' systems structure of the *robot* systems definition version 1 and b) analysis' systems behavior of the *robot* systems definition version 1.

19-1-2-1-1 Analysis' Systems Structure of the Robot Systems Definition Version 1

The entire analysis' systems structure of the *robot* systems definition version 1 includes: a) *Analysis' AHD*, b) *Analysis' FD*, c) *Analysis' COD* and d) *Analysis' CCD* of the *robot* systems definition version 1.

We first draw the analysis' AHD of the *robot* systems definition version 1. As shown in Figure 19-6, *Robot* is composed of *Head* and *Limb*. In the figure, *Robot* is an aggregated system while *Head* and *Limb* are non-aggregated systems.

186

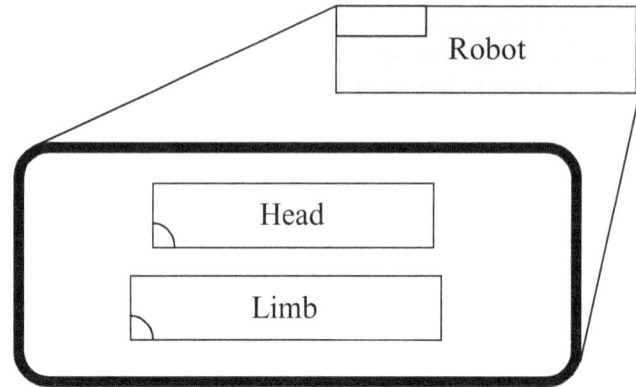

Figure 19-6   Analysis' AHD of the *Robot* Systems Definition Version 1

Figure 19-7 shows the analysis' FD of the *robot* systems definition version 1. In the figure, *Technology_SubLayer_2* contains the *Head* component; *Technology_SubLayer_1* contains the *Limb* component.

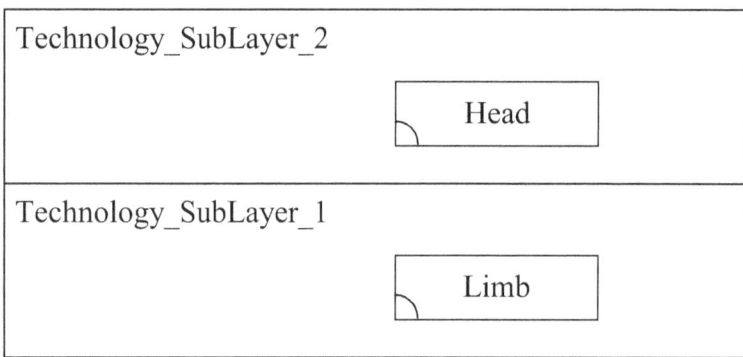

Figure 19-7   Analysis' FD of the *Robot* Systems Definition Version 1

Figure 19-8 shows the analysis' COD of the *robot* systems definition version 1. In the figure, component *Head* has two operations: *Receive_Write_Signal* and *Receive_Walk_Signal*; component *Limb* has two operation: *Move_Hand* and *Move_Leg*.

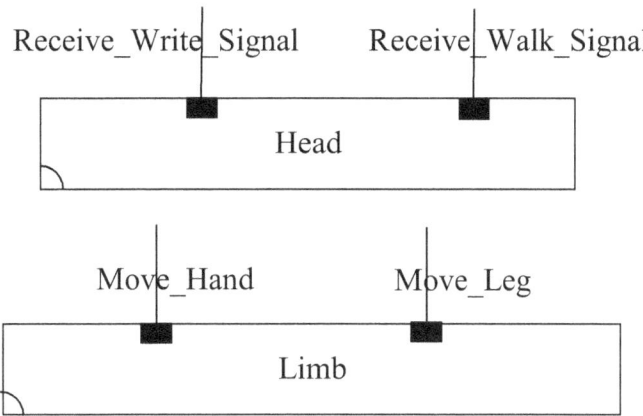

Figure 19-8   Analysis' COD of the *Robot* Systems Definition Version 1

Figure 19-9 shows the analysis' CCD of the *robot* systems definition version 1. In the figure, actor *Remote_Controller* has two connections with the *Head* component; component *Head* also has two connections with the *Limb* component.

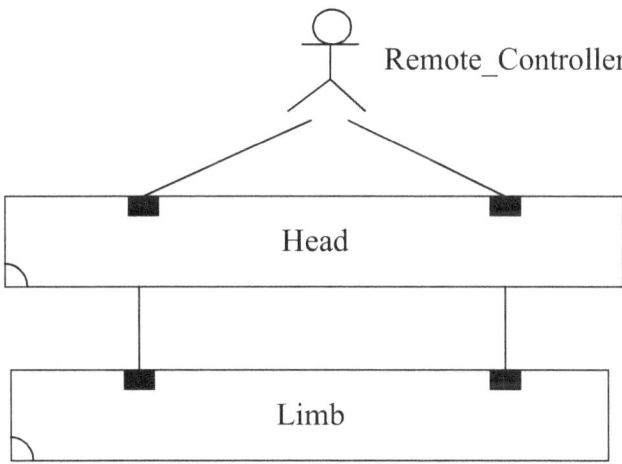

Figure 19-9   Analysis' CCD of the *Robot* Systems Definition Version 1

19-1-2-1-2 Analysis' Systems Behavior of the Robot Systems Definition Version 1

The entire analysis' systems behavior of the *robot* systems definition version 1 includes: a) *Analysis' SBCD* and b) *Analysis' IFD* of the *robot* systems definition version 1.

Figure 19-10 shows the analysis' SBCD of the *robot* systems definition version 1 in which interactions among the *Remote_Controller* actor and the *Head, Limb* components shall draw forth the *Writing* and *Walking* behaviors.

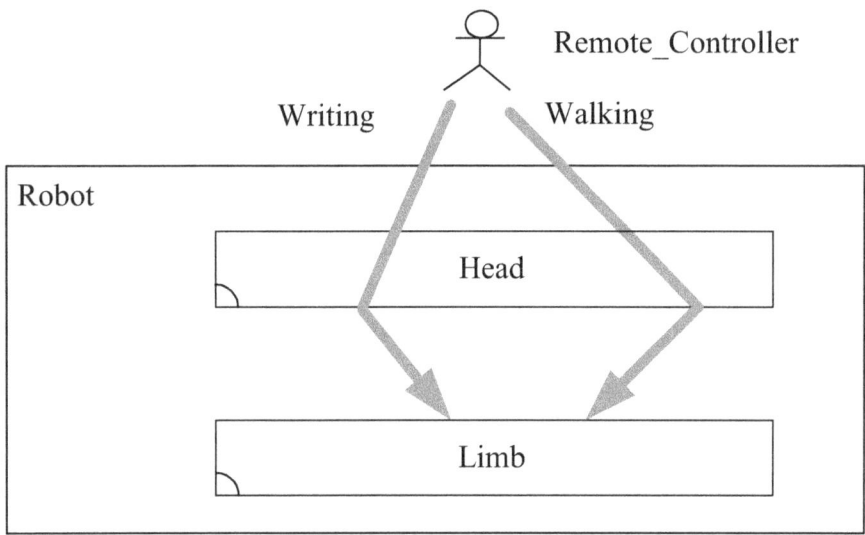

Figure 19-10   Analysis' SBCD of the *Robot* Systems Definition Version 1

The overall analysis' behavior of the *robot* systems definition version 1 includes the *Writing* and *Walking* behaviors. In other words, behaviors *Writing* and *Walking* together provide the overall analysis' behavior of the *robot* systems definition version 1.

Be noticed that the *Writing* and *Walking* behaviors are mutually independent of each other. They tend to be executed concurrently [Hoar85, Miln89, Miln99].

The overall behavior of the *robot* systems definition version 1 includes two individual behaviors: *Writing* and *Walking*. Each individual behavior is represented by an execution path. We use an interaction flow diagram (IFD) to define each one of these execution paths. Figure 19-11 shows the analysis' IFD of the *robot* systems definition version 1 *Writing* behavior. First, actor *Remote_Controller* interacts with the *Head* component through the operation call interaction *Receive_Write_Signal*. Finally, component *Head* interacts with the *Limb* component through the *Move_Hand* operation call interaction.

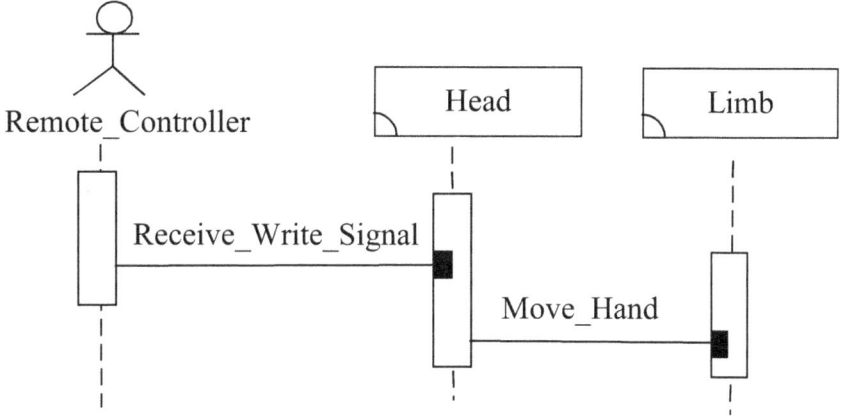

Figure 19-11　　Analysis' IFD of
the *Robot* Systems Definition Version 1 *Writing* Behavior

Figure 19-12 shows the analysis' IFD of the *robot* systems definition version 1 *Walking* behavior. First, actor *Remote_Controller* interacts with the *Head* component through the *Receive_Walk_Signal* operation call interaction. Finally, component *Head* interacts with the *Limb* component through the *Move_Leg* operation call interaction.

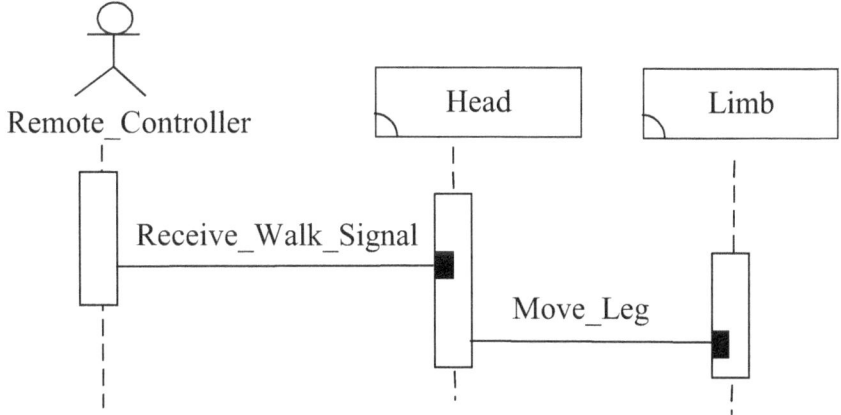

Figure 19-12　　Analysis' IFD of
the *Robot* Systems Definition Version 1 *Walking* Behavior

19-1-2-2 Design View of the Robot Systems Definition Version 1

The design view of the *robot* systems definition version 1 is shown in Figure 19-13.

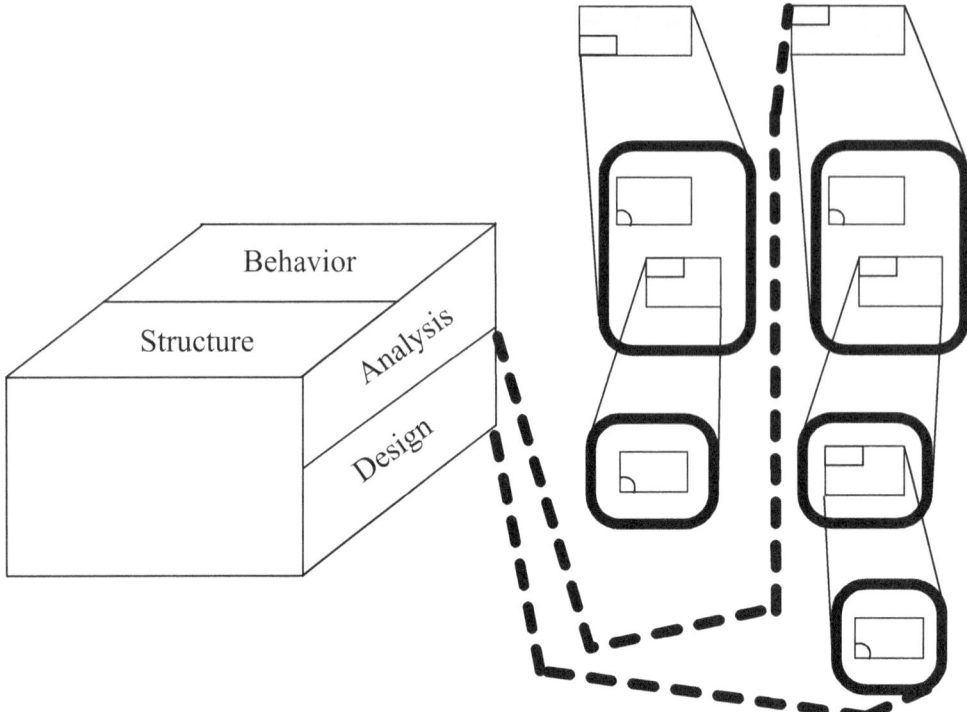

Figure 19-13   Design View of the *Robot* Systems Definition Version 1

The design view of the *robot* systems definition version 1 consists of: a) design's systems structure of the *robot* systems definition version 1 and b) design's systems behavior of the *robot* systems definition version 1.

19-1-2-2-1 Design's Systems Structure of the Robot Systems Definition Version 1

The entire design's systems structure of the *robot* systems definition version 1 includes: a) *Design's AHD*, b) *Design's FD*, c) *Design's COD* and d) *Design's CCD* of the *robot* systems definition version 1.

We first draw the design's AHD of the *robot* systems definition version 1. As shown in Figure 19-14, *Robot* is composed of *Head* and *Limb*; *Limb* is composed of *Hands* and *Legs*. In the figure, *Robot* and *Limb* are aggregated systems while *Head*, *Hands* and *Legs* are non-aggregated systems.

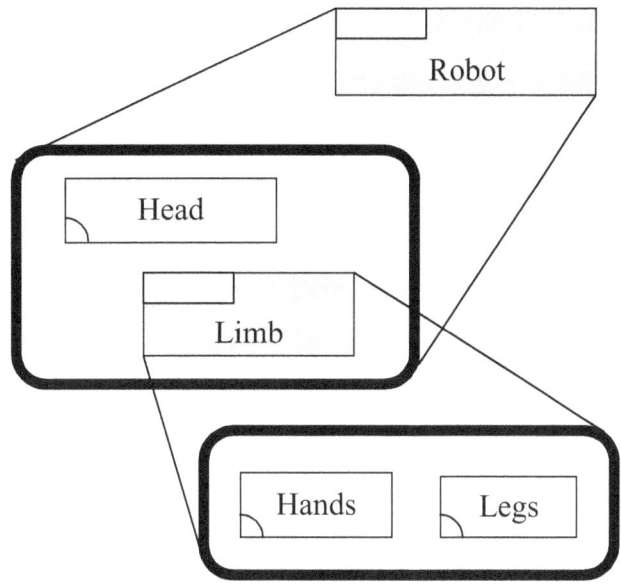

Figure 19-14   Design's AHD of the *Robot* Systems Definition Version 1

Figure 19-15 shows the design's FD of the *robot* systems definition version 1. In the figure, *Technology_SubLayer_2* contains the *Head* component; *Technology_SubLayer_1* contains the *Hands* and *Legs* components.

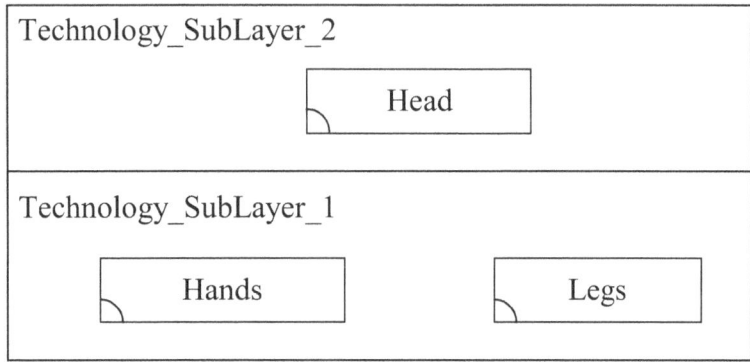

Figure 19-15   Design's FD of the *Robot* Systems Definition Version 1

Figure 19-16 shows the design's COD of the *robot* systems definition version 1. In the figure, component *Head* has two operations: *Receive_Write_Signal* and *Receive_Walk_Signal*; component *Hands* has one operation: *Move_Hand*; component *Legs* has one operation: *Move_Leg*.

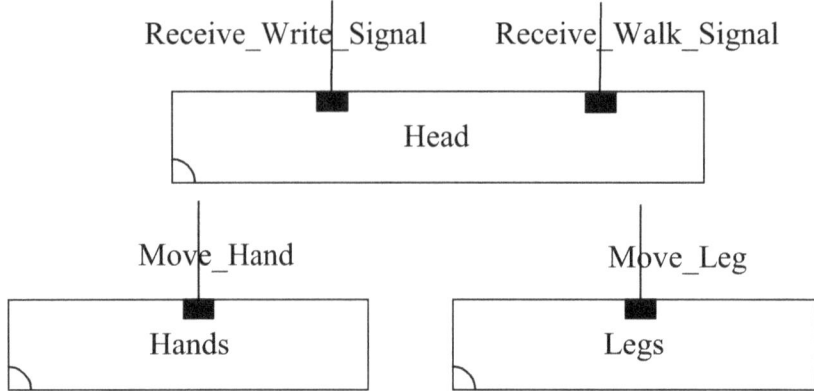

Figure 19-16   Design's COD of the *Robot* Systems Definition Version 1

Figure 19-17 shows the design's CCD of the *robot* systems definition version 1. In the figure, actor *Remote_Controller* has a connection with the *Head* component; component *Head* has a connection with each one of the *Hands* and *Legs* components.

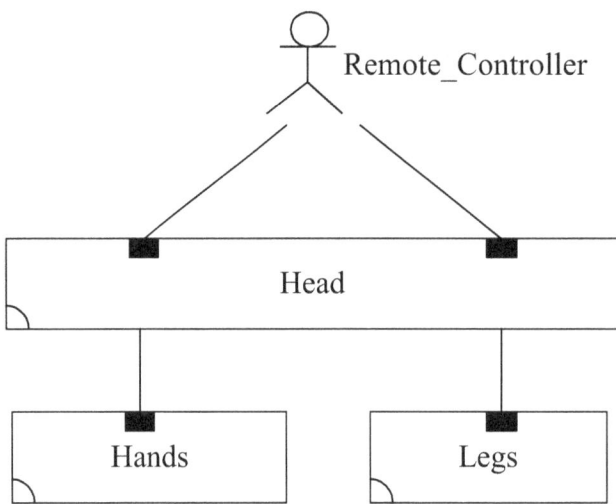

Figure 19-17   Design's CCD of the *Robot* Systems Definition Version 1

19-1-2-2-2 Design's Systems Behavior of the Robot Systems Definition Version 1

The entire design's systems behavior of the *robot* systems definition version 1 includes: a) *Design's SBCD* and b) *Design's IFD* of the *robot* systems definition version 1.

Figure 19-18 shows the design's SBCD of the *robot* systems definition version 1 in which interactions among the *Remote_Controller* actor and the *Head*, *Hand*s, *Legs* components shall draw forth the *Writing* and *Walking* behaviors.

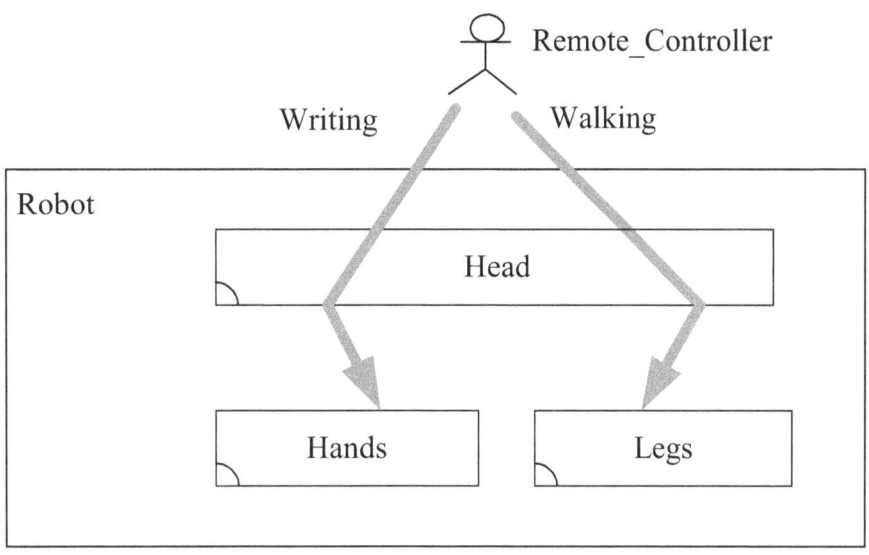

Figure 19-18   Design's SBCD of the *Robot* Systems Definition Version 1

The overall design's behavior of the *robot* systems definition version 1 includes the *Writing* and *Walking* behaviors. In other words, behaviors *Writing* and *Walking* together provide the overall design's behavior of the *robot* systems definition version 1.

Be noticed that the *Writing* and *Walking* behaviors are mutually independent of each other. They tend to be executed concurrently [Hoar85, Miln89, Miln99].

The overall design's behavior of the *robot* systems definition version 1 includes two individual behaviors: *Writing* and *Walking*. Each individual behavior is represented by an execution path. We use an interaction flow diagram (IFD) to define each one of these execution paths. Figure 19-19 shows the design's IFD of the *robot* systems definition version 1 *Writing* behavior. First, actor *Remote_Controller* interacts with the *Head* component through the operation call interaction *Receive_Write_Signal*. Finally, component *Head* interacts with the *Hands* component through the *Move_Hand* operation call interaction.

194

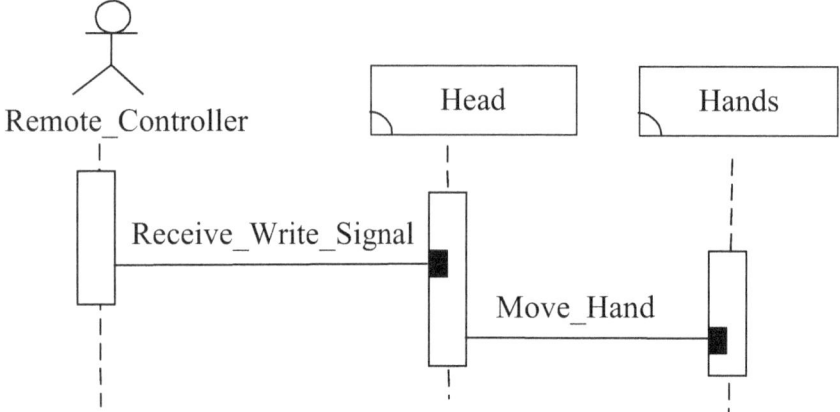

Figure 19-19   Design's IFD of
the *Robot* Systems Definition Version 1 *Writing* Behavior

Figure 19-20 shows the design's IFD of the *robot* systems definition version 1 *Walking* behavior. First, actor *Remote_Controller* interacts with the *Head* component through the *Receive_Walk_Signal* operation call interaction. Finally, component *Head* interacts with the *Legs* component through the *Move_Leg* operation call interaction.

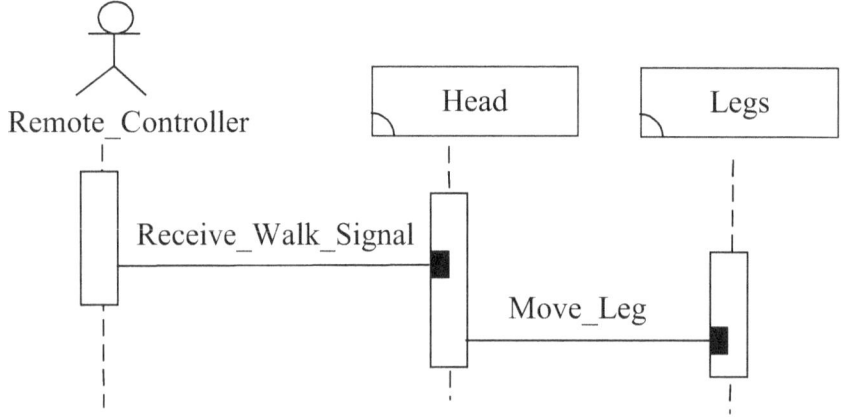

Figure 19-20   Design's IFD of
the *Robot* Systems Definition Version 1 *Walking* Behavior

19-2 Robot Systems Definition Strategy/Version 2

The *robot* motivation model shown in Figure 19-21, being a higher-order system, has the *robot* systems definition strategy 2 as its input and the *robot* systems definition version 2 as its output.

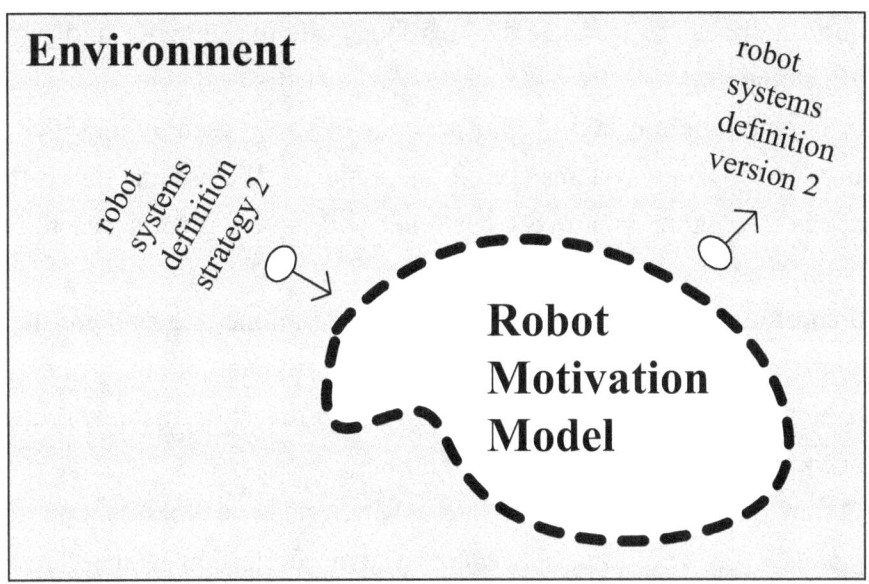

Figure 19-21   *Robot* Motivation Model is a Higher-Order System

The *robot* systems definition strategy 2 mapped to the *robot* systems definition version 2 can be represented as an ordered pair, as shown in Figure 19-22.

(*Robot* Systems Definition Strategy 2   ⟶   *Robot* Systems Definition Version 2)

Figure 19-22   An Ordered Pair

19-2-1 Robot Systems Definition Strategy 2

Here, we use (a) goal drivers, (b) goal assumptions, (c) goal constraints and (d) SWOT analysis, to illustrate the strategic means of the "robot systems definition strategy 2".

Goal drivers are up from the policy considerations, the goal driver is kind of wahy we want to have this robot systems definition version 2. The goal drivers of robot systems definition version 2 are: currently, a robot system can only write and walk is losing its appeal to customers; the robot system can write and jump shall attract more customers.

Goal assumptions are taking into account of those assumptions that have a positive impact on the robot systems definition version 2. We assume that if the robot system's price is affordable, then every customer will have a great desire to buy one. This is the major goal assumption of this strategy.

Goal constraints are up from the policy considerations, the goal constraints are related to those restrictions which have a negative impact on the robot systems definition version 2. If the company is short of fund to invest in this new project, then this would become the goal constraint of this strategy.

SWOT analysis is to analyze the internal strengths, weaknesses, opportunities, and threats, and so for executing this robot systems definition strategy 2. Being a leader of robot systems manufacturer, it should be trivial for the company to produce the robot system which is able to write and jump. This is the internal strength of this company. However, kind of bulky makes the company may not react fast enough to carry out this new project. This is the internal weakness of this company.

19-2-2 Robot Systems Definition Version 2

Using the SBC multi-level (hierarchical) view, an architect goes through: a) analysis view and b) design view for the robot systems definition version 2 as shown in Figure 19-23.

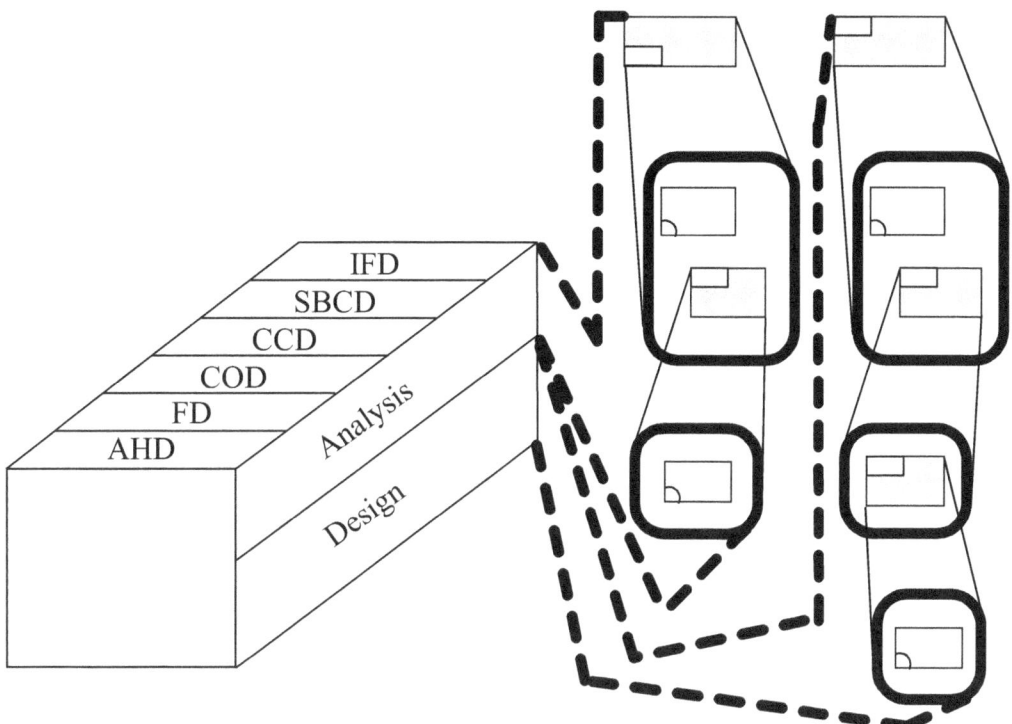

Figure 19-23   *Robot* Systems Definition Version 2

19-2-2-1 Analysis View of the Robot Systems Definition Version 2

The analysis view of the *robot* systems definition version 2 is shown in Figure 19-24.

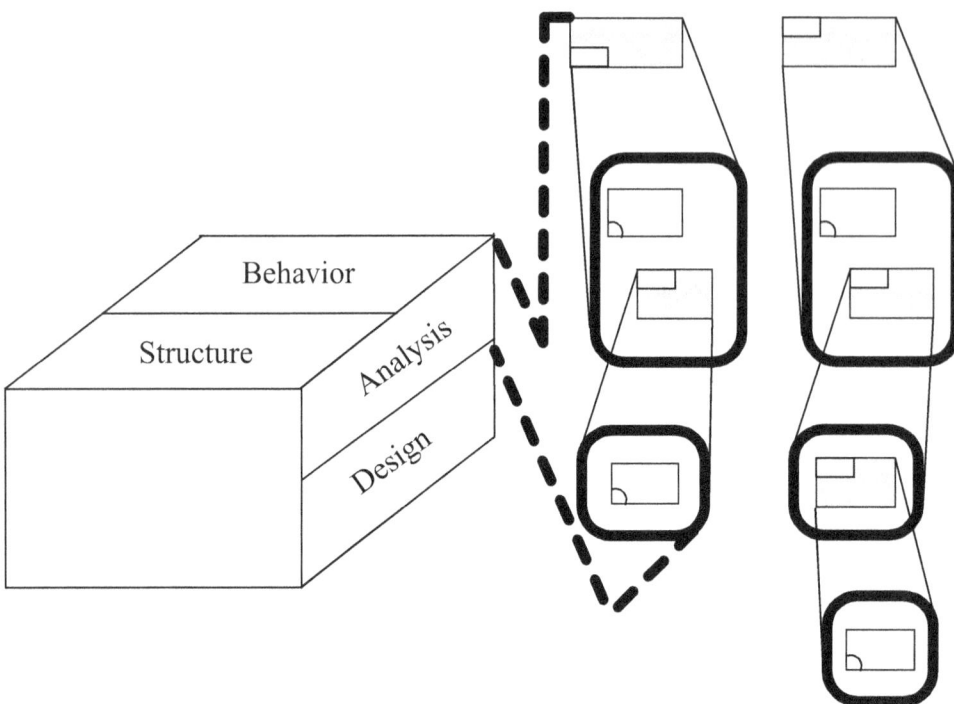

Figure 19-24   Analysis View of the *Robot* Systems Definition Version 2

The analysis view of the *robot* systems definition version 2 consists of: a) analysis' systems structure of the *robot* systems definition version 2 and b) analysis' systems behavior of the *robot* systems definition version 2.

19-2-2-1-1 Analysis' Systems Structure of the Robot Systems Definition Version 2

The entire analysis' systems structure of the *robot* systems definition version 2 includes: a) *Analysis' AHD*, b) *Analysis' FD*, c) *Analysis' COD* and d) *Analysis' CCD* of the *robot* systems definition version 2.

We first draw the analysis' AHD of the *robot* systems definition version 2. As shown in Figure 19-25, *Robot* is composed of *Head* and *Limb*. In the figure, *Robot* is an aggregated system while *Head* and *Limb* are non-aggregated systems.

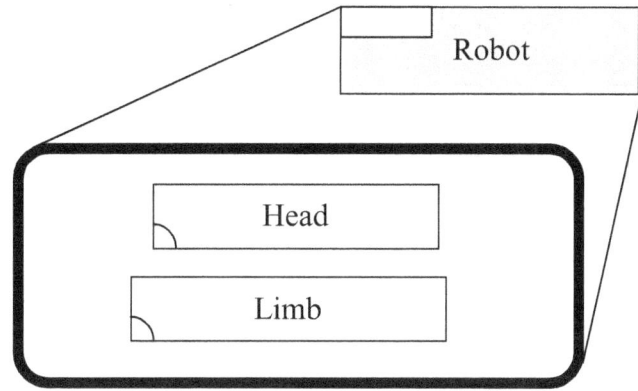

Figure 19-25   Analysis' AHD of the *Robot* Systems Definition Version 2

Figure 19-26 shows the analysis' FD of the *robot* systems definition version 2. In the figure, *Technology_SubLayer_2* contains the *Head* component; *Technology_SubLayer_1* contains the *Limb* component.

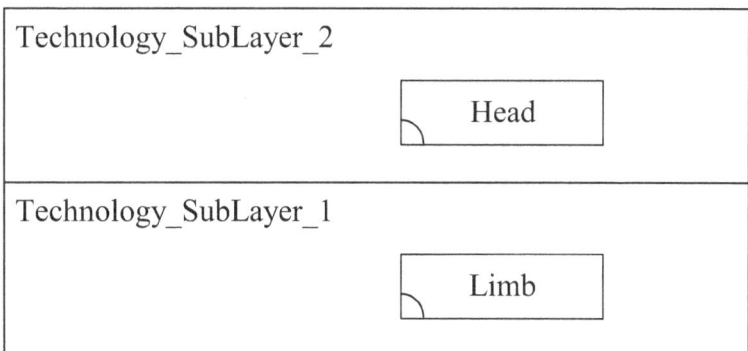

Figure 19-26   Analysis' FD of the *Robot* Systems Definition Version 2

Figure 19-27 shows the analysis' COD of the *robot* systems definition version 2. In the figure, component *Head* has two operations: *Receive_Write_Signal* and *Receive_Jump_Signal*; component *Limb* has two operation: *Move_Hand* and *Move_Spring_Leg*.

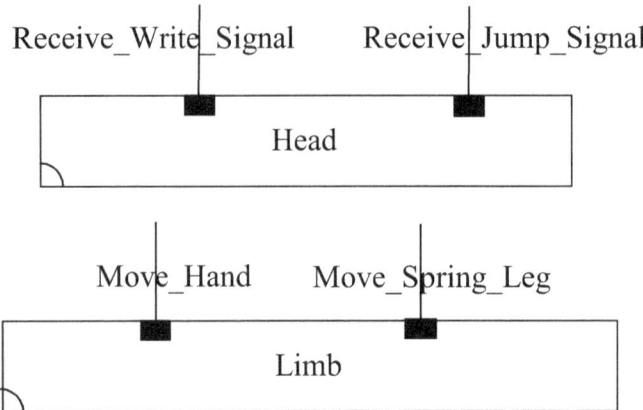

Figure 19-27   Analysis' COD of the *Robot* Systems Definition Version 2

Figure 19-28 shows the analysis' CCD of the *robot* systems definition version 2. In the figure, actor *Remote_Controller* has two connections with the *Head* component; component *Head* also has two connections with the *Limb* component.

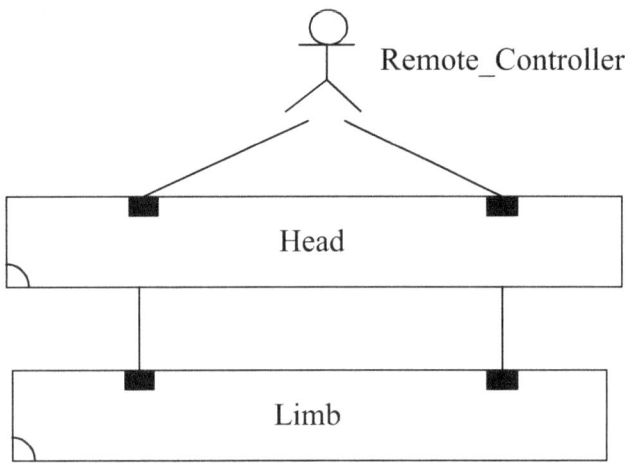

Figure 19-28   Analysis' CCD of the *Robot* Systems Definition Version 2

19-2-2-1-2 Analysis' Systems Behavior of the Robot Systems Definition Version 2

The entire analysis' systems behavior of the *robot* systems definition version 2 includes: a) *Analysis' SBCD* and b) *Analysis' IFD* of the *robot* systems definition version 2.

Figure 19-29 shows the analysis' SBCD of the *robot* systems definition version 2 in which interactions among the *Remote_Controller* actor and the *Head, Limb* components shall draw forth the *Writing* and *Jumping* behaviors.

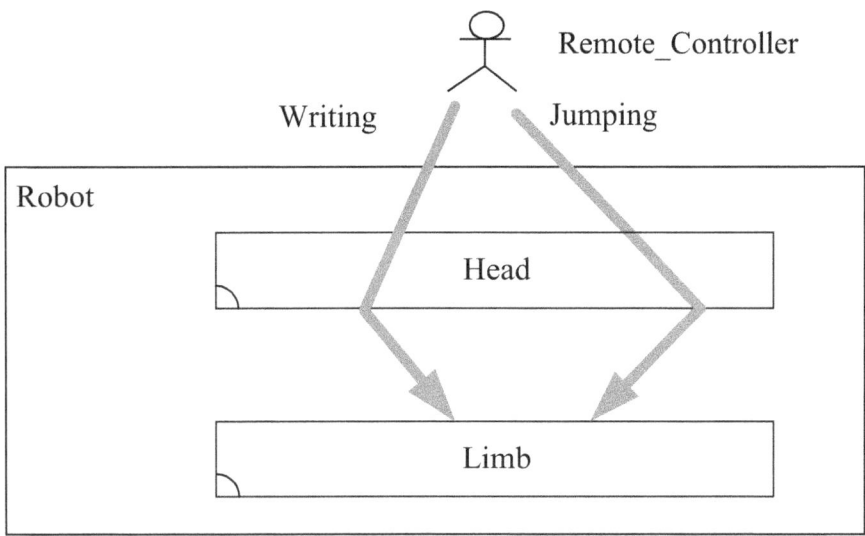

Figure 19-29   Analysis' SBCD of the *Robot* Systems Definition Version 2

The overall analysis' behavior of the *robot* systems definition version 2 includes the *Writing* and *Jumping* behaviors. In other words, behaviors *Writing* and *Jumping* together provide the overall analysis' behavior of the *robot* systems definition version 2.

Be noticed that the *Writing* and *Jumping* behaviors are mutually independent of each other. They tend to be executed concurrently [Hoar85, Miln89, Miln99].

The overall behavior of the *robot* systems definition version 2 includes two individual behaviors: *Writing* and *Jumping*. Each individual behavior is represented by an execution path. We use an interaction flow diagram (IFD) to define each one of these execution paths. Figure 19-30 shows the analysis' IFD of the *robot* systems definition version 2 *Writing* behavior. First, actor *Remote_Controller* interacts with the *Head* component through the operation call interaction *Receive_Write_Signal*. Finally, component *Head* interacts with the *Limb* component through the *Move_Hand* operation call interaction.

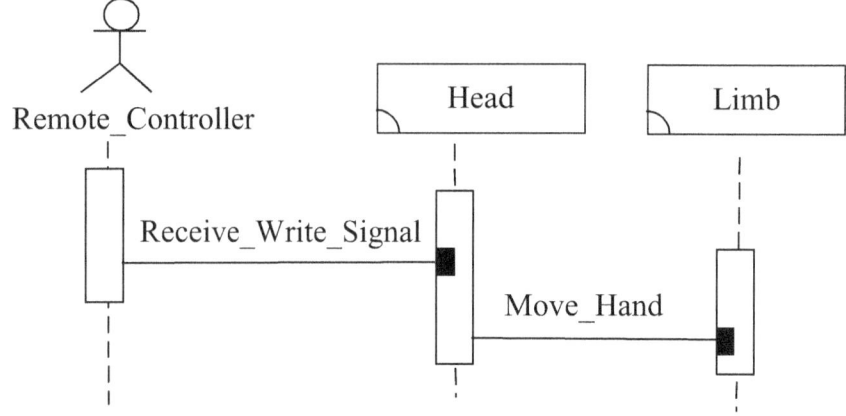

Figure 19-30   Analysis' IFD of
the *Robot* Systems Definition Version 2 *Writing* Behavior

Figure 19-31 shows the analysis' IFD of the *robot* systems definition version 2 *Jumping* behavior. First, actor *Remote_Controller* interacts with the *Head* component through the *Receive_Jump_Signal* operation call interaction. Finally, component *Head* interacts with the *Limb* component through the *Move_Spring_Leg* operation call interaction.

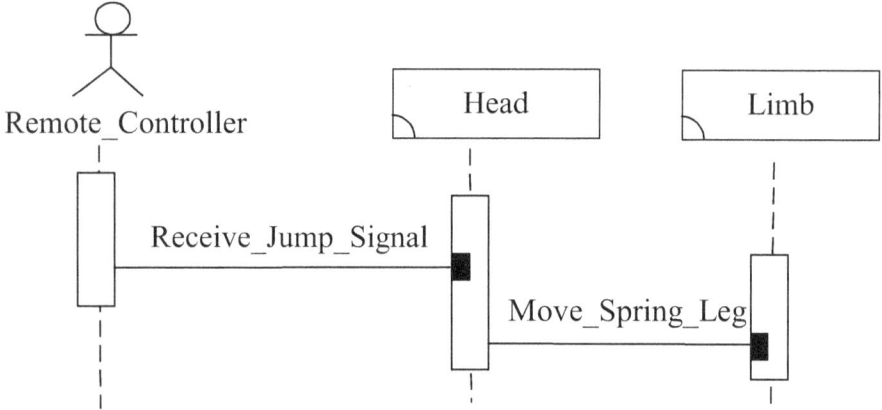

Figure 19-31   Analysis' IFD of
the *Robot* Systems Definition Version 2 *Jumping* Behavior

19-2-2-2 Design View of the Robot Systems Definition Version 2

The design view of the *robot* systems definition version 2 is shown in Figure 19-32.

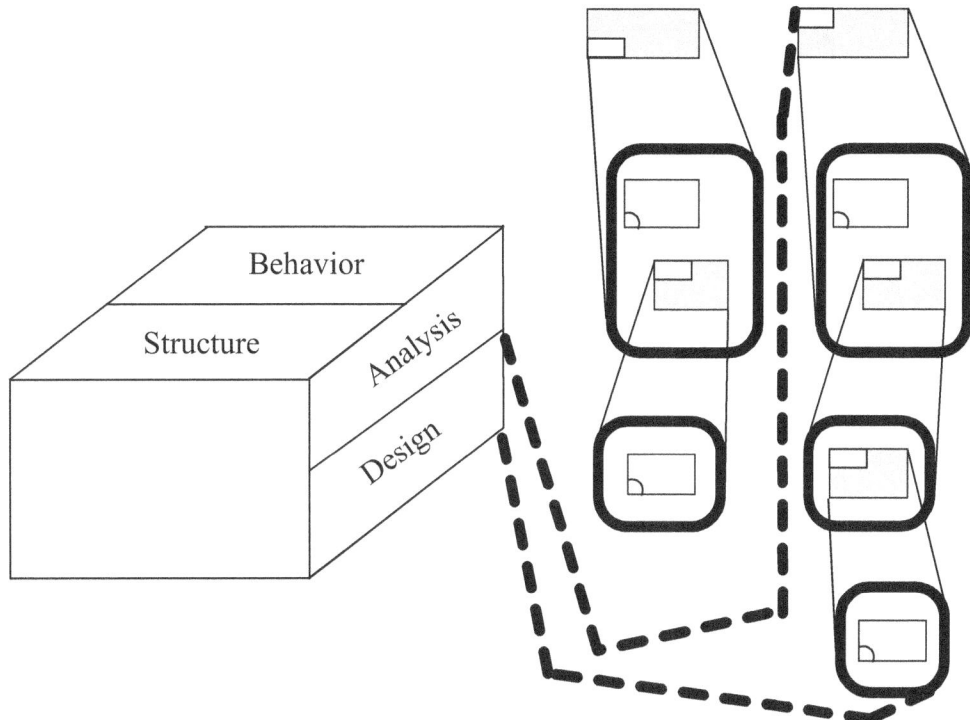

Figure 19-32   Design View of the *Robot* Systems Definition Version 2

The design view of the *robot* systems definition version 2 consists of: a) design's systems structure of the *robot* systems definition version 2 and b) design's systems behavior of the *robot* systems definition version 2.

19-2-2-2-1 Design's Systems Structure of the Robot Systems Definition Version 2

The entire design's systems structure of the *robot* systems definition version 2 includes: a) *Design's AHD*, b) *Design's FD*, c) *Design's COD* and d) *Design's CCD* of the *robot* systems definition version 2.

We first draw the design's AHD of the *robot* systems definition version 2. As shown in Figure 19-33, *Robot* is composed of *Head* and *Limb*; *Limb* is composed of *Hands* and *Spring_Legs*. In the figure, *Robot* and *Limb* are aggregated systems while *Head*, *Hands* and *Spring_Legs* are non-aggregated systems.

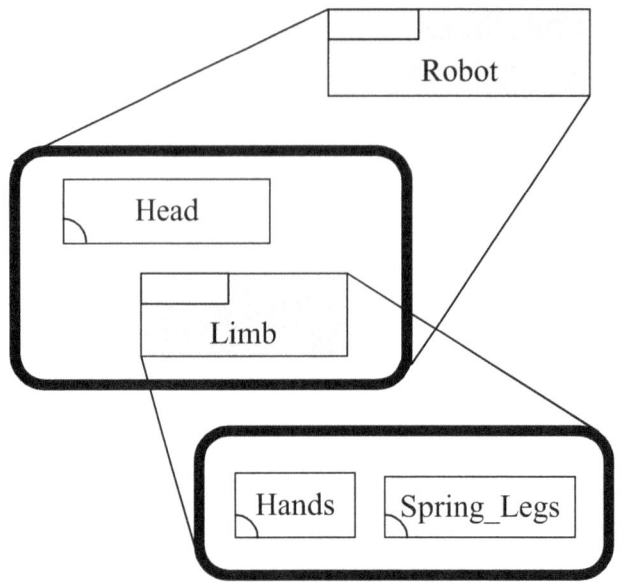

Figure 19-33   Design's AHD of the *Robot* Systems Definition Version 2

Figure 19-34 shows the design's FD of the *robot* systems definition version 2. In the figure, *Technology_SubLayer_2* contains the *Head* component; *Technology_SubLayer_1* contains the *Hands* and *Spring_Legs* components.

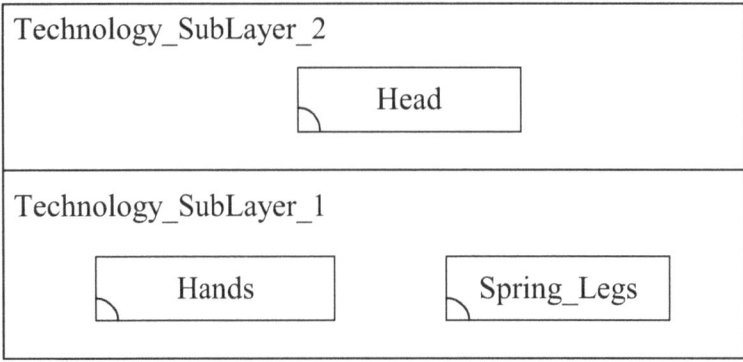

Figure 19-34   Design's FD of the *Robot* Systems Definition Version 2

Figure 19-35 shows the design's COD of the *robot* systems definition version 2. In the figure, component *Head* has two operations: *Receive_Write_Signal* and *Receive_Jump_Signal*; component *Hands* has one operation: *Move_Hand*; component *Spring_Legs* has one operation: *Move_Spring_Leg*.

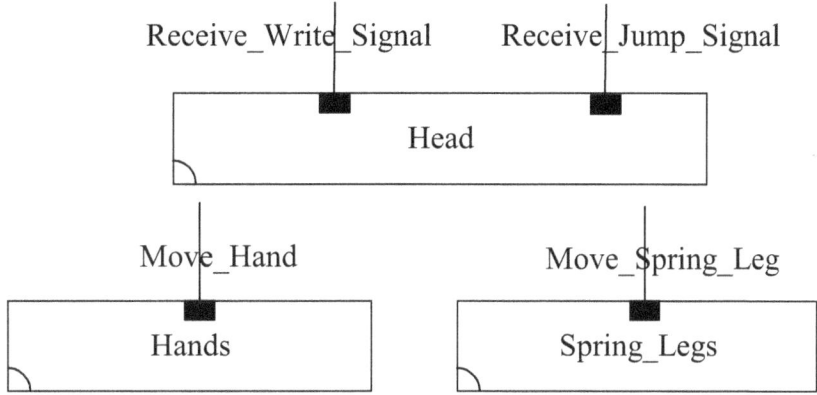

Figure 19-35   Design's COD of the *Robot* Systems Definition Version 2

Figure 19-36 shows the design's CCD of the *robot* systems definition version 2. In the figure, actor *Remote_Controller* has a connection with the *Head* component; component *Head* has a connection with each one of the *Hands* and *Spring_Legs* components.

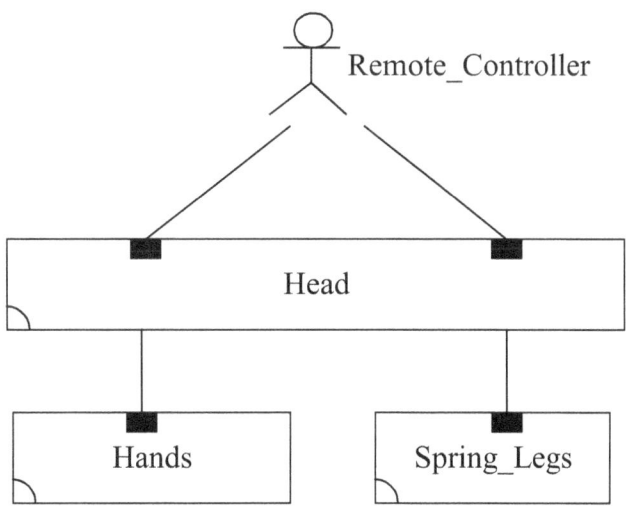

Figure 19-36   Design's CCD of the *Robot* Systems Definition Version 2

19-2-2-2-2 Design's Systems Behavior of the Robot Systems Definition Version 2

The entire design's systems behavior of the *robot* systems definition version 2 includes: a) *Design's SBCD* and b) *Design's IFD* of the *robot* systems definition version 2.

Figure 19-37 shows the design's SBCD of the *robot* systems definition version 2 in which interactions among the *Remote_Controller* actor and the *Head, Hand*s, *Spring_Legs* components shall draw forth the *Writing* and *Jumping* behaviors.

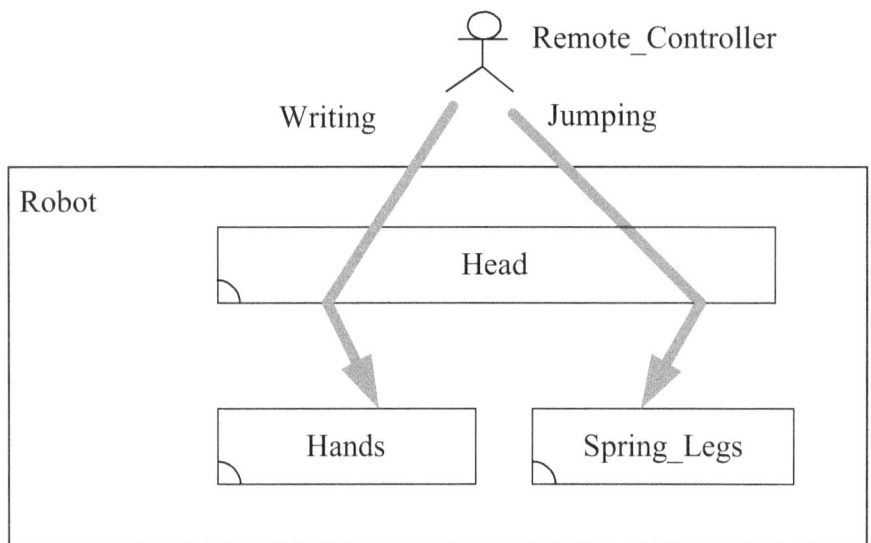

Figure 19-37   Design's SBCD of the *Robot* Systems Definition Version 2

The overall design's behavior of the *robot* systems definition version 2 includes the *Writing* and *Jumping* behaviors. In other words, behaviors *Writing* and *Jumping* together provide the overall design's behavior of the *robot* systems definition version 2.

Be noticed that the *Writing* and *Jumping* behaviors are mutually independent of each other. They tend to be executed concurrently [Hoar85, Miln89, Miln99].

The overall design's behavior of the *robot* systems definition version 2 includes two individual behaviors: *Writing* and *Jumping*. Each individual behavior is represented by an execution path. We use an interaction flow diagram (IFD) to define each one of these execution paths. Figure 19-38 shows the design's IFD of the *robot* systems definition version 2 *Writing* behavior. First, actor *Remote_Controller* interacts with the *Head* component through the operation call interaction *Receive_Write_Signal*. Finally, component *Head* interacts with the *Hands* component through the *Move_Hand* operation call interaction.

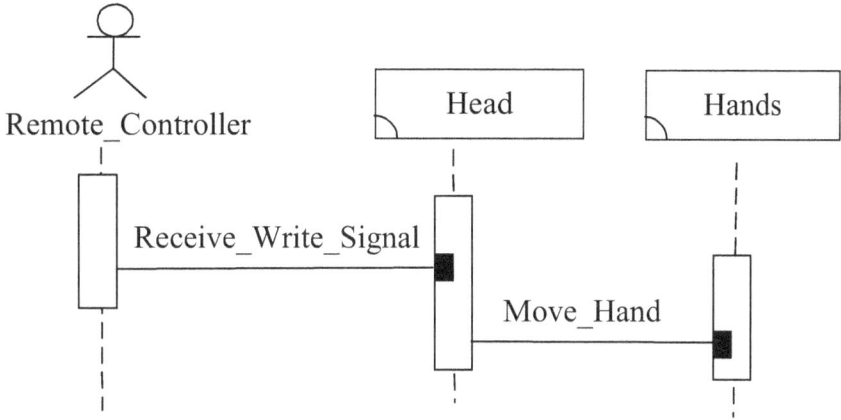

Figure 19-38   Design's IFD of
the *Robot* Systems Definition Version 2 *Writing* Behavior

Figure 19-39 shows the design's IFD of the *robot* systems definition version 2 *Jumping* behavior. First, actor *Remote_Controller* interacts with the *Head* component through the *Receive_Jump_Signal* operation call interaction. Finally, component *Head* interacts with the *Spring_Legs* component through the *Move_Spring_Leg* operation call interaction.

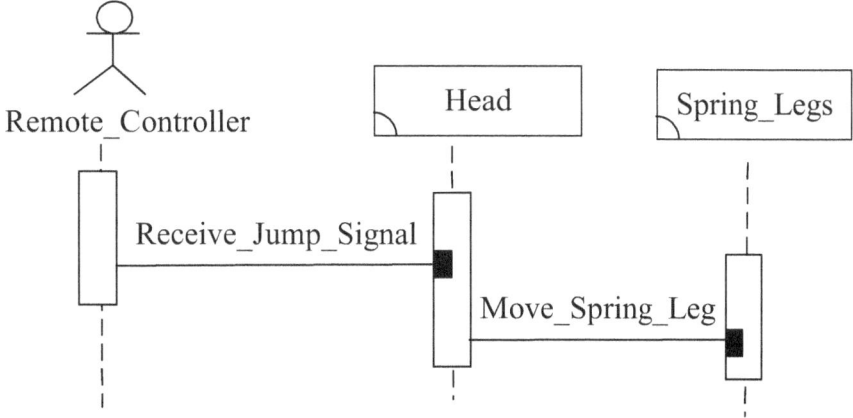

Figure 19-39   Design's IFD of
the *Robot* Systems Definition Version 2 *Jumping* Behavior

19-3 Robot Systems Definition Strategy/Version 3

The *robot* motivation model shown in Figure 19-40, being a higher-order system, has the *robot* systems definition strategy 3 as its input and the *robot* systems definition version 3 as its output.

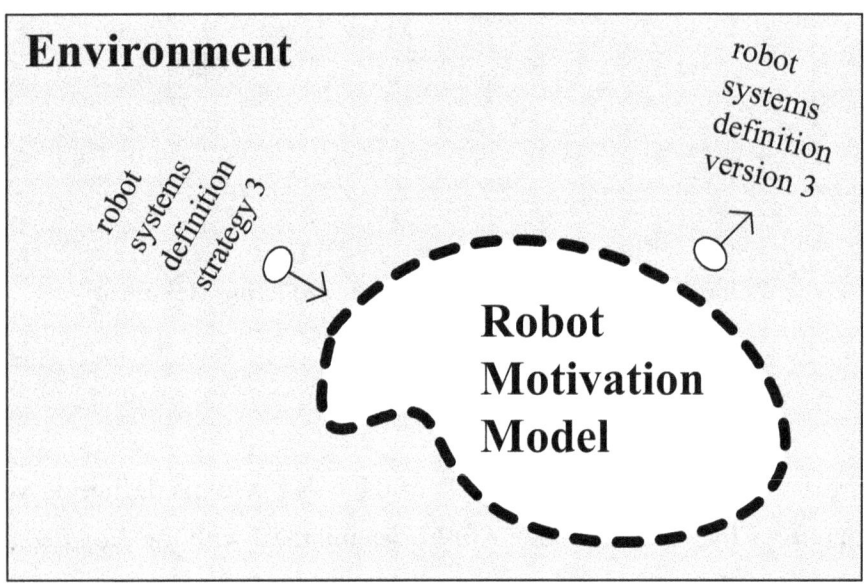

Figure 19-40   *Robot* Motivation Model is a Higher-Order System

The *robot* systems definition strategy 3 mapped to the *robot* systems definition version 3 can be represented as an ordered pair, as shown in Figure 19-41.

(*Robot* Systems Definition Strategy 3 ⟶ *Robot* Systems Definition Version 3)

Figure 19-41   An Ordered Pair

19-3-1 Robot Systems Definition Strategy 3

Here, we use (a) goal drivers, (b) goal assumptions, (c) goal constraints and (d) SWOT analysis, to illustrate the strategic means of the "*robot* systems definition strategy 3".

Goal drivers are up from the policy considerations, the goal driver is kind of wahy we want to have this *robot* systems definition version 3. The goal drivers of *robot* systems definition version 3 are: at present, a robot system can only write and jump is losing its business; only the robot system can write and fly will expand the company's business.

Goal assumptions are taking into account of those assumptions that have a positive impact on the *robot* systems definition version 3. We assume that if the robot system's price is affordable, then every customer will have a great desire to buy one. This is the major goal assumption of this strategy.

Goal constraints are up from the policy considerations, the goal constraints are related to those restrictions which have a negative impact on the *robot* systems definition version 3. If the company is short of fund to invest in this new project, then this would become the goal constraint of this strategy.

SWOT analysis is to analyze the internal strengths, weaknesses, opportunities and threats, and so for executing this *robot* systems definition strategy 3. Being a leader of robot systems manufacturer, it should be trivial for the company to produce the robot system which is able to write and fly. This is the internal strength of this company. However, kind of bulky makes the company may not react fast enough to carry out this new project. This is the internal weakness of this company.

19-3-2 Robot Systems Definition Version 3

Using the SBC multi-level (hierarchical) view, an architect goes through: a) analysis view and b) design view for the *robot* systems definition version 3 as shown in Figure 19-42.

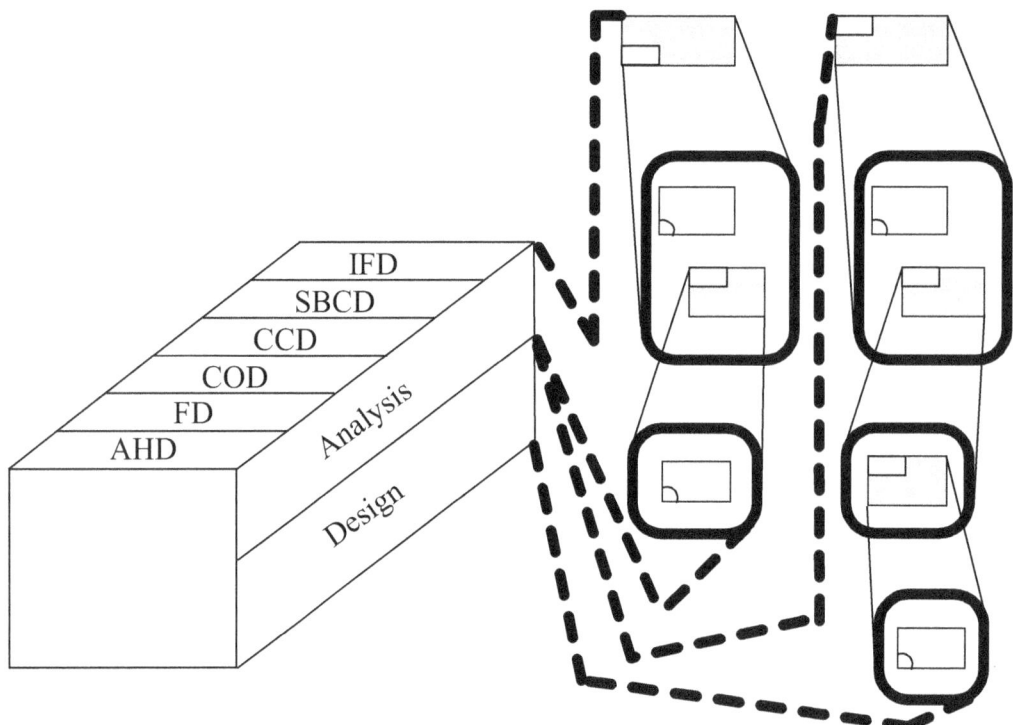

Figure 19-42   *Robot* Systems Definition Version 3

19-3-2-1 Analysis View of the Robot Systems Definition Version 3

The analysis view of the *robot* systems definition version 3 is shown in Figure 19-43.

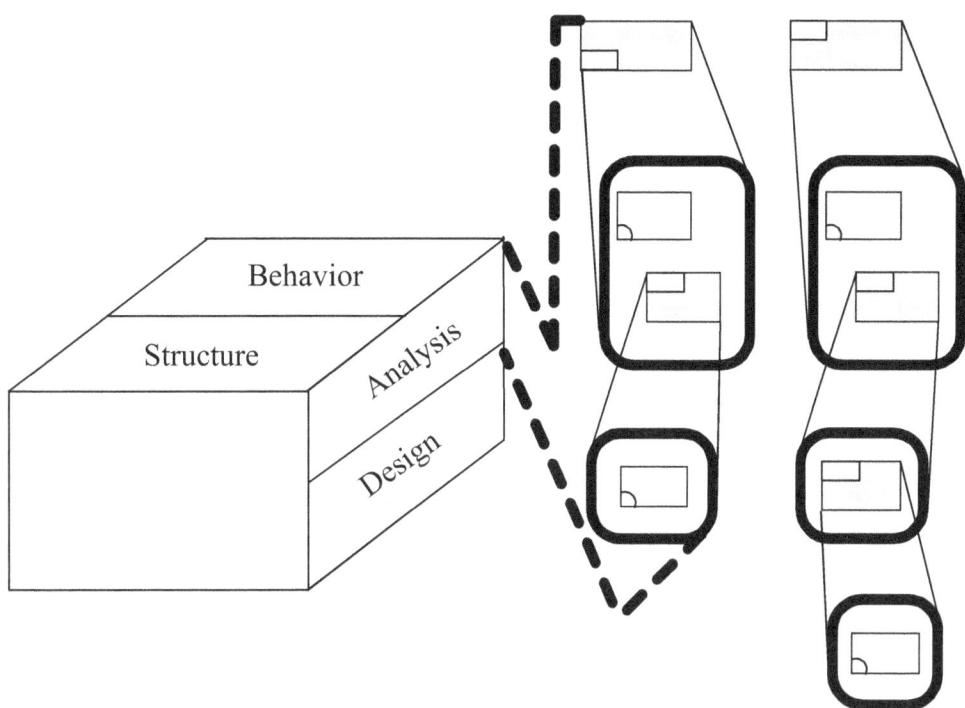

Figure 19-43   Analysis View of the *Robot* Systems Definition Version 3

The analysis view of the *robot* systems definition version 3 consists of: a) analysis' systems structure of the *robot* systems definition version 3 and b) analysis' systems behavior of the *robot* systems definition version 3.

19-3-2-1-1 Analysis' Systems Structure of the Robot Systems Definition Version 3

The entire analysis' systems structure of the *robot* systems definition version 3 includes: a) *Analysis' AHD*, b) *Analysis' FD*, c) *Analysis' COD* and d) *Analysis' CCD* of the *robot* systems definition version 3.

We first draw the analysis' AHD of the *robot* systems definition version 3. As shown in Figure 19-44, *Robot* is composed of *Head* and *Limb*. In the figure, *Robot* is an aggregated system while *Head* and *Limb* are non-aggregated systems.

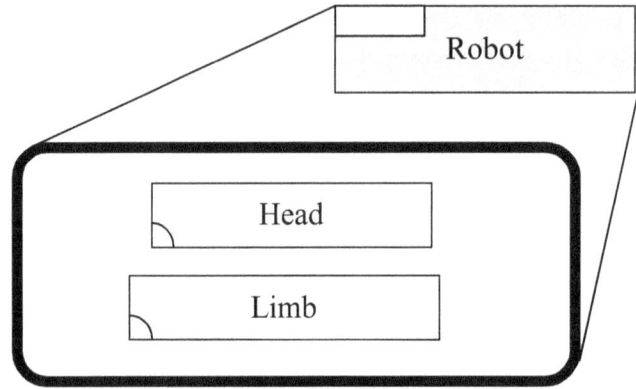

Figure 19-44   Analysis' AHD of the *Robot* Systems Definition Version 3

Figure 19-45 shows the analysis' FD of the *robot* systems definition version 3. In the figure, *Technology_SubLayer_2* contains the *Head* component; *Technology_SubLayer_1* contains the *Limb* component.

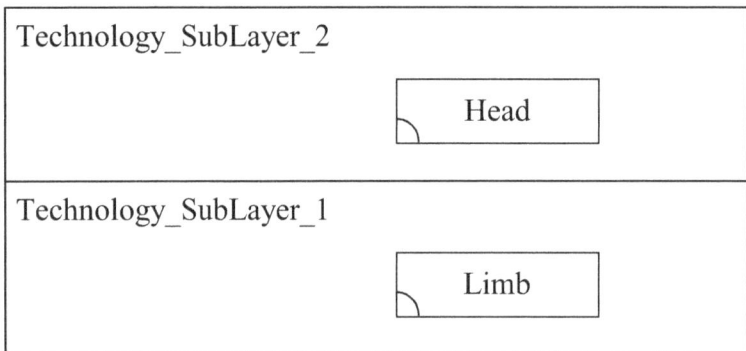

Figure 19-45   Analysis' FD of the *Robot* Systems Definition Version 3

Figure 19-46 shows the analysis' COD of the *robot* systems definition version 3. In the figure, component *Head* has two operations: *Receive_Write_Signal* and *Receive_Fly_Signal*; component *Limb* has two operation: *Move_Hand* and *Ignite_Jet_Thruster*.

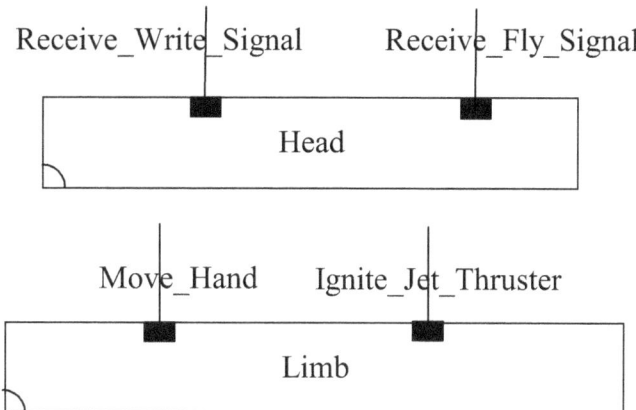

Figure 19-46   Analysis' COD of the *Robot* Systems Definition Version 3

Figure 19-47 shows the analysis' CCD of the *robot* systems definition version 3. In the figure, actor *Remote_Controller* has two connections with the *Head* component; component *Head* also has two connections with the *Limb* component.

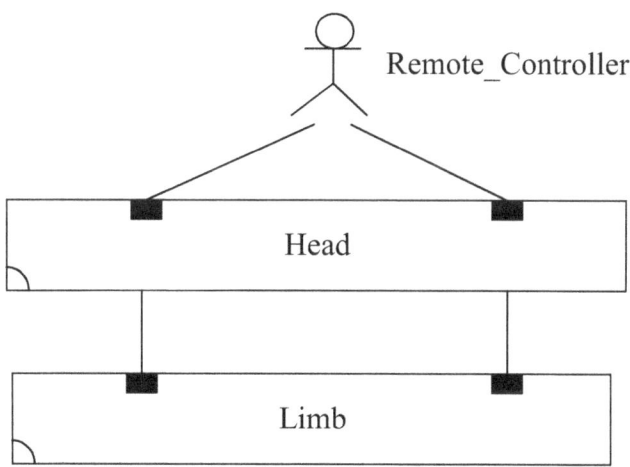

Figure 19-47   Analysis' CCD of the *Robot* Systems Definition Version 3

19-3-2-1-2 Analysis' Systems Behavior of the Robot Systems Definition Version 3

The entire analysis' systems behavior of the *robot* systems definition version 3 includes: a) *Analysis' SBCD* and b) *Analysis' IFD* of the *robot* systems definition version 3.

Figure 19-48 shows the analysis' SBCD of the *robot* systems definition version 3 in which interactions among the *Remote_Controller* actor and the *Head, Limb* components shall draw forth the *Writing* and *Flying* behaviors.

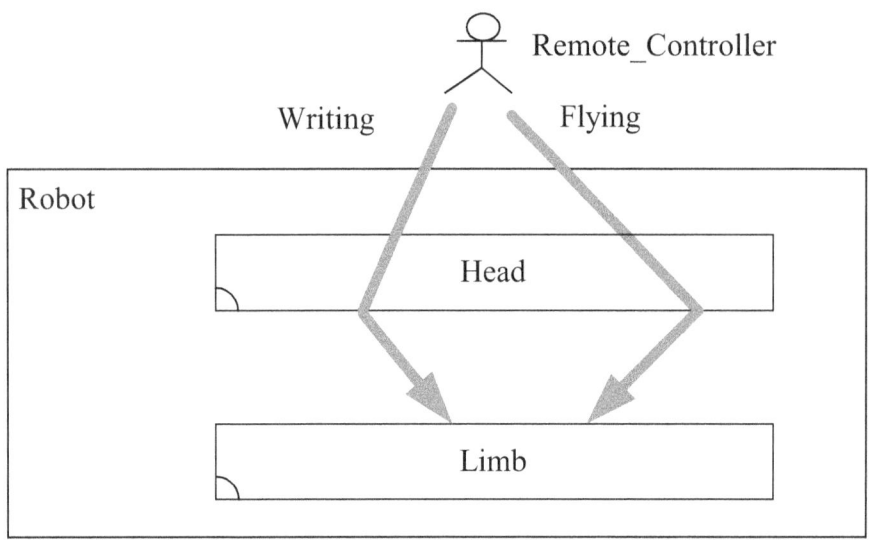

Figure 19-48   Analysis' SBCD of the *Robot* Systems Definition Version 3

The overall analysis' behavior of the *robot* systems definition version 3 includes the *Writing* and *Flying* behaviors. In other words, behaviors *Writing* and *Flying* together provide the overall analysis' behavior of the *robot* systems definition version 3.

Be noticed that the *Writing* and *Flying* behaviors are mutually independent of each other. They tend to be executed concurrently [Hoar85, Miln89, Miln99].

The overall behavior of the *robot* systems definition version 3 includes two individual behaviors: *Writing* and *Flying*. Each individual behavior is represented by an execution path. We use an interaction flow diagram (IFD) to define each one of these execution paths. Figure 19-49 shows the analysis' IFD of the *robot* systems definition version 3 *Writing* behavior. First, actor *Remote_Controller* interacts with the *Head* component through the operation call interaction *Receive_Write_Signal*. Finally, component *Head* interacts with the *Limb* component through the *Move_Hand* operation call interaction.

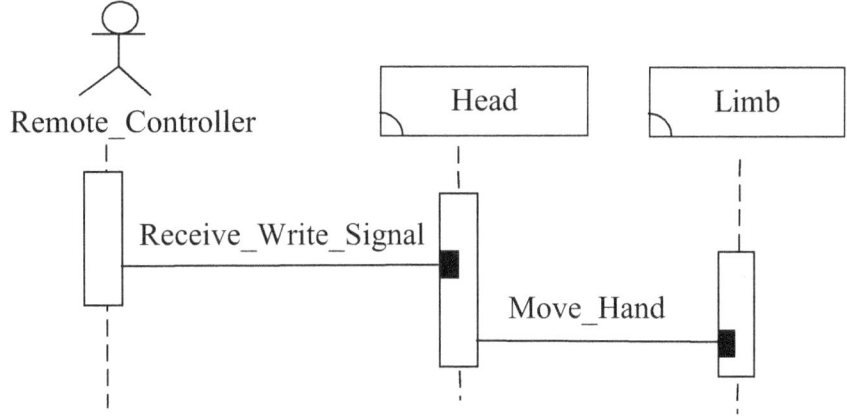

Figure 19-49   Analysis' IFD of
the *Robot* Systems Definition Version 3 *Writing* Behavior

Figure 19-50 shows the analysis' IFD of the *robot* systems definition version 3 *Flying* behavior. First, actor *Remote_Controller* interacts with the *Head* component through the *Receive_Fly_Signal* operation call interaction. Finally, component *Head* interacts with the *Limb* component through the *Ignite_Jet_Thruster* operation call interaction.

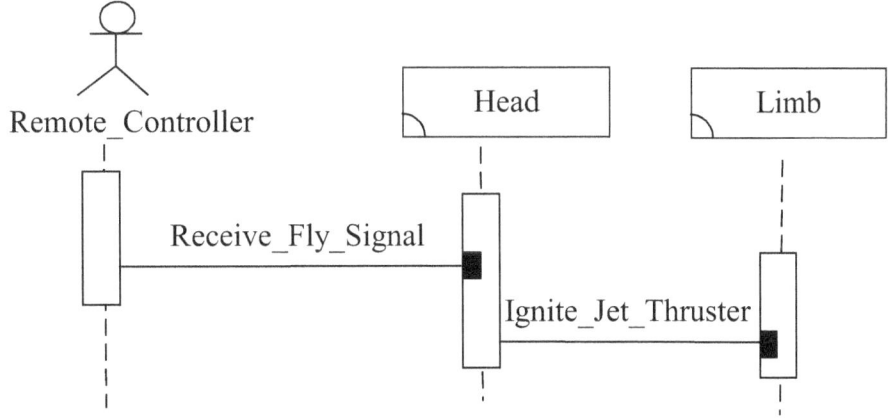

Figure 19-50   Analysis' IFD of
the *Robot* Systems Definition Version 3 *Flying* Behavior

19-3-2-2 Design View of the Robot Systems Definition Version 3

The design view of the *robot* systems definition version 3 is shown in Figure 19-51.

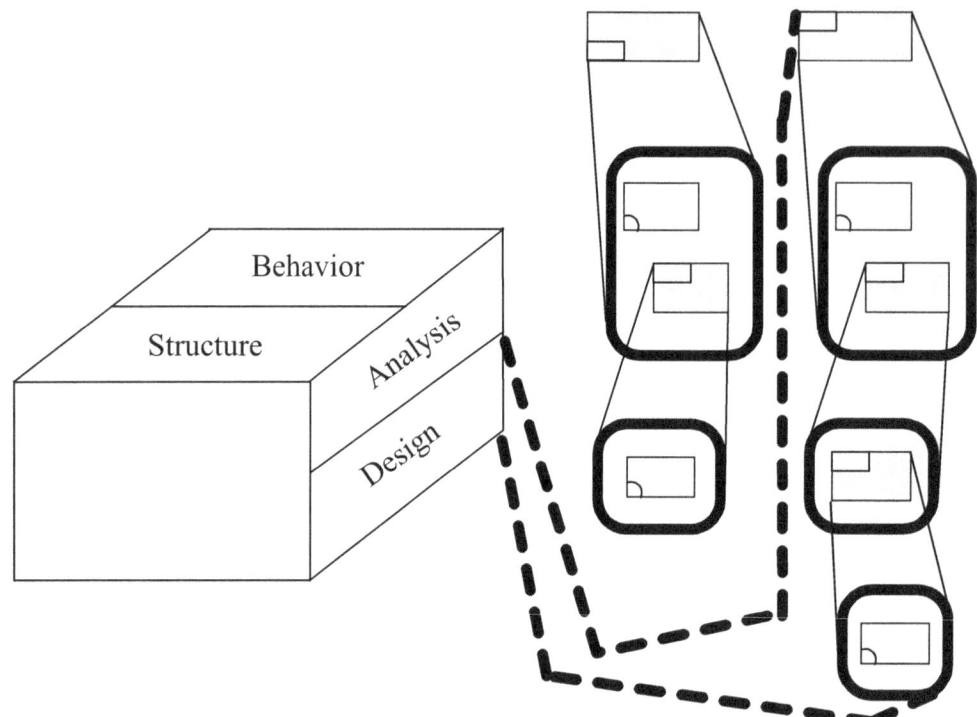

Figure 19-51   Design View of the *Robot* Systems Definition Version 3

The design view of the *robot* systems definition version 3 consists of: a) design's systems structure of the *robot* systems definition version 3 and b) design's systems behavior of the *robot* systems definition version 3.

19-3-2-2-1 Design's Systems Structure of the Robot Systems Definition Version 3

The entire design's systems structure of the *robot* systems definition version 3 includes: a) *Design's AHD*, b) *Design's FD*, c) *Design's COD* and d) *Design's CCD* of the *robot* systems definition version 3.

We first draw the design's AHD of the *robot* systems definition version 3. As shown in Figure 19-52, *Robot* is composed of *Head* and *Limb*; *Limb* is composed of *Hands* and *Jet_Thrusters*. In the figure, *Robot* and *Limb* are aggregated systems while *Head*, *Hands* and *Jet_Thrusters* are non-aggregated systems.

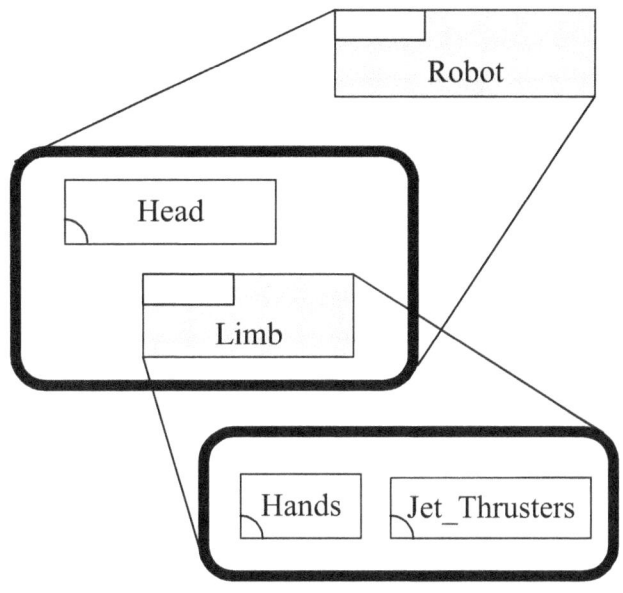

Figure 19-52   Design's AHD of the *Robot* Systems Definition Version 3

Figure 19-53 shows the design's FD of the *robot* systems definition version 3. In the figure, *Technology_SubLayer_2* contains the *Head* component; *Technology_SubLayer_1* contains the *Hands* and *Jet_Thrusters* components.

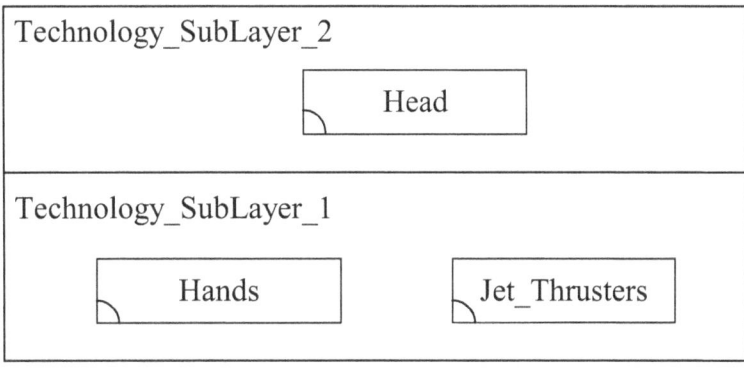

Figure 19-53   Design's FD of the *Robot* Systems Definition Version 3

Figure 19-54 shows the design's COD of the *robot* systems definition version 3. In the figure, component *Head* has two operations: *Receive_Write_Signal* and *Receive_Fly_Signal*; component *Hands* has one operation: *Move_Hand*; component *Jet_Thrusters* has one operation: *Ignite_Jet_Thruster*.

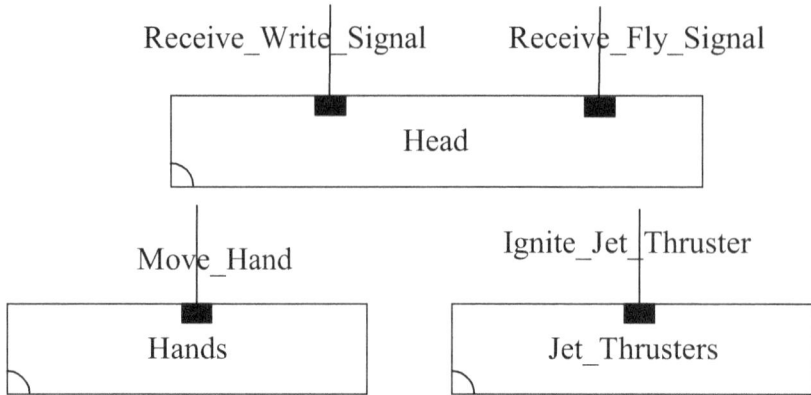

Figure 19-54   Design's COD of the *Robot* Systems Definition Version 3

Figure 19-55 shows the design's CCD of the *robot* systems definition version 3. In the figure, actor *Remote_Controller* has a connection with the *Head* component; component *Head* has a connection with each one of the *Hands* and *Jet_Thrusters* components.

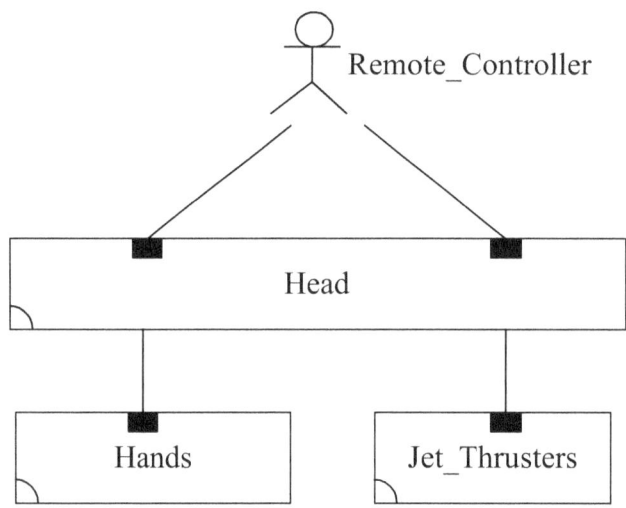

Figure 19-55   Design's CCD of the *Robot* Systems Definition Version 3

19-3-2-2-2 Design's Systems Behavior of the Robot Systems Definition Version 3

The entire design's systems behavior of the *robot* systems definition version 3 includes: a) *Design's SBCD* and b) *Design's IFD* of the *robot* systems definition version 3.

Figure 19-56 shows the design's SBCD of the *robot* systems definition version 3 in which interactions among the *Remote_Controller* actor and the *Head, Hands, Jet_Thrusters* components shall draw forth the *Writing* and *Flying* behaviors.

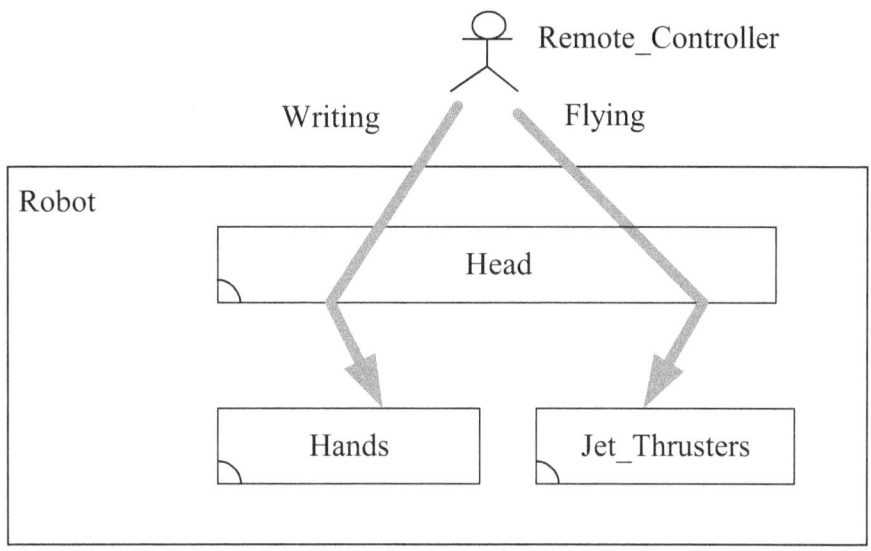

Figure 19-56   Design's SBCD of the *Robot* Systems Definition Version 3

The overall design's behavior of the *robot* systems definition version 3 includes the *Writing* and *Flying* behaviors. In other words, behaviors *Writing* and *Flying* together provide the overall design's behavior of the *robot* systems definition version 3.

Be noticed that the *Writing* and *Flying* behaviors are mutually independent of each other. They tend to be executed concurrently [Hoar85, Miln89, Miln99].

The overall design's behavior of the *robot* systems definition version 3 includes two individual behaviors: *Writing* and *Flying*. Each individual behavior is represented by an execution path. We use an interaction flow diagram (IFD) to define each one of these execution paths. Figure 19-57 shows the design's IFD of the *robot* systems definition version 3 *Writing* behavior. First, actor *Remote_Controller* interacts with the *Head* component through the operation call interaction *Receive_Write_Signal*. Finally, component *Head* interacts with the *Hands* component through the *Move_Hand* operation call interaction.

220

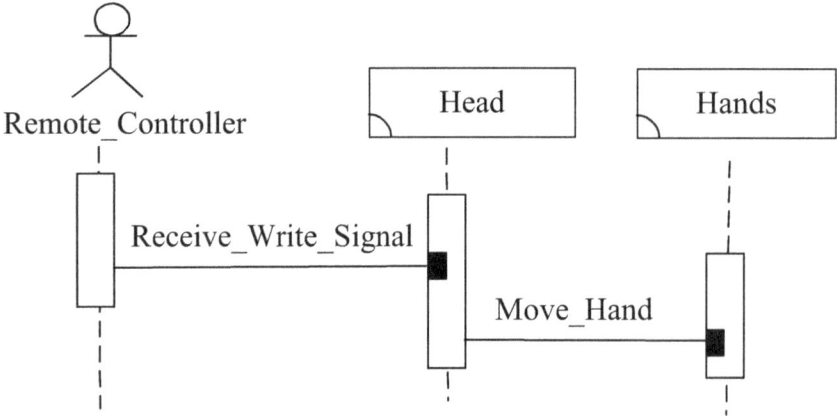

Figure 19-57   Design's IFD of
the *Robot* Systems Definition Version 3 *Writing* Behavior

Figure 19-58 shows the design's IFD of the *robot* systems definition version 3 *Flying* behavior. First, actor *Remote_Controller* interacts with the *Head* component through the *Receive_Fly_Signal* operation call interaction. Finally, component *Head* interacts with the *Jet_Thrusters* component through the *Ignite_Jet_Thruster* operation call interaction.

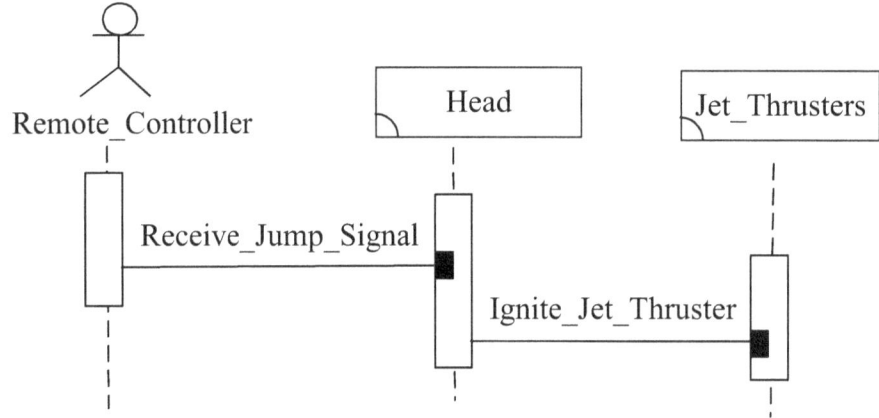

Figure 19-58   Design's IFD of
the *Robot* Systems Definition Version 3 *Flying* Behavior

Chapter 20: General Systems Theory 2.0 Defining the Purchasing Management

This chapter demonstrates how to achieve general systems theory 2.0 (general architectural theory) defining the *purchasing management*, through the application of SBC architecture.

Based on the SBC architecture, we define the *purchasing management* as shown in Figure 20-1.

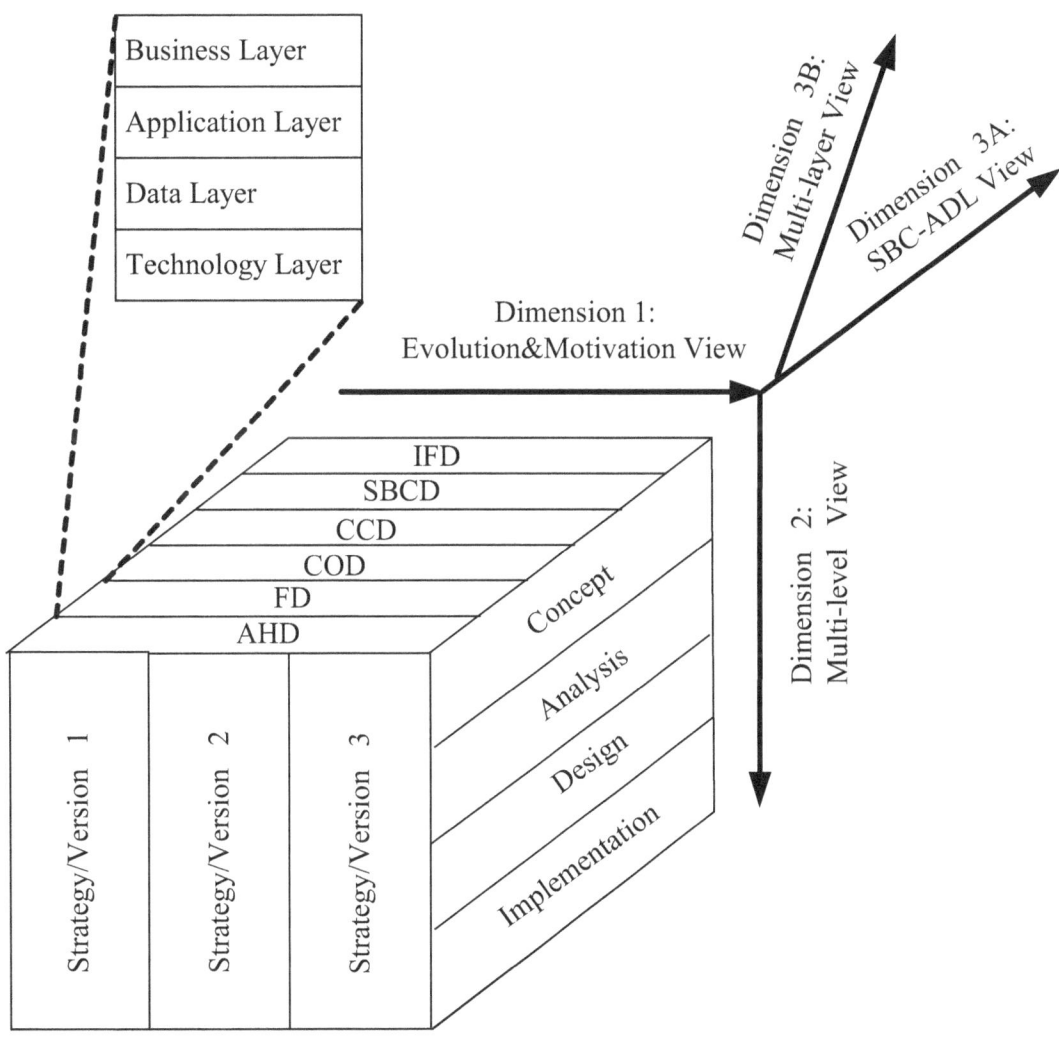

Figure 20-1   General Systems Theory 2.0 Defining the *Purchasing Management*

Using the SBC evolution&motivation view, we shall go through: a) purchasing management systems definition strategy/version 1, b) purchasing management systems definition strategy/version 2 and c) purchasing management systems definition strategy/version 3, to accomplish general systems theory 2.0 (general architectural theory) systems definition of the *purchasing management*.

20-1 Purchasing Management Systems Definition Strategy/Version 1

The *purchasing management* motivation model shown in Figure 20-2, being a higher-order system, has the *purchasing management* systems definition strategy 1 as its input and the *purchasing management* systems definition version 1 as its output.

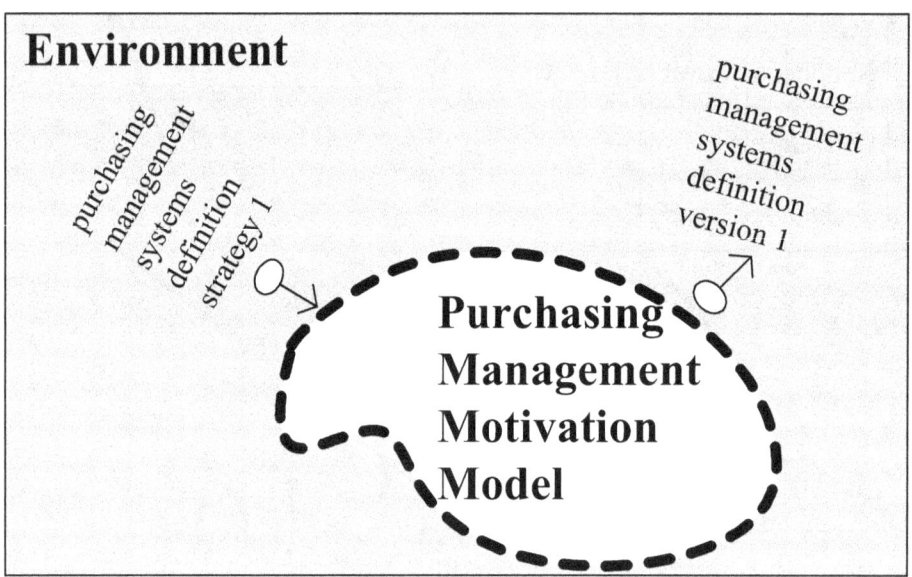

Figure 20-2   *Purchasing Management* Motivation Model is a Higher-Order System

The *purchasing management* systems definition strategy 1 mapped to the *purchasing management* systems definition version 1 can be represented as an ordered pair, as shown in Figure 20-3.

```
┌─────────────────────────────────────────────────────────────────┐
│                                                                   │
│   (Purchasing Management              Purchasing Management        │
│   Systems Definition Strategy 1  ──►  Systems Definition Version 1)│
│                                                                   │
└─────────────────────────────────────────────────────────────────┘
```

Figure 20-3   An Ordered Pair

20-1-1 Purchasing Management Systems Definition Strategy 1

Here, we use (a) goal drivers, (b) goal assumptions, (c) goal constraints and (d) SWOT analysis, to illustrate the strategic means of the "*purchasing management systems definition strategy 1*".

Goal drivers are up from the policy considerations, the goal driver is kind of why we want to have this *purchasing management* systems definition version 1. The goal drivers of *purchasing management* systems definition version 1 are: nowadays, a purchasing management system with the *Quotation, Purchase_Order and Purchase* behaviors is very attractive to most managers.

Goal assumptions are taking into account of those assumptions that have a positive impact on the *purchasing management* systems definition version 1. We assume that if the cost of installing the purchasing management system is affordable, then every manager will have a great desire to install one. This is the major goal assumption of this strategy.

Goal constraints are up from the policy considerations, the goal constraints are related to those restrictions which have a negative impact on the *purchasing management* systems definition version 1. If the company is short of fund to invest in this new installation, then this would become the goal constraint of this strategy.

SWOT analysis is to analyze the internal strengths, weaknesses, opportunities and threats, and so for executing this *purchasing management* systems definition strategy 1. Being a big company, it should be trivial for the company to install this purchasing management system with the *Quotation, Purchase_Order and Purchase* behaviors. This is the internal strength of this company. However, kind of bulky makes the company may not react fast enough to carry out this new installation. This is the internal weakness of this company.

224

20-1-2 Purchasing Management Systems Definition Version 1

Using the SBC multi-level (hierarchical) view, an architect goes through: a) concept view, b) analysis view, c) design view and d) implementation view for the *purchasing management* systems definition version 1 as shown in Figure 20-4.

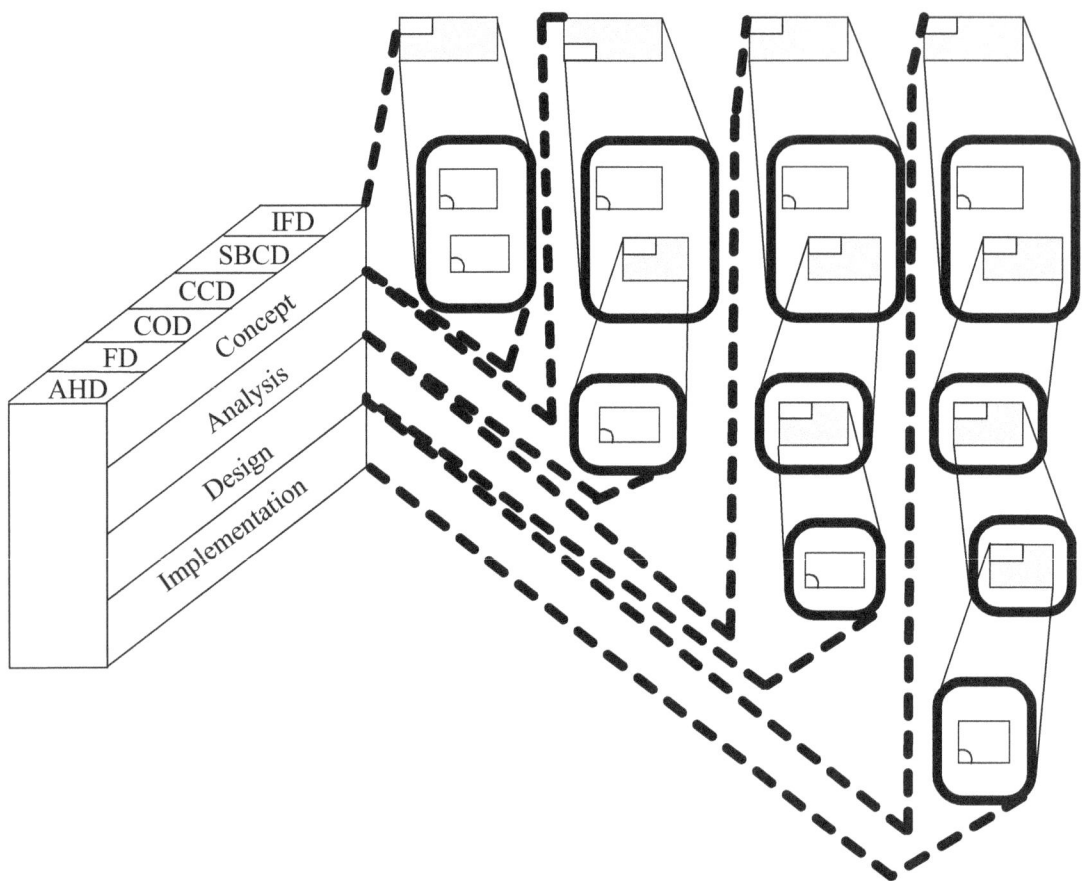

Figure 20-4   *Purchasing Management* Systems Definition Version 1

20-1-2-1 Concept View of the Purchasing Management Systems Definition Version 1

The concept view of the *purchasing management* systems definition version 1 is shown in Figure 20-5.

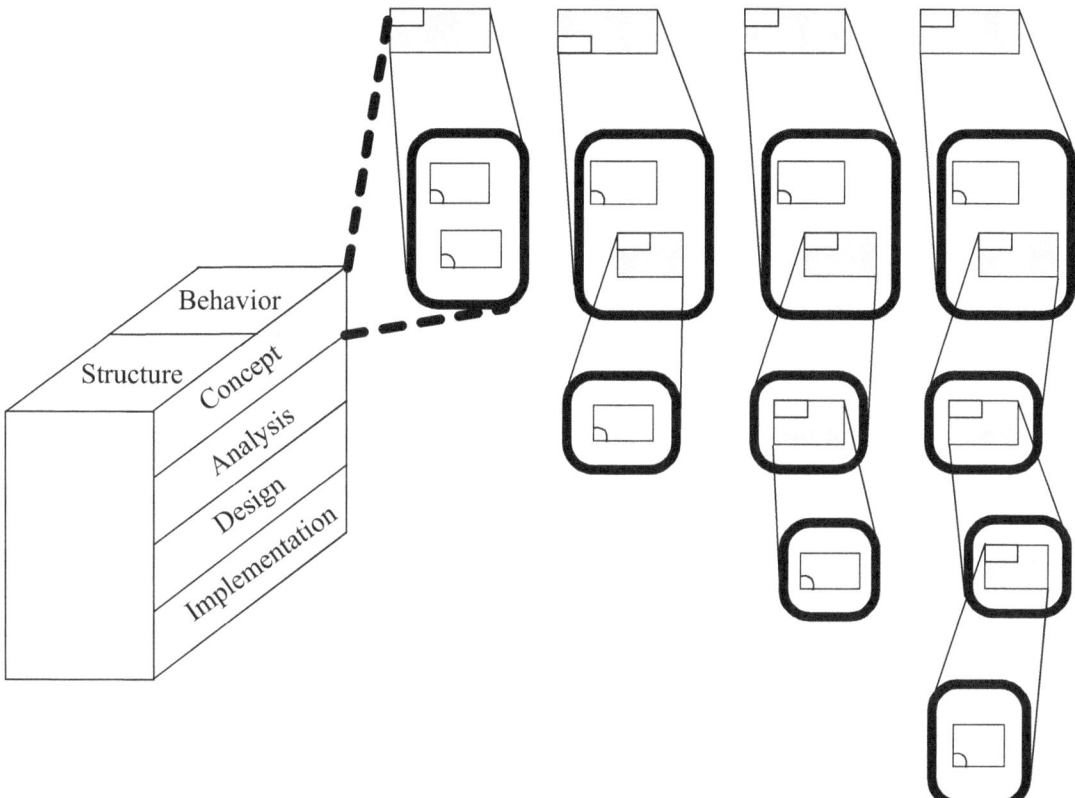

Figure 20-5   Concept View of the *Purchasing Management* Systems Definition Version 1

The concept view of the *purchasing management* systems definition version 1 consists of: a) concept's systems structure of the *purchasing management* systems definition version 1 and b) concept's systems behavior of the *purchasing management* systems definition version 1.

20-1-2-1-1 Concept's Systems Structure of the Purchasing Management Systems Definition Version 1

The entire concept's systems structure of the *purchasing management* systems definition version 1 includes: a) *Concept's AHD*, b) *Concept's FD*, c) *Concept's COD* and d) *Concept's CCD* of the *purchasing management* systems definition version 1.

We first draw the concept's AHD of the *purchasing management* systems definition version 1. As shown in Figure 20-6, *Purchasing Management* is composed of *Purchasing_Department* and *PM_Subsystem_3*. In the figure, *Purchasing Management* is an aggregated system while *Purchasing_Department* and *PM_Subsystem_3* are non-aggregated systems.

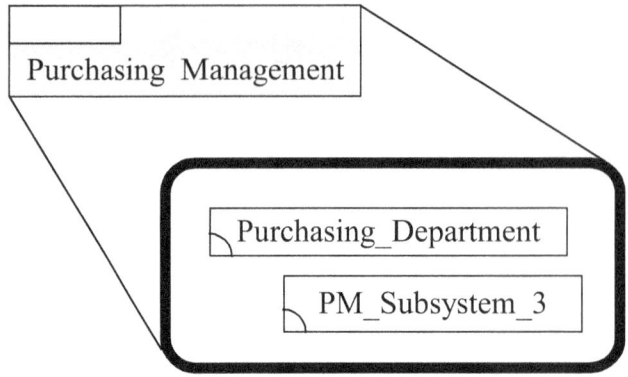

Figure 20-6   Concept's AHD of the *Purchasing Management*
Systems Definition Version 1

Figure 20-7 shows the concept's FD of the *purchasing management* systems definition version 1. In the figure, *Business_Layer* contains the *Purchasing_Department* and *PM_Subsystem_3* components.

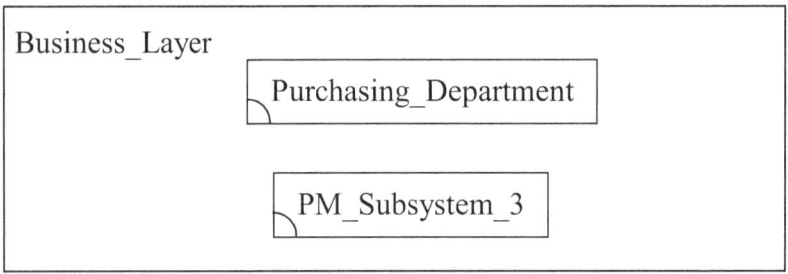

Figure 20-7   Concept's FD of the *Purchasing Management*
Systems Definition Version 1

Figure 20-8 shows the concept's COD of the *purchasing management* systems definition version 1. In the figure, component *Purchasing_Department* has three operations: *Quotation_Verify*, *Purchase_Order_Verify* and *Purchase_Verify*; component *PM_Subsystem_3* has three operations: *Quotation_Processing_Button_Click, Purchase_Order_Processing_Button_Click and Purchase_Processing_Button_Click*.

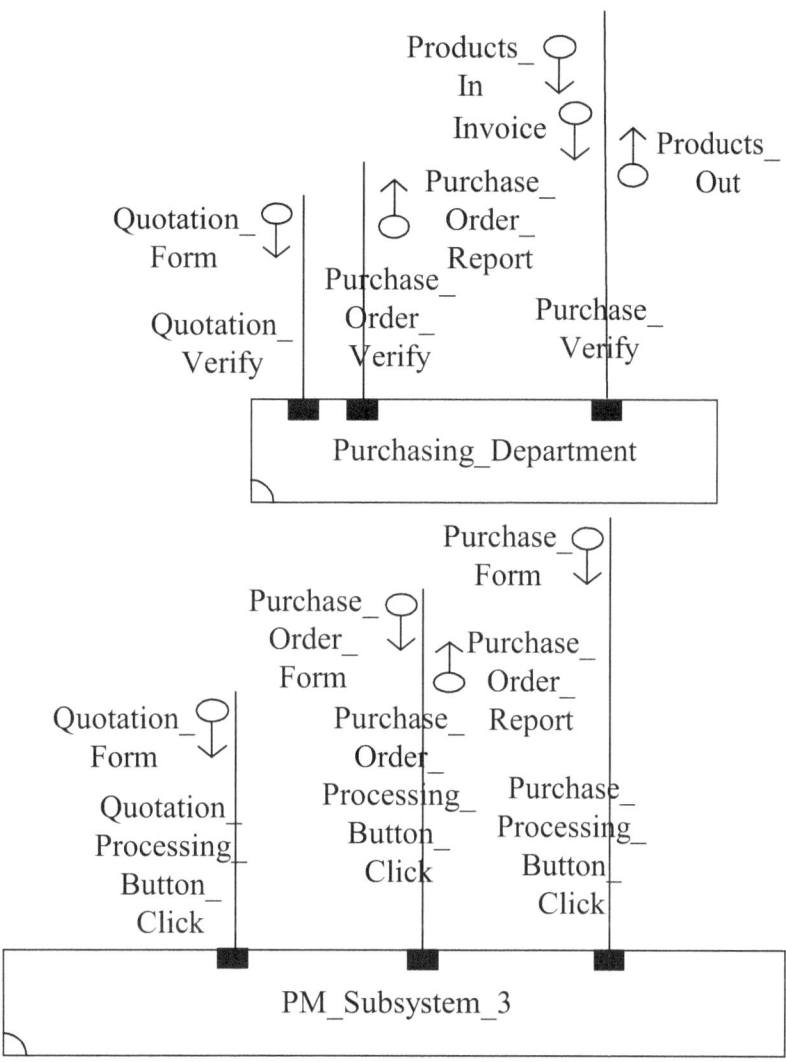

Figure 20-8   Concept's COD of the *Purchasing Management*
Systems Definition Version 1

The operation formula of *Quotation_Verify* is *Quotation_Verify(In Quotation_Form)*. The operation formula of *Purchase_Order_Verify* is *Purchase_Order_Verify(Out Purchase_Order_Report)*. The operation formula of *Purchase_Verify* is *Purchase_Verify(In Products_In, Invoice; Out Products_Out)*. The operation formula of *Quotation_Processing_Button_Click* is *Quotation_Processing_Button_Click(In Quotation_Form)*. The operation formula of *Purchase_Order_Processing_Button_Click* is *Purchase_Order_Processing_Button_Click(In Purchase_Order_Form; Out Purchase_Order_Report)*. The operation formula of *Purchase_Processing_Button_Click* is *Purchase_Processing_Button_Click(In*

228

Purchase_Form).

Figure 20-9 shows the composite data type specification of the *Quotation_Form* input parameter occurring in the *Quotation_Verify(In Quotation_Form)* and *Quotation_Processing_Button_Click(In Quotation_Form)* operation formulas.

Parameter	*Quotation_Form*
Data Type	TABLE of Date : Text SupplierName: Text ProductNo : Text Quantity : Integer UnitPrice : Real Total : Real End TABLE ;
Instances	**Quotation Form** Date: 2011/10/25 SupplierName : Johnson Corp. ProductNo Quantity UnitPrice A00001(Pen) 300 100.00 _A00002(Mouse)____400____200.00__ _A00003(Camera)___500____300.00__ Total : 260,000.00

Figure 20-9   Composite Data Type Specification

Figure 20-10 shows the composite data type specification of the *Purchase_Order_Report* output parameter occurring in the *Purchase_Order_Verify(Out Purchase_Order_Report)* and *Purchase_Order_Processing_Button_Click(In Purchase_Order_Form; Out Purchase_Order_Report)* operation formulas.

Parameter	*Purchase_Order_Report*

Data Type	TABLE of
	  Date : Text
	  SupplierName: Text
	  ProductNo : Text
	  Quantity : Integer
	End TABLE ;

Instances	Date : 20111118 SupplierName : Johnson Corp. ProductNo / Quantity: A00001(Pen) — 300 A00002(Mouse) — 400 A00003(Camera) — 500

Figure 20-10   Composite Data Type Specification

Figure 20-11 shows the primitive data type specification of the *Products_In* input parameter occurring in the *Purchase_Verify(In Products_In, Invoice; Out Products_Out)* operation formula.

Parameter	Data Type	Instances
Products_In	Physical Object	Pen, Mouse, Camera

Figure 20-11   Primitive Data Type Specification

Figure 20-12 shows the composite data type specification of the *Invoice* input parameter occurring in the *Purchase_Verify(In Products_In, Invoice; Out Products_Out)* operation formula.

Parameter	*Invoice*
Data Type	TABLE of 　Date : Text 　SupplierName: Text 　ProductNo : Text 　Quantity : Integer 　UnitPrice : Real 　Total : Real End TABLE ;
Instances	Date : 20111130 SupplierName :　Johnson Corp. | ProductNo	Quantity	UnitPrice | |---|---|---| | A00001(Pen) | 300 | 100.00 | | A00002(Mouse) | 400 | 200.00 | | A00003(Camera) | 500 | 300.00 | Total : 260,000.00

Figure 20-12　　Composite Data Type Specification

Figure 20-13 shows the primitive data type specification of the *Products_Out* output parameter occurring in the *Purchase_Verify(In Products_In, Invoice; Out Products_Out)* operation formula.

Parameter	Data Type	Instances
Products_Out	Physical Object	Pen, Mouse, Camera

Figure 20-13　　Primitive Data Type Specification

Figure 20-14 shows the composite data type specification of the *Purchase_Order_Form* input parameter occurring in the *Purchase_Order_Processing_Button_Click(In Purchase_Order_Form; Out Purchase_Order_Report)* operation formula.

Parameter	*Purchase_Order_Form*

Data Type	TABLE of
	  Date : Text
	  SupplierName: Text
	  ProductNo : Text
	  Quantity : Integer
	End TABLE ;

Instances	**Purchase Order Form** Date: 2011/11/18 SupplierName : Johnson Corp. ProductNo Quantity __A00001(Pen) ____300 ___ __A00002(Mouse)___400___ __A00003(Camera)___500___

Figure 20-14   Composite Data Type Specification

Figure 20-15 shows the composite data type specification of the *Purchase_Form* input parameter occurring in the *Purchase_Processing_Button_Click(In Purchase_Form)* operation formula.

Parameter	*Purchase_Form*

Data Type	TABLE of
	   Date : Text
	   SupplierName: Text
	   ProductNo : Text
	   Quantity : Integer
	   UnitPrice : Real
	   ReturnQuantity : Integer
	   Total : Real
	End TABLE ;

Instances	**Purchase Form** Date: 2011/12/12 SupplierName : Johnson Corp. ProductNo Quantity UnitPrice ReturnQuantity A00001(Pen) 300 100.00 0 A00002(Mouse) 390 200.00 10 A00003(Camera)500 300.00 0 Total : 258,000.00

Figure 20-15   Composite Data Type Specification

Figure 20-16 shows the concept's CCD of the *purchasing management* systems definition version 1. In the figure, actor *Supplier* has three connections with the *Purchasing_Department* component; component *Purchasing_Department* also has three connections with the *PM_Subsystem_3* component.

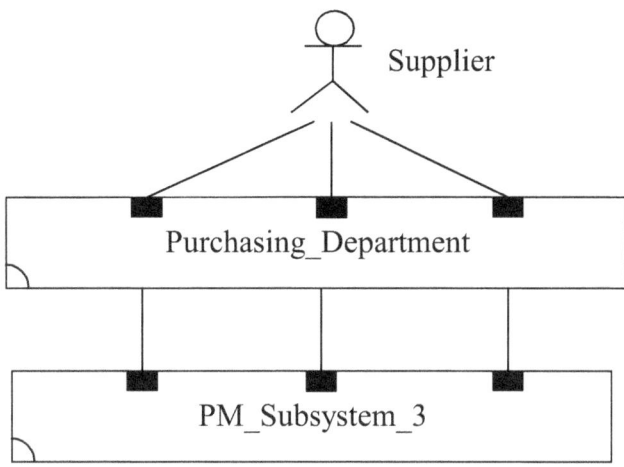

Figure 20-16   Concept's CCD of the *Purchasing Management*
Systems Definition Version 1

20-1-2-1-2 Concept's Systems Behavior of the Purchasing Management Systems Definition Version 1

The entire concept's systems behavior of the *purchasing management* systems definition version 1 includes: a) *Concept's SBCD* and b) *Concept's IFD* of the *purchasing management* systems definition version 1.

Figure 20-17 shows the concept's SBCD of the *purchasing management* systems definition version 1 in which interactions among the *Supplier* actor and the *Purchasing_Department*, *PM_Subsystem_3* components shall draw forth the *Quotation, Purchase_Order and Purchase* behaviors.

234

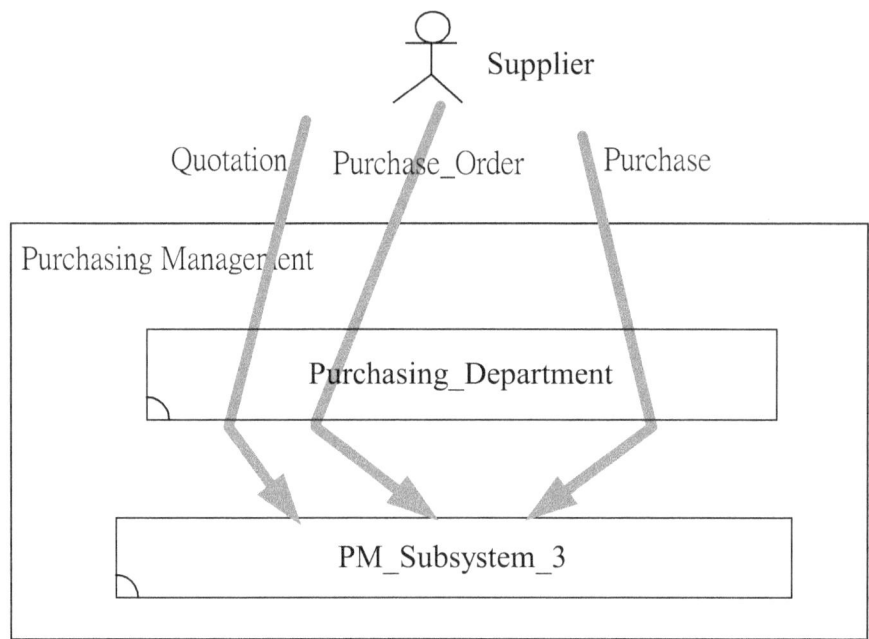

Figure 20-17   Concept's SBCD of the *Purchasing Management*
Systems Definition Version 1

The overall concept's behavior of the *purchasing management* systems definition version 1 includes the *Quotation, Purchase_Order and Purchase* behaviors. In other words, the *Quotation, Purchase_Order and Purchase* behaviors together provide the overall concept's behavior of the *purchasing management* systems definition version 1.

Be noticed that the *Quotation, Purchase_Order and Purchase* behaviors are mutually independent of each other. They tend to be executed concurrently [Hoar85, Miln89, Miln99].

The overall concept's behavior of the *purchasing management* systems definition version 1 includes three individual behaviors: *Quotation, Purchase_Order and Purchase*. Each individual behavior is represented by an execution path. We use an interaction flow diagram (IFD) to define each one of these execution paths. Figure 20-18 shows the concept's IFD of the *purchasing management* systems definition version 1 *Quotation* behavior. First, actor *Supplier* interacts with the *Purchasing_Department* component through the *Quotation_Verify* operation call interaction, carrying the *Quotation_Form* input parameter. Finally, component *Purchasing_Department* interacts with the *PM_Subsystem_3* component through the *Quotation_Processing_Button_Click* operation call interaction, carrying the *Quotation_Form* input parameter.

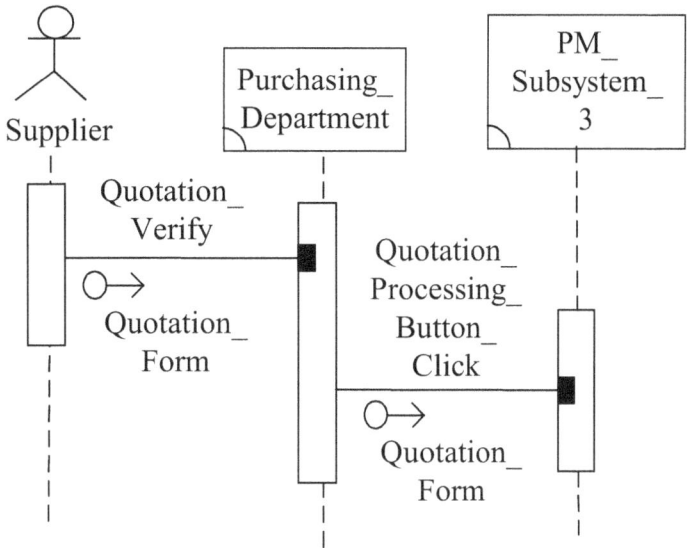

Figure 20-18   Concept's IFD of the *Purchasing Management*
Systems Definition Version 1 *Quotation* Behavior

Figure 20-19 shows the concept's IFD of the *purchasing management* systems definition version 1 *Purchase_Order* behavior. First, actor *Supplier* interacts with the *Purchasing_Department* component through the *Purchase_Order_Verify* operation call interaction. Next, component *Purchasing_Department* interacts with the *PM_Subsystem_3* component through the *Purchase_Order_Processing_Button_Click* operation call interaction, carrying the *Purchase_Order_Form* input parameter and *Purchase_Order_Report* output parameter. Finally, actor *Supplier* interacts with the *Purchasing_Department* component through the *Purchase_Order_Verify* operation return interaction, carrying the *Purchase_Order_Report* output parameter.

236

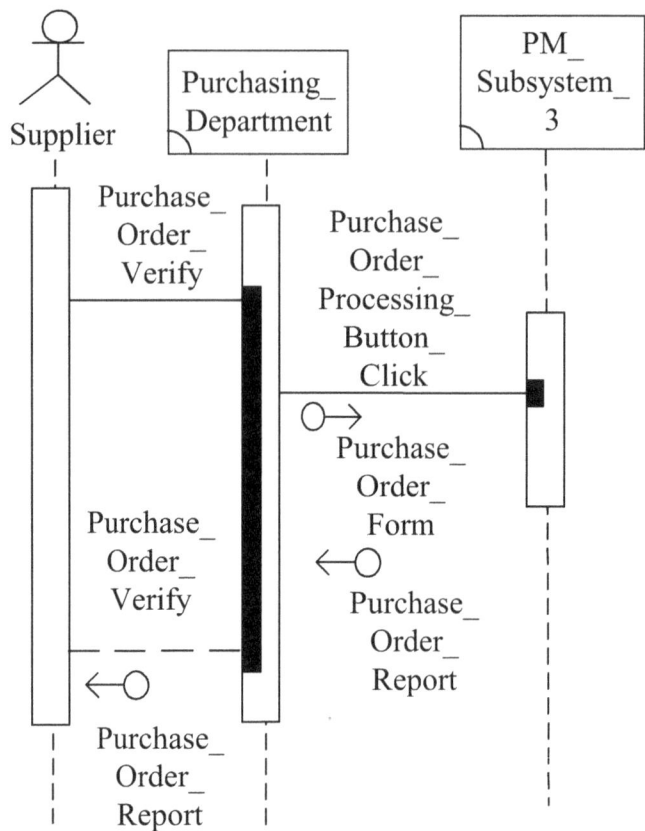

Figure 20-19   Concept's IFD of the *Purchasing Management*
Systems Definition Version 1 *Purchase_Order* Behavior

Figure 20-20 shows the concept's IFD of the *purchasing management* systems definition version 1 *Purchase* behavior. First, actor *Supplier* interacts with the *Purchasing_Department* component through the *Purchase_Verify* operation call interaction, carrying the *Products_In* and *Invoice* input parameters. Next, component *Purchasing_Department* interacts with the *PM_Subsystem_3* component through the *Purchase_Processing_Button_Click* operation call interaction, carrying the *Purchase_Form* input parameter. Finally, actor *Supplier* interacts with the *Purchasing_Department* component through the *Purchase_Verify* operation return interaction, carrying the *Products_Out* output parameter.

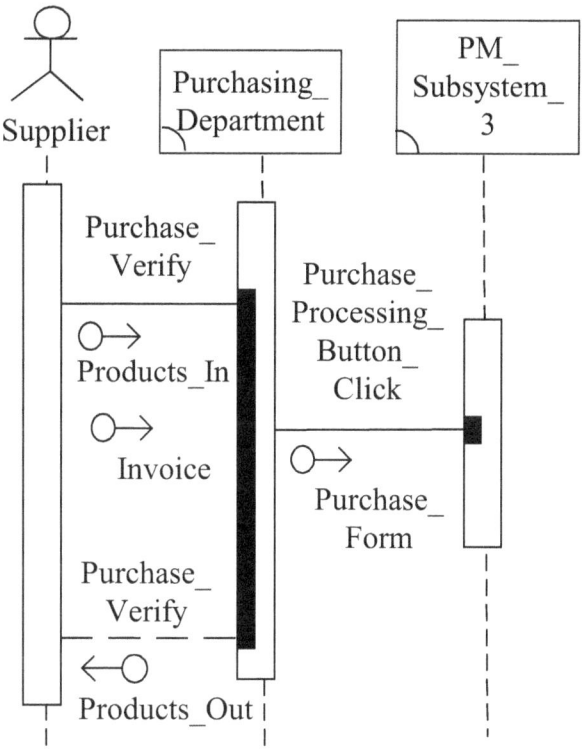

Figure 20-20   Concept's IFD of the *Purchasing Management*
Systems Definition Version 1 *Purchase* Behavior

20-1-2-2 Analysis View of the Purchasing Management Systems Definition Version 1

The analysis view of the *purchasing management* systems definition version 1 is shown in Figure 20-21.

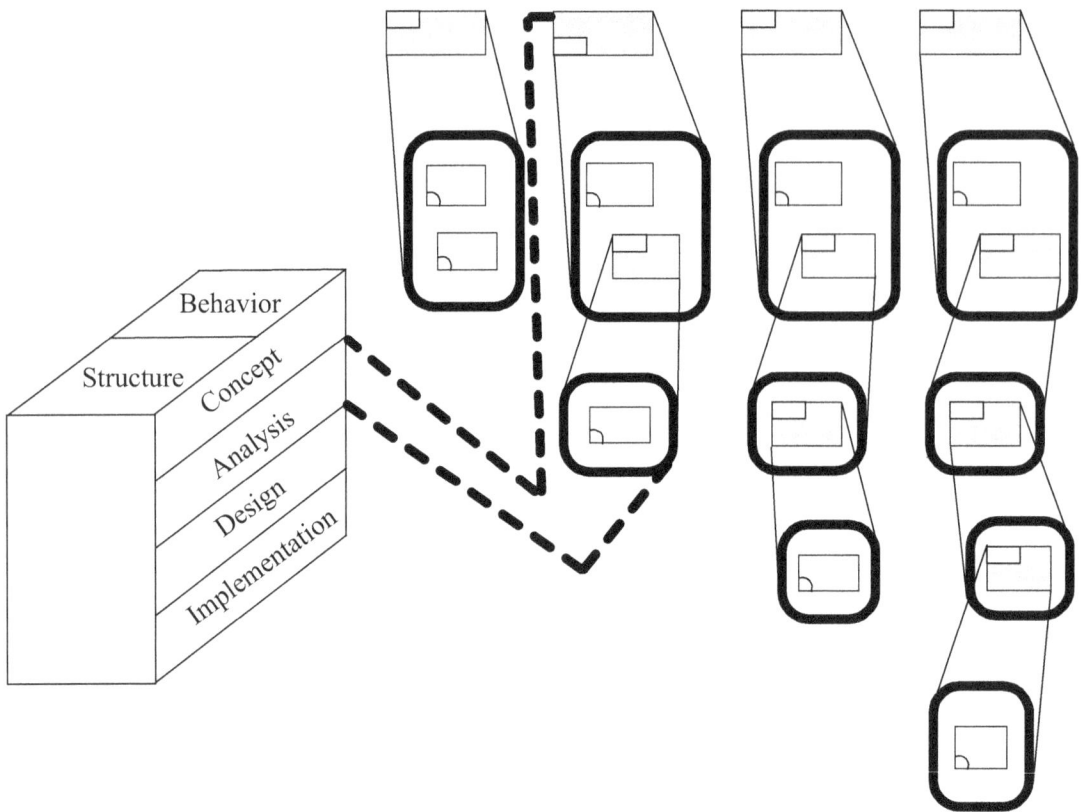

Figure 20-21   Analysis View of the *Purchasing Management* Systems Definition Version 1

The analysis view of the *purchasing management* systems definition version 1 consists of: a) analysis' systems structure of the *purchasing management* systems definition version 1 and b) analysis' systems behavior of the *purchasing management* systems definition version 1.

20-1-2-2-1 Analysis' Systems Structure of the Purchasing Management Systems Definition Version 1

The entire analysis' systems structure of the *purchasing management* systems definition version 1 includes: a) *Analysis' AHD*, b) *Analysis' FD*, c) *Analysis' COD* and d) *Analysis' CCD* of the *purchasing management* systems definition version 1.

We first draw the analysis' AHD of the *purchasing management* systems definition version 1. As shown in Figure 20-22, *Purchasing Management* is composed of *Purchasing_Department* and *PM_Subsystem_3*; *PM_Subsystem_3* is composed of *Quotation_Processing_GUI*, *Purchase_Order_Processing_GUI*, *Purchase_Processing_GUI* and *PM_Subsystem_2*. In the figure, *Purchasing Management* and *PM_Subsystem_3* are aggregated systems while

Purchasing_Department,   *Quotation_Processing_GUI,*
Purchase_Order_Processing_GUI, Purchase_Processing_GUI and *PM_Subsystem_2*
are non-aggregated systems.

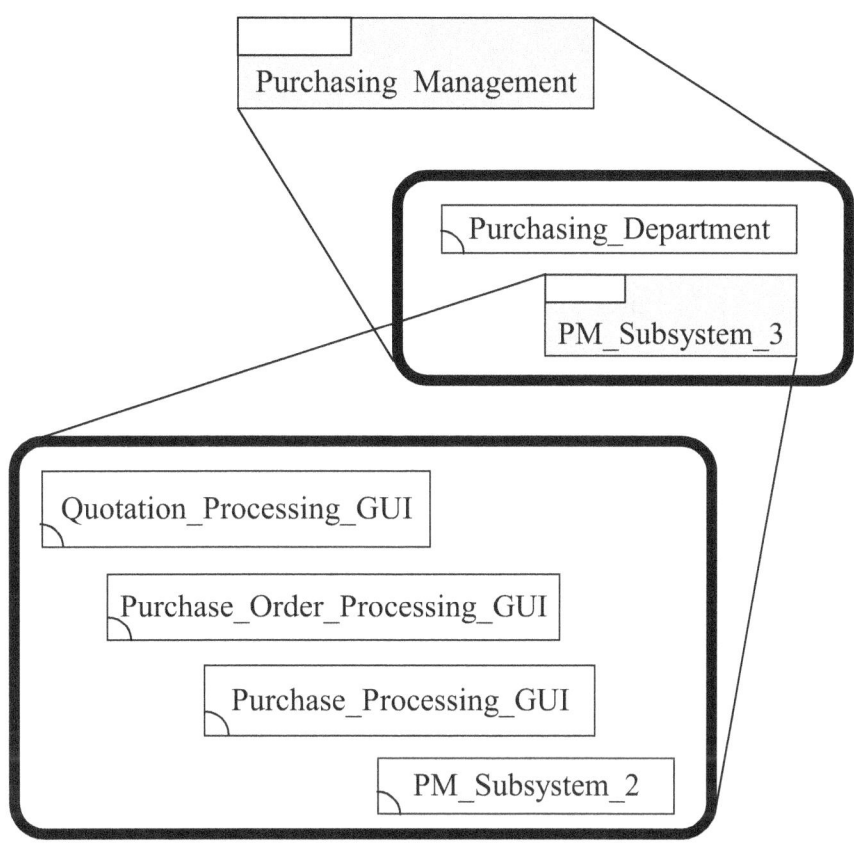

Figure 20-22   Analysis' AHD of the *Purchasing Management*
Systems Definition Version 1

Figure 20-23 shows the analysis' FD of the *purchasing management* systems
definition version 1. In the figure, *Business_Layer* contains the
Purchasing_Department component; *Application_Layer* contains the
Quotation_Processing_GUI,   *Purchase_Order_Processing_GUI,*
Purchase_Processing_GUI and *PM_Subsystem_2* components.

240

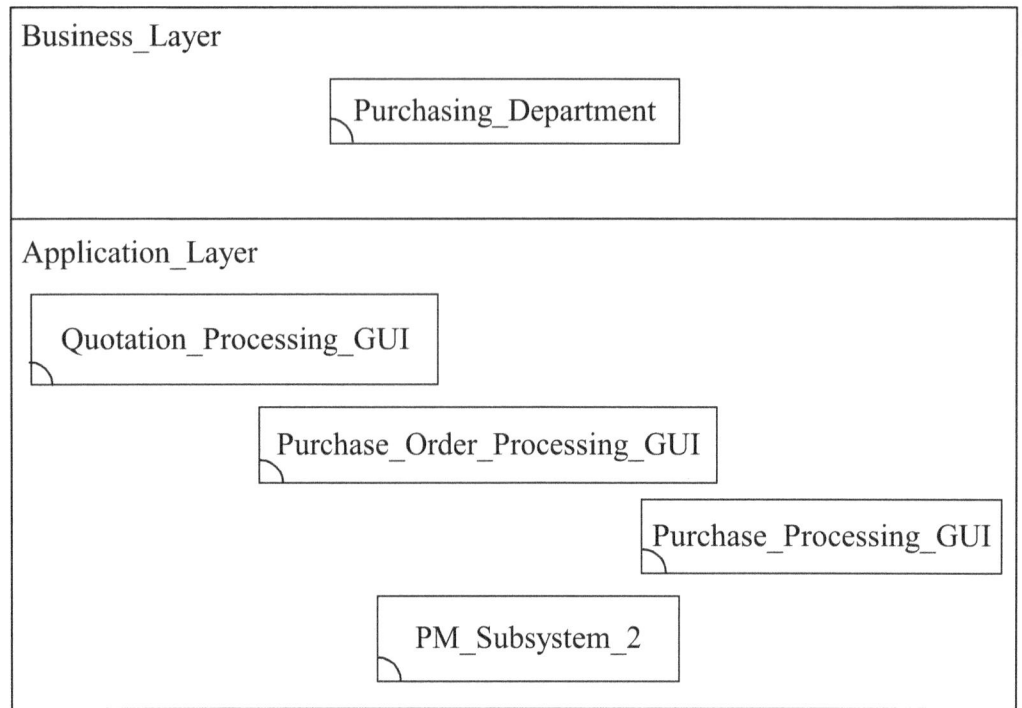

Figure 20-23   Analysis' FD of the *Purchasing Management*
Systems Definition Version 1

Figure 20-24 shows the analysis' COD of the *purchasing management* systems definition version 1. In the figure, component *Purchasing_Department* has three operations: *Quotation_Verify*, *Purchase_Order_Verify* and *Purchase_Verify*; component *Quotation_Processing_GUI* has one operation: *Quotation_Processing_Button_Click*; component *Purchase_Order_Processing_GUI* has one operation: *Purchase_Order_Processing_Button_Click*; component *Purchase_Processing_GUI* has one operation: *Purchase_Processing_Button_Click*; component *PM_Subsystem_2* has three operations: *SQL_Quotation_Insert*, *SQL_Purchase_Order_Insert* and *SQL_Purchase_Insert*.

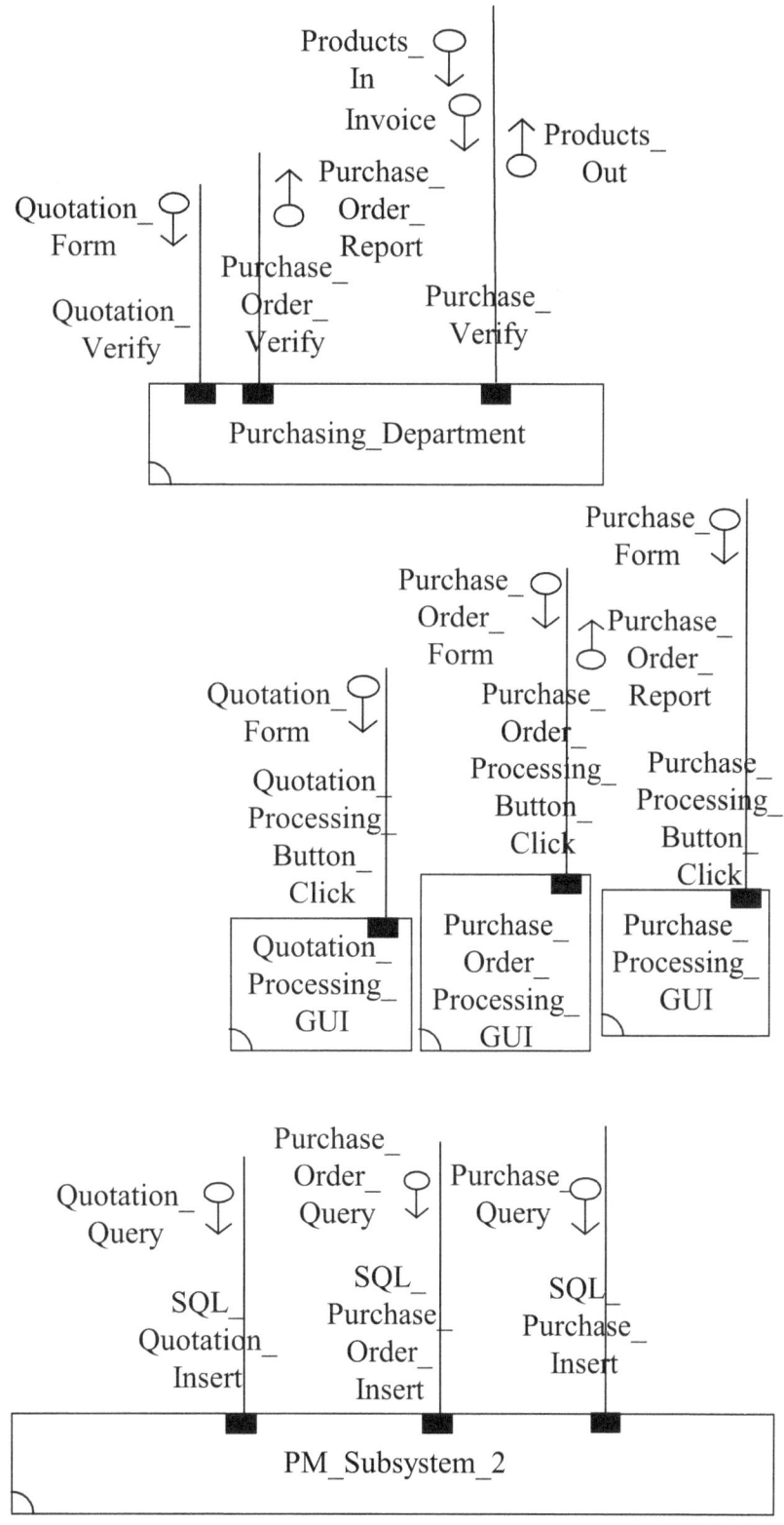

Figure 20-24   Analysis' COD of the *Purchasing Management*
Systems Definition Version 1

The operation formula of *Quotation_Verify* is *Quotation_Verify(In Quotation_Form)*. The operation formula of *Purchase_Order_Verify* is *Purchase_Order_Verify(Out Purchase_Order_Report)*. The operation formula of *Purchase_Verify* is *Purchase_Verify(In Products_In, Invoice; Out Products_Out)*. The operation formula of *Quotation_Processing_Button_Click* is *Quotation_Processing_Button_Click(In Quotation_Form)*. The operation formula of *Purchase_Order_Processing_Button_Click* is *Purchase_Order_Processing_Button_Click(In Purchase_Order_Form; Out Purchase_Order_Report)*. The operation formula of *Purchase_Processing_Button_Click* is *Purchase_Processing_Button_Click(In Purchase_Form)*. The operation formula of *SQL_Quotation_Insert* is *SQL_Quotation_Insert(In Quotation_Query)*. The operation formula of *SQL_Purchase_Order_Insert* is *SQL_Purchase_Order_Insert (In Purchase_Order_Query)*. The operation formula of *SQL_Purchase_Insert* is *SQL_Purchase_Insert (In Purchase_Query)*.

Figure 20-25 shows the composite data type specification of the *Quotation_Form* input parameter occurring in the *Quotation_Verify(In Quotation_Form)* and *Quotation_Processing_Button_Click(In Quotation_Form)* operation formulas.

Parameter	*Quotation_Form*
Data Type	TABLE of 　Date : Text 　SupplierName: Text 　ProductNo : Text 　Quantity : Integer 　UnitPrice : Real 　Total : Real End TABLE ;
Instances	**Quotation Form** Date: 2011/10/25 SupplierName : Johnson Corp. ProductNo　　　Quantity　UnitPrice A 0 0 0 0 1 (P e n)　3 0 0　1 0 0 . 0 0 _A00002(Mouse)____400____200.00__ _A00003(Camera)___500____300.00__ Total : 260,000.00

Figure 20-25　　Composite Data Type Specification

Figure 20-26 shows the composite data type specification of the *Purchase_Order_Report* output parameter occurring in the *Purchase_Order_Verify(Out Purchase_Order_Report)* and *Purchase_Order_Processing_Button_Click(In Purchase_Order_Form; Out Purchase_Order_Report)* operation formulas.

Parameter	*Purchase_Order_Report*

Data Type	TABLE of
	  Date : Text
	  SupplierName: Text
	  ProductNo : Text
	  Quantity : Integer
	End TABLE ;

Instances	Date : 20111118 SupplierName : Johnson Corp. <table><tr><th>ProductNo</th><th>Quantity</th></tr><tr><td>A00001(Pen)</td><td>300</td></tr><tr><td>A00002(Mouse)</td><td>400</td></tr><tr><td>A00003(Camera)</td><td>500</td></tr></table>

Figure 20-26   Composite Data Type Specification

Figure 20-27 shows the primitive data type specification of the *Products_In* input parameter occurring in the *Purchase_Verify(In Products_In, Invoice; Out Products_Out)* operation formula.

Parameter	Data Type	Instances
Products_In	Physical Object	Pen, Mouse, Camera

Figure 20-27   Primitive Data Type Specification

Figure 20-28 shows the composite data type specification of the *Invoice* input parameter occurring in the *Purchase_Verify(In Products_In, Invoice; Out Products_Out)* operation formula.

Parameter	*Invoice*

Data Type	TABLE of
	  Date : Text
	  SupplierName: Text
	  ProductNo : Text
	  Quantity : Integer
	  UnitPrice : Real
	  Total : Real
	End TABLE ;

Instances	Date : 20111130 SupplierName : Johnson Corp. ProductNo / Quantity / UnitPrice: A00001(Pen) — 300 — 100.00 A00002(Mouse) — 400 — 200.00 A00003(Camera) — 500 — 300.00 Total : 260,000.00

Figure 20-28   Composite Data Type Specification

Figure 20-29 shows the primitive data type specification of the *Products_Out* output parameter occurring in the *Purchase_Verify(In Products_In, Invoice; Out Products_Out)* operation formula.

Parameter	Data Type	Instances
Products_Out	Physical Object	Pen, Mouse, Camera

Figure 20-29   Primitive Data Type Specification

Figure 20-30 shows the composite data type specification of the *Purchase_Order_Form* input parameter occurring in the *Purchase_Order_Processing_Button_Click(In Purchase_Order_Form; Out Purchase_Order_Report)* operation formula.

Parameter	*Purchase_Order_Form*

Data Type	TABLE of
	  Date : Text
	  SupplierName: Text
	  ProductNo : Text
	  Quantity : Integer
	End TABLE ;

Instances	**Purchase Order Form** Date: 2011/11/18 SupplierName : Johnson Corp. ProductNo Quantity __A00001(Pen) ____300___ __A00002(Mouse)____400___ __A00003(Camera)___500___

Figure 20-30   Composite Data Type Specification

Figure 20-31 shows the composite data type specification of the *Purchase_Form* input parameter occurring in the *Purchase_Processing_Button_Click(In Purchase_Form)* operation formula.

Parameter	*Purchase_Form*

Data Type	TABLE of
	  Date : Text
	  SupplierName: Text
	  ProductNo : Text
	  Quantity : Integer
	  UnitPrice : Real
	  ReturnQuantity : Integer
	  Total : Real
	End TABLE ;

Instances	**Purchase Form** Date: 2011/12/12 SupplierName : Johnson Corp. ProductNo Quantity UnitPrice ReturnQuantity A00001(Pen) 300 100.00 0 A00002(Mouse) 390 200.00 10 A00003(Camera)500 300.00 0 Total : 258,000.00

Figure 20-31   Composite Data Type Specification

Figure 20-32 shows the composite data type specification of the *Quotation_Query* input parameter occurring in the *SQL_Quotation_Insert(In Quotation_Query)* operation formula.

248

Parameter	*Quotation_Query*
Data Type	TABLE of Date : Text SupplierName: Text ProductNo : Text Quantity : Integer UnitPrice : Real Total : Real End TABLE ;
Instances	<table><tr><th>Date</th><th>SupplierName</th><th>Total</th></tr><tr><td>20111025</td><td>Johnson Corp.</td><td>260,000.00</td></tr></table> <table><tr><th>ProductNo</th><th>Quantity</th><th>UnitPrice</th></tr><tr><td>A00001(Pen)</td><td>300</td><td>100.00</td></tr><tr><td>A00002(Mouse)</td><td>400</td><td>200.00</td></tr><tr><td>A00003(Camera)</td><td>500</td><td>300.00</td></tr></table>

Figure 20-32   Composite Data Type Specification

Figure 20-33 shows the composite data type specification of the *Purchase_Order_Query* input parameter occurring in the *SQL_Purchase_Order_Insert(In Purchase_Order_Query)* operation formula.

Parameter	*Purchase_Order_Query*
Data Type	TABLE of Date : Text SupplierName: Text ProductNo : Text Quantity : Integer End TABLE ;
Instances	

Date	SupplierName
20111118	Johnson Corp.

ProductNo	Quantity
A00001(Pen)	300
A00002(Mouse)	400
A00003(Camera)	500

Figure 20-33   Composite Data Type Specification

Figure 20-34 shows the composite data type specification of the *Purchase_Query* input parameter occurring in the *SQL_Purchase_Insert (In Purchase_Query)* operation formula.

250

Parameter	*Purchase_Query*

Data Type	TABLE of
	  Date : Text
	  SupplierName: Text
	  ProductNo : Text
	  Quantity : Integer
	  UnitPrice : Real
	  ReturnQuantity : Integer
	  Total : Real
	End TABLE ;

Instances	

Date	SupplierName	Total
20111212	Johnson Corp.	258,000.00

ProductNo	Quantity	UnitPrice	ReturnQuantity
A00001(Pen)	300	100.00	0
A00002(Mouse)	390	200.00	10
A00003(Camera)	500	300.00	0

Figure 20-34   Composite Data Type Specification

Figure 20-35 shows the analysis' CCD of the *purchasing management* systems definition version 1. In the figure, actor *Supplier* has three connections with the *Purchasing_Department* component; component *Purchasing_Department* has a connection with each one of the *Quotation_Processing_GUI*, *Purchase_Order_Processing_GUI* and *Purchase_Processing_GUI* components; each one of the *Quotation_Processing_GUI*, *Purchase_Order_Processing_GUI* and *Purchase_Processing_GUI* components has a connection with the *PM_Subsystem_2* component.

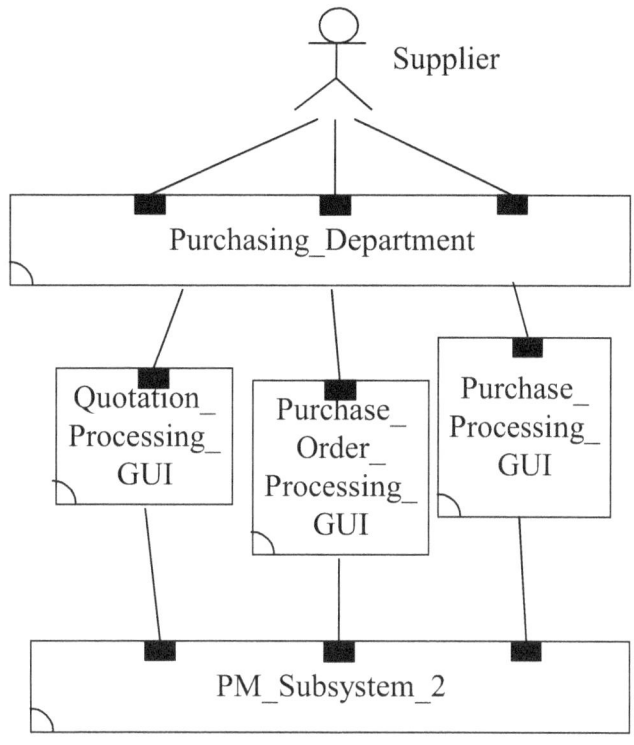

Figure 20-35   Analysis' CCD of the *Purchasing Management Systems* Definition Version 1

20-1-2-2-2 Analysis' Systems Behavior of the Purchasing Management Systems Definition Version 1

The entire analysis' systems behavior of the *purchasing management* systems definition version 1 includes: a) *Analysis' SBCD* and b) *Analysis' IFD* of the *purchasing management* systems definition version 1.

Figure 20-36 shows the analysis' SBCD of the *purchasing management* systems definition version 1 in which interactions among the *Supplier* actor and the *Purchasing_Department*,   *Quotation_Processing_GUI*, *Purchase_Order_Processing_GUI*,   *Purchase_Processing_GUI*,   *PM_Subsystem_2* components shall draw forth the *Quotation, Purchase_Order* and *Purchase* behaviors.

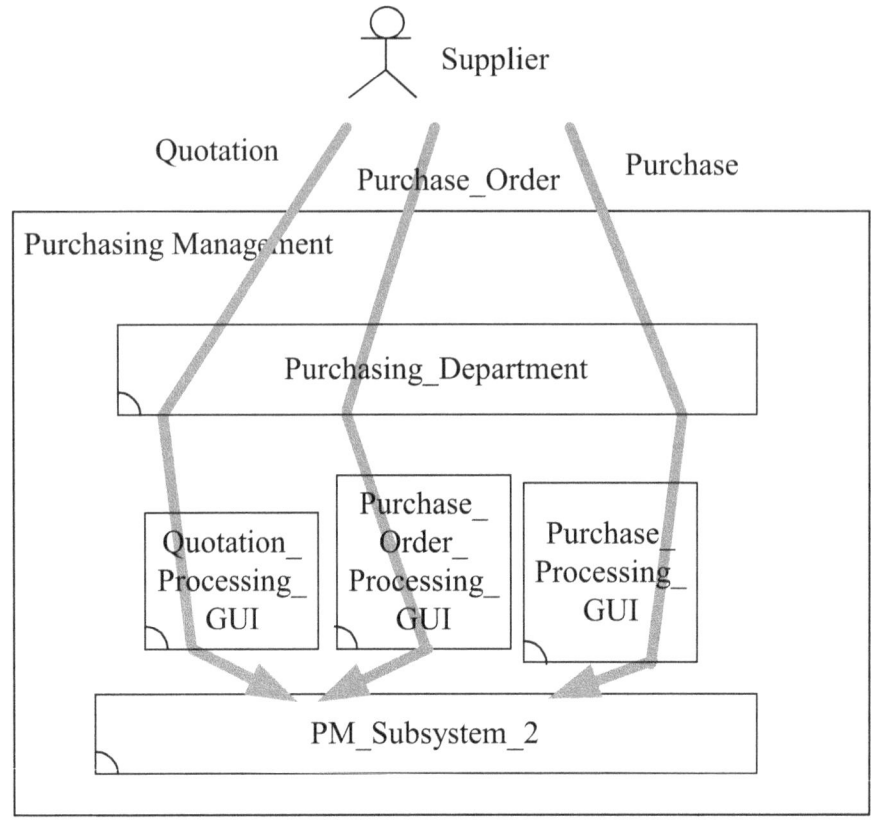

Figure 20-36   Analysis' SBCD of the *Purchasing Management*
Systems Definition Version 1

The overall analysis' behavior of the *purchasing management* systems definition version 1 includes the *Quotation, Purchase_Order and Purchase* behaviors. In other words, the *Quotation, Purchase_Order and Purchase* behaviors together provide the overall analysis' behavior of the *purchasing management* systems definition version 1.

Be noticed that the *Quotation, Purchase_Order and Purchase* behaviors are mutually independent of each other. They tend to be executed concurrently [Hoar85, Miln89, Miln99].

The overall analysis' behavior of the *purchasing management* systems definition version 1 includes three individual behaviors: *Quotation, Purchase_Order and Purchase*. Each individual behavior is represented by an execution path. We use an interaction flow diagram (IFD) to define each one of these execution paths. Figure 20-37 shows the analysis' IFD of the *purchasing management* systems definition version 1 *Quotation* behavior. First, actor *Supplier* interacts with the *Purchasing_Department* component through the *Quotation_Verify* operation call interaction, carrying the *Quotation_Form* input parameter. Next, component *Purchasing_Department* interacts with the *Quotation_Processing_GUI* component through the *Quotation_Processing_Button_Click* operation call interaction, carrying the *Quotation_Form* input parameter. Finally, component *Quotation_Processing_GUI* interacts with the *PM_Subsystem_2* component through the *SQL_Quotation_Insert* operation call interaction, carrying the *Quotation_Query* input parameter.

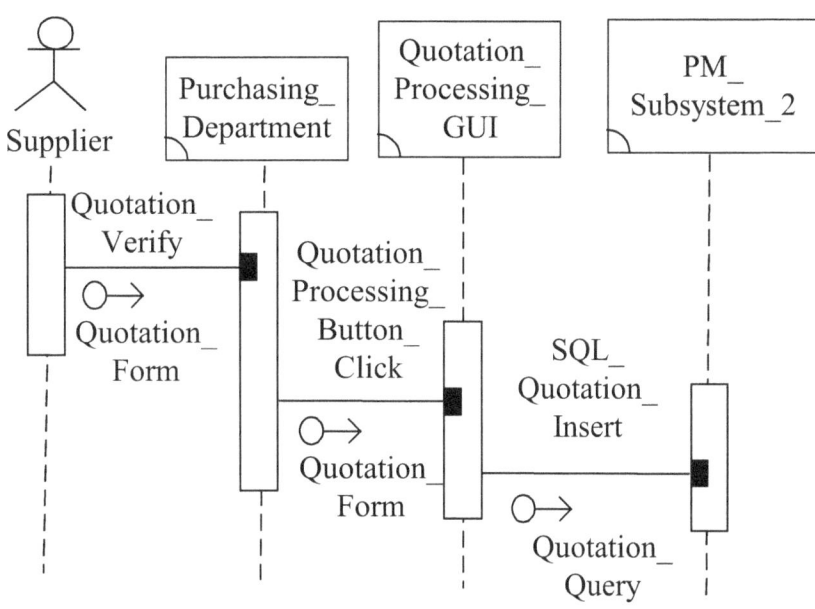

Figure 20-37   Analysis' IFD of the *Purchasing Management*
Systems Definition Version 1 *Quotation* Behavior

Figure 20-38 shows the analysis' IFD of the *purchasing management* systems definition version 1 *Purchase_Order* behavior. First, actor *Supplier* interacts with the *Purchasing_Department* component through the *Purchase_Order_Verify* operation call interaction. Next, component *Purchasing_Department* interacts with the *Purchase_Order_Processing_GUI* component through the *Purchase_Order_Processing_Button_Click* operation call interaction, carrying the *Purchase_Order_Form* input parameter. Continuingly, component *Purchase_Order_Processing_GUI* interacts with the *PM_Subsystem_2* component through the *SQL_Purchase_Order_Insert* operation call interaction, carrying the *Purchase_Order_Query* input parameter. Continuingly, component *Purchasing_Department* interacts with the *Purchase_Order_Processing_GUI* component through the *Purchase_Order_Processing_Button_Click* operation return interaction, carrying the *Purchase_Order_Report* output parameter. Finally, actor *Supplier* interacts with the *Purchasing_Department* component through the

Purchase_Order_Verify operation return interaction, carrying the *Purchase_Order_Report* output parameter.

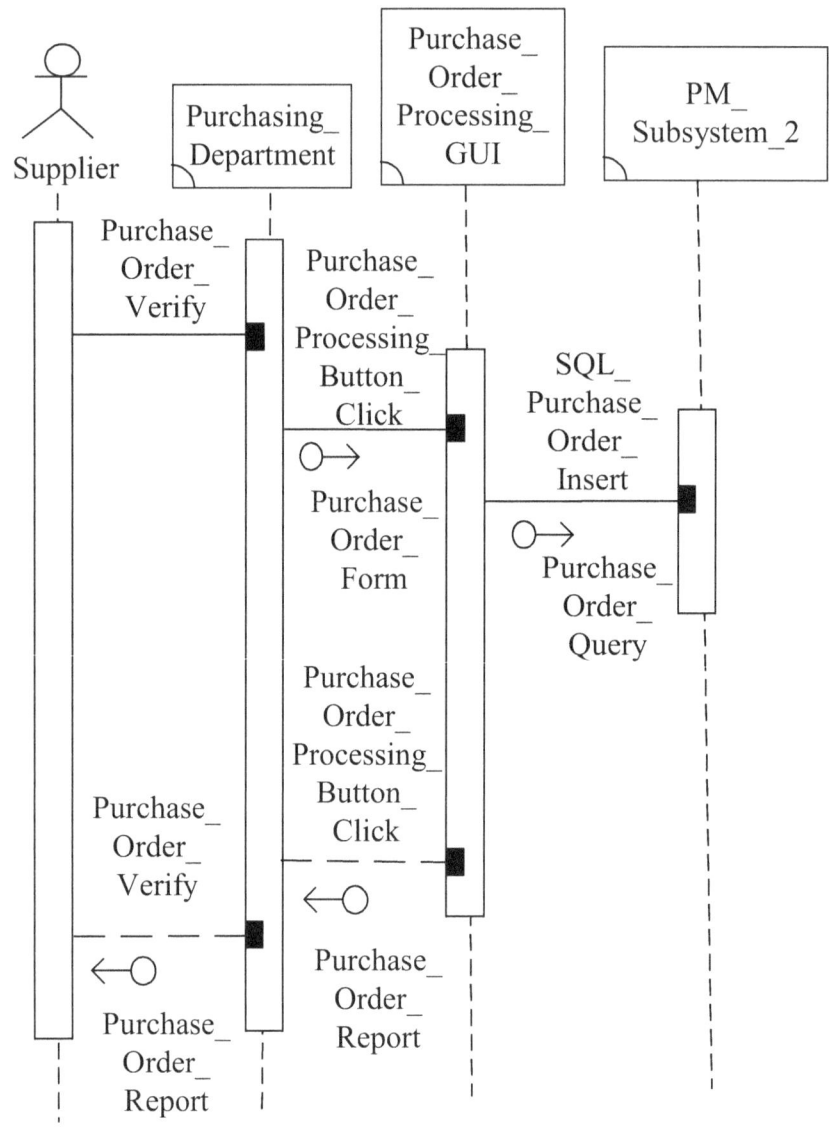

Figure 20-38   Analysis' IFD of the *Purchasing Management*
Systems Definition Version 1 *Purchase_Order* Behavior

Figure 20-39 shows the analysis' IFD of the *purchasing management* systems definition version 1 *Purchase* behavior. First, actor *Supplier* interacts with the *Purchasing_Department* component through the *Purchase_Verify* operation call interaction, carrying the *Products_In* and *Invoice* input parameters. Next, component *Purchasing_Department* interacts with the *Purchase_Processing_GUI* component through the *Purchase_Processing_Button_Click* operation call interaction, carrying the *Purchase_Form* input parameter. Continuingly, component

Purchase_Processing_GUI interacts with the *PM_Subsystem_2* component through the *SQL_Purchase_Insert* operation call interaction, carrying the *Purchase_Query* input parameter. Finally, actor *Supplier* interacts with the *Purchasing_Department* component through the *Purchase_Verify* operation return interaction, carrying the *Products_Out* output parameter.

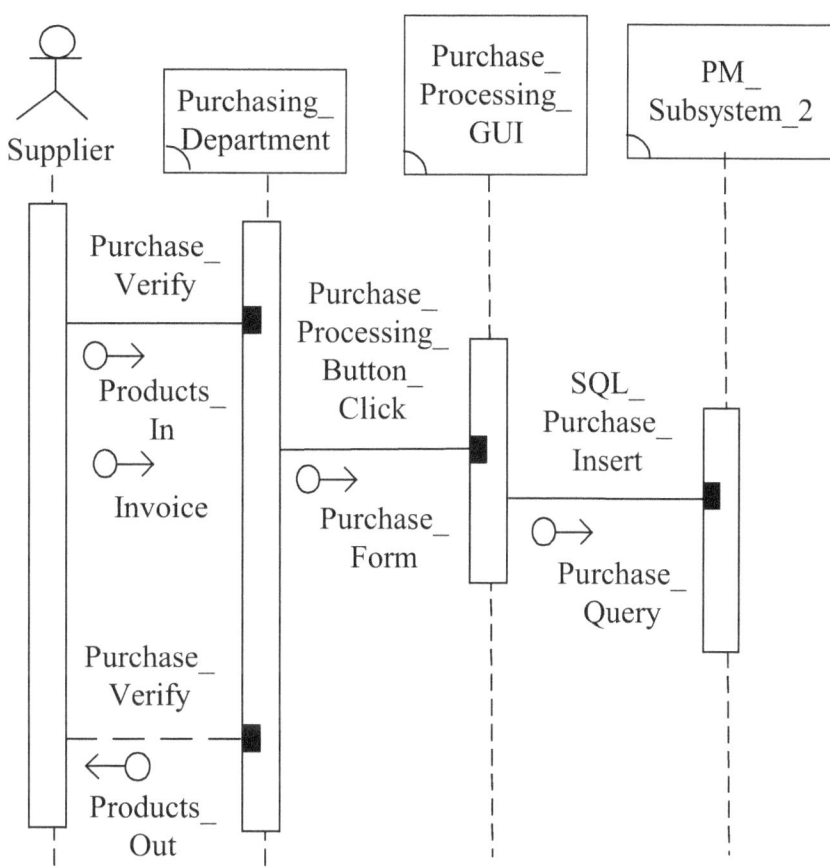

Figure 20-39   Analysis' IFD of the *Purchasing Management*
Systems Definition Version 1 *Purchase* Behavior

20-1-2-3 Design View of the Purchasing Management Systems Definition Version 1

The design view of the *purchasing management* systems definition version 1 is shown in Figure 20-40.

256

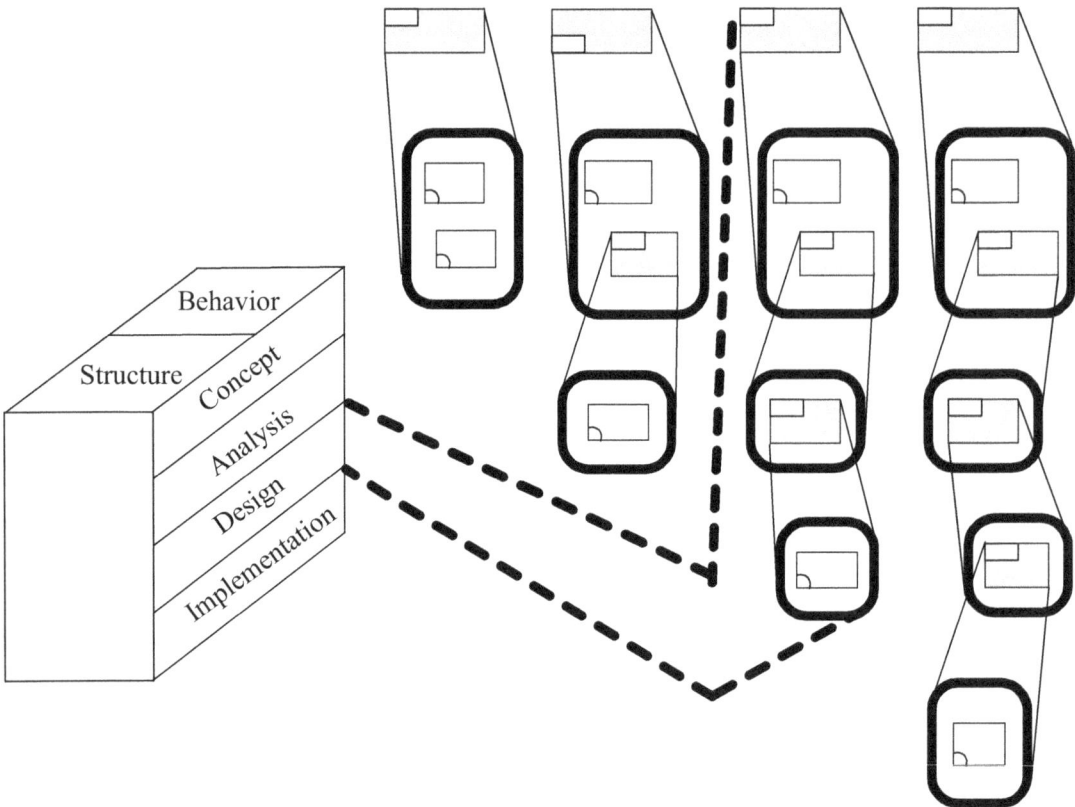

Figure 20-40   Design View of the *Purchasing Management* Systems Definition Version 1

The design view of the *purchasing management* systems definition version 1 consists of: a) design's systems structure of the *purchasing management* systems definition version 1 and b) design's systems behavior of the *purchasing management* systems definition version 1.

20-1-2-3-1 Design's Systems Structure of the Purchasing Management Systems Definition Version 1

The entire design's systems structure of the *purchasing management* systems definition version 1 includes: a) *Design's AHD*, b) *Design's FD*, c) *Design's COD* and d) *Design's CCD* of the *purchasing management* systems definition version 1.

We first draw the design's AHD of the *purchasing management* systems definition version 1. As shown in Figure 20-41, *Purchasing Management* is composed of *Purchasing_Department* and *PM_Subsystem_3*; *PM_Subsystem_3* is composed of *Quotation_Processing_GUI*, *Purchase_Order_Processing_GUI*, *Purchase_Processing_GUI* and *PM_Subsystem_2*; *PM_Subsystem_2* is composed of *Purchasing_Database* and *PM_Subsystem_1*. In the figure, *Purchasing Management,*

PM_Subsystem_3 and PM_Subsystem_2 are aggregated systems while *Purchasing_Department,* *Quotation_Processing_GUI,* *Purchase_Order_Processing_GUI,* *Purchase_Processing_GUI,* *Purchasing_Database* and *PM_Subsystem_1* are non-aggregated systems.

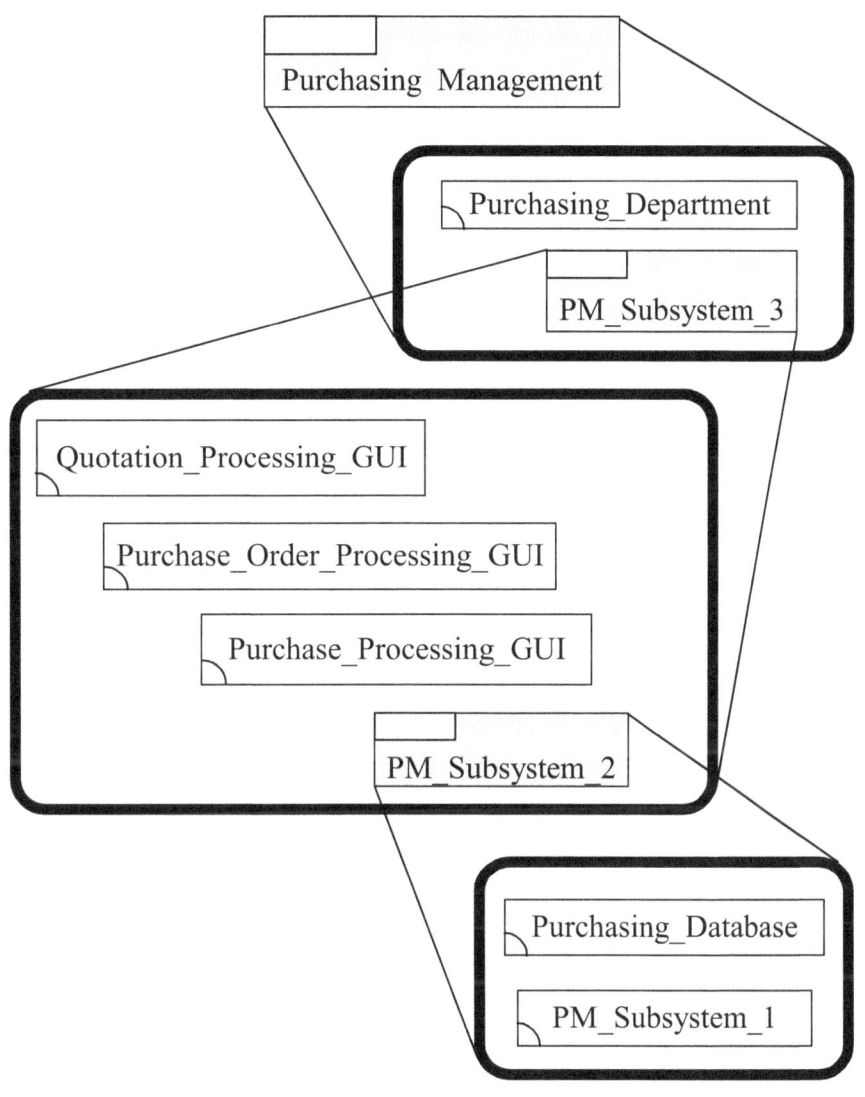

Figure 20-41   Design's AHD of the *Purchasing Management* Systems Definition Version 1

Figure 20-42 shows the design's FD of the *purchasing management* systems definition version 1. In the figure, *Business_Layer* contains the *Purchasing_Department* component; *Application_Layer* contains the *Quotation_Processing_GUI,* *Purchase_Order_Processing_GUI* and *Purchase_Processing_GUI* components; *Data_Layer* contains the *Purchasing_Database* and *PM_Subsystem_1* components.

258

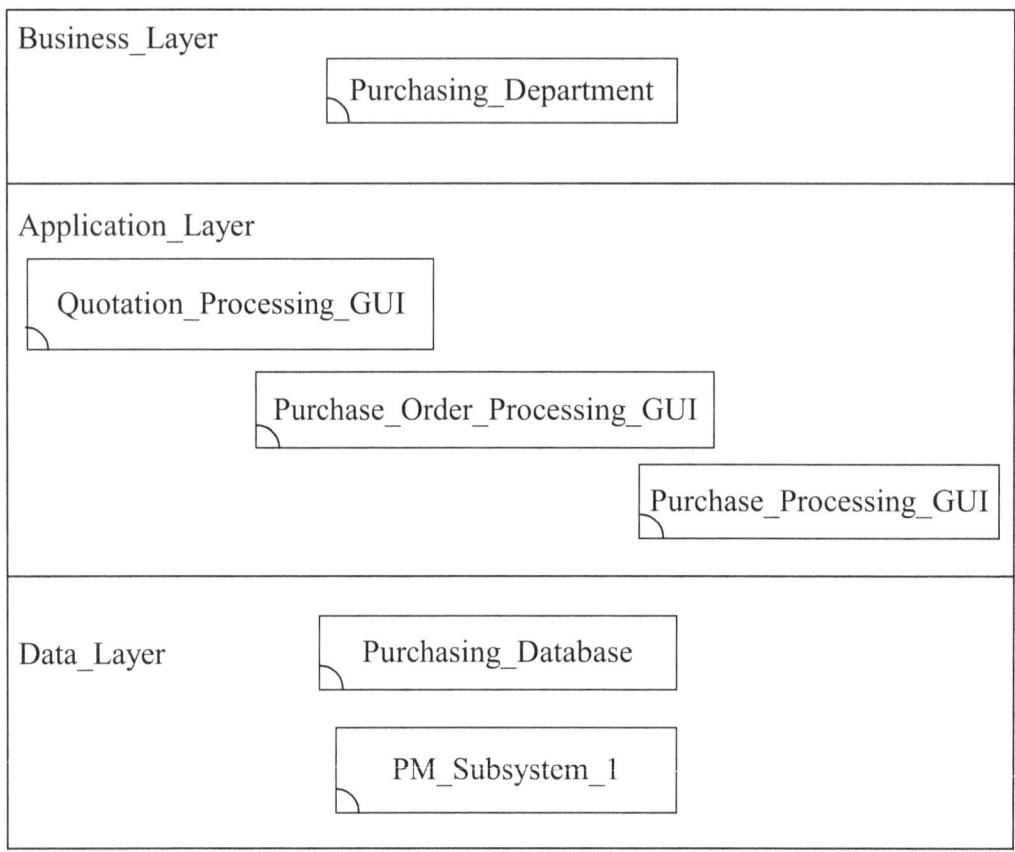

Figure 20-42   Design's FD of the *Purchasing Management*
Systems Definition Version 1

Figure 20-43 shows the design's COD of the *purchasing management* systems definition version 1. In the figure, component *Purchasing_Department* has three operations: *Quotation_Verify*, *Purchase_Order_Verify* and *Purchase_Verify*; component *Quotation_Processing_GUI* has one operation: *Quotation_Processing_Button_Click*; component *Purchase_Order_Processing_GUI* has one operation: *Purchase_Order_Processing_Button_Click*; component *Purchase_Processing_GUI* has one operation: *Purchase_Processing_Button_Click*; component *Purchasing_Database* has three operations: *SQL_Quotation_Insert*, *SQL_Purchase_Order_Insert* and *SQL_Purchase_Insert*; component *PM_Subsystem_1* has one operation: *Infrastructure_Resources_Share*.

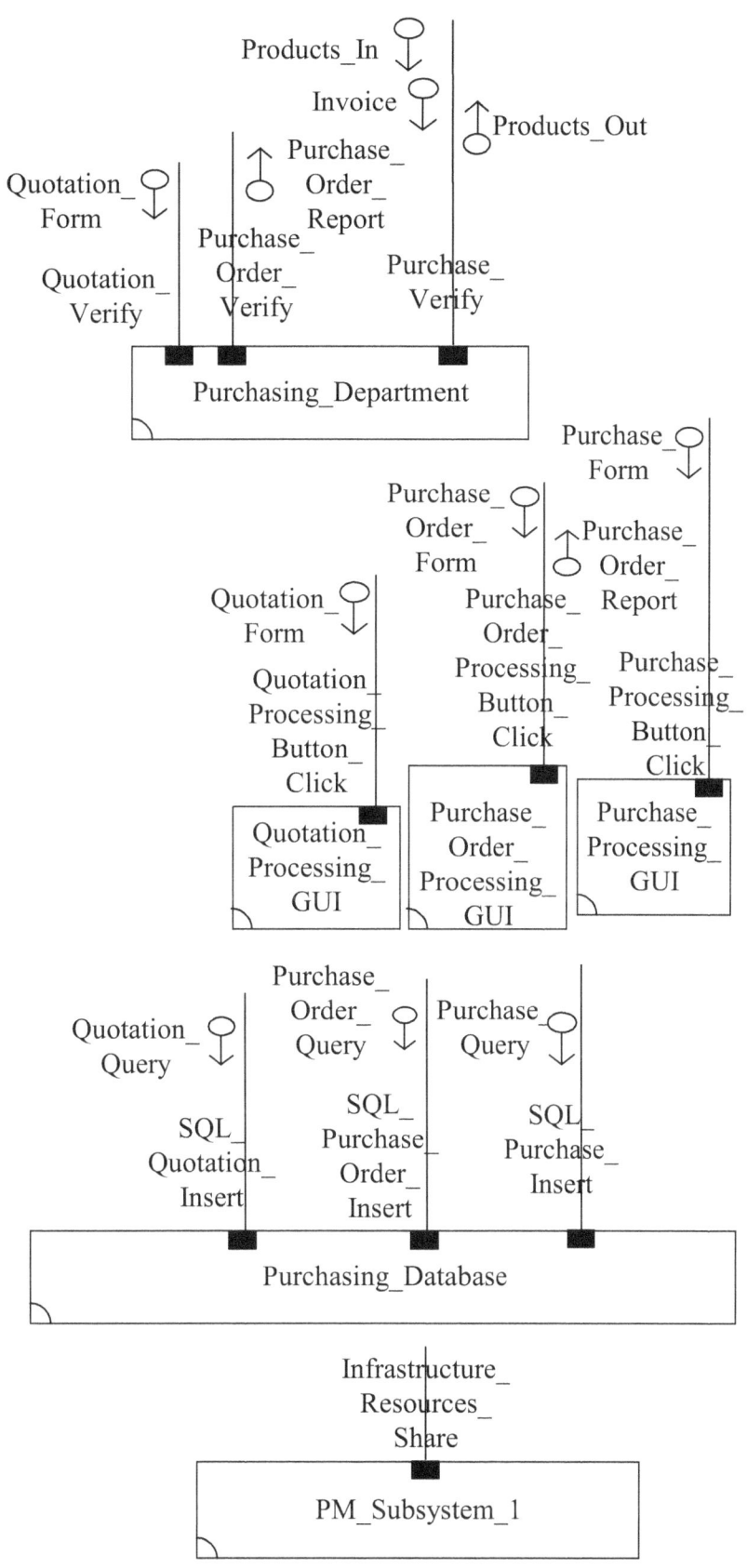

Figure 20-43   Design's COD of the *Purchasing Management Systems Definition Version 1*

The operation formula of *Quotation_Verify* is *Quotation_Verify(In Quotation_Form)*. The operation formula of *Purchase_Order_Verify* is *Purchase_Order_Verify(Out Purchase_Order_Report)*. The operation formula of *Purchase_Verify* is *Purchase_Verify(In Products_In, Invoice; Out Products_Out)*. The operation formula of *Quotation_Processing_Button_Click* is *Quotation_Processing_Button_Click(In Quotation_Form)*. The operation formula of *Purchase_Order_Processing_Button_Click* is *Purchase_Order_Processing_Button_Click(In Purchase_Order_Form; Out Purchase_Order_Report)*. The operation formula of *Purchase_Processing_Button_Click* is *Purchase_Processing_Button_Click(In Purchase_Form)*. The operation formula of *SQL_Quotation_Insert* is *SQL_Quotation_Insert(In Quotation_Query)*. The operation formula of *SQL_Purchase_Order_Insert* is *SQL_Purchase_Order_Insert (In Purchase_Order_Query)*. The operation formula of *SQL_Purchase_Insert* is *SQL_Purchase_Insert (In Purchase_Query)*. The operation formula of *Infrastructure_Resources_Share* is *Infrastructure_Resources_Share*.

Figure 20-44 shows the composite data type specification of the *Quotation_Form* input parameter occurring in the *Quotation_Verify(In Quotation_Form)* and *Quotation_Processing_Button_Click(In Quotation_Form)* operation formulas.

Parameter	*Quotation_Form*

Data Type	TABLE of
	  Date : Text
	  SupplierName: Text
	  ProductNo : Text
	  Quantity : Integer
	  UnitPrice : Real
	  Total : Real
	End TABLE ;

Instances	**Quotation Form**
	
	Date: 2011/10/25
	
	SupplierName : Johnson Corp.
	
	  ProductNo Quantity UnitPrice
	  A 0 0 0 0 1 (P e n) 3 0 0 1 0 0 . 0 0
	  _A00002(Mouse)____400____200.00__
	  _A00003(Camera)___500____300.00__
	
	               Total : 260,000.00

Figure 20-44   Composite Data Type Specification

Figure 20-45 shows the composite data type specification of the *Purchase_Order_Report* output parameter occurring in the *Purchase_Order_Verify(Out Purchase_Order_Report)* and *Purchase_Order_Processing_Button_Click(In Purchase_Order_Form; Out Purchase_Order_Report)* operation formulas.

Parameter	*Purchase_Order_Report*

Data Type	TABLE of
	  Date : Text
	  SupplierName: Text
	  ProductNo : Text
	  Quantity : Integer
	End TABLE ;

Instances	Date : 20111118 SupplierName : Johnson Corp. <table><tr><th>ProductNo</th><th>Quantity</th></tr><tr><td>A00001(Pen)</td><td>300</td></tr><tr><td>A00002(Mouse)</td><td>400</td></tr><tr><td>A00003(Camera)</td><td>500</td></tr></table>

Figure 20-45   Composite Data Type Specification

Figure 20-46 shows the primitive data type specification of the *Products_In* input parameter occurring in the *Purchase_Verify(In Products_In, Invoice; Out Products_Out)* operation formula.

Parameter	Data Type	Instances
Products_In	Physical Object	Pen, Mouse, Camera

Figure 20-46   Primitive Data Type Specification

Figure 20-47 shows the composite data type specification of the *Invoice* input parameter occurring in the *Purchase_Verify(In Products_In, Invoice; Out Products_Out)* operation formula.

Parameter	*Invoice*

Data Type	TABLE of
	  Date : Text
	  SupplierName: Text
	  ProductNo : Text
	  Quantity : Integer
	  UnitPrice : Real
	  Total : Real
	End TABLE ;

Instances	Date : 20111130 SupplierName : Johnson Corp. | ProductNo	Quantity	UnitPrice
---	---	---
A00001(Pen)	300	100.00
A00002(Mouse)	400	200.00
A00003(Camera)	500	300.00	 Total : 260,000.00

Figure 20-47   Composite Data Type Specification

Figure 20-48 shows the primitive data type specification of the *Products_Out* output parameter occurring in the *Purchase_Verify(In Products_In, Invoice; Out Products_Out)* operation formula.

Parameter	Data Type	Instances
Products_Out	Physical Object	Pen, Mouse, Camera

Figure 20-48   Primitive Data Type Specification

Figure 20-49 shows the composite data type specification of the *Purchase_Order_Form* input parameter occurring in the *Purchase_Order_Processing_Button_Click(In Purchase_Order_Form; Out Purchase_Order_Report)* operation formula.

264

Parameter	*Purchase_Order_Form*
Data Type	TABLE of Date : Text SupplierName: Text ProductNo : Text Quantity : Integer End TABLE ;
Instances	**Purchase Order Form** Date: 2011/11/18 SupplierName : Johnson Corp. ProductNo Quantity __A00001(Pen) ____300 ___ __A00002(Mouse)____400___ __A00003(Camera)___500___

Figure 20-49   Composite Data Type Specification

Figure 20-50 shows the composite data type specification of the *Purchase_Form* input parameter occurring in the *Purchase_Processing_Button_Click(In Purchase_Form)* operation formula

Parameter	*Purchase_Form*

Data Type	TABLE of
	  Date : Text
	  SupplierName: Text
	  ProductNo : Text
	  Quantity : Integer
	  UnitPrice : Real
	  ReturnQuantity : Integer
	  Total : Real
	End TABLE ;

Instances	**Purchase Form** Date: 2011/12/12 SupplierName : Johnson Corp. ProductNo Quantity UnitPrice ReturnQuantity A00001(Pen) 300 100.00 0 A00002(Mouse) 390 200.00 10 A00003(Camera)500 300.00 0 Total : 258,000.00

Figure 20-50   Composite Data Type Specification

Figure 20-51 shows the composite data type specification of the *Quotation_Query* input parameter occurring in the *SQL_Quotation_Insert(In Quotation_Query)* operation formula.

266

Parameter	*Quotation_Query*

Data Type	TABLE of
	  Date : Text
	  SupplierName: Text
	  ProductNo : Text
	  Quantity : Integer
	  UnitPrice : Real
	  Total : Real
	End TABLE ;

Instances	<table><tr><th>Date</th><th>SupplierName</th><th>Total</th></tr><tr><td>20111025</td><td>Johnson Corp.</td><td>260,000.00</td></tr></table> <table><tr><th>ProductNo</th><th>Quantity</th><th>UnitPrice</th></tr><tr><td>A00001(Pen)</td><td>300</td><td>100.00</td></tr><tr><td>A00002(Mouse)</td><td>400</td><td>200.00</td></tr><tr><td>A00003(Camera)</td><td>500</td><td>300.00</td></tr></table>

Figure 20-51   Composite Data Type Specification

Figure 20-52 shows the composite data type specification of the *Purchase_Order_Query* input parameter occurring in the *SQL_Purchase_Order_Insert(In Purchase_Order_Query)* operation formula.

Parameter	*Purchase_Order_Query*

Data Type	TABLE of
	  Date : Text
	  SupplierName: Text
	  ProductNo : Text
	  Quantity : Integer
	End TABLE ;

Instances	

Instances table:

Date	SupplierName
20111118	Johnson Corp.

ProductNo	Quantity
A00001(Pen)	300
A00002(Mouse)	400
A00003(Camera)	500

Figure 20-52   Composite Data Type Specification

Figure 20-53 shows the composite data type specification of the *Purchase_Query* input parameter occurring in the *SQL_Purchase_Insert (In Purchase_Query)* operation formula.

268

Parameter	*Purchase_Query*

Data Type	TABLE of
	  Date : Text
	  SupplierName: Text
	  ProductNo : Text
	  Quantity : Integer
	  UnitPrice : Real
	  ReturnQuantity : Integer
	  Total : Real
	End TABLE ;

Instances	<table><tr><th>Date</th><th>SupplierName</th><th>Total</th></tr><tr><td>20111212</td><td>Johnson Corp.</td><td>258,000.00</td></tr></table> <table><tr><th>ProductNo</th><th>Quantity</th><th>UnitPrice</th><th>ReturnQuantity</th></tr><tr><td>A00001(Pen)</td><td>300</td><td>100.00</td><td>0</td></tr><tr><td>A00002(Mouse)</td><td>390</td><td>200.00</td><td>10</td></tr><tr><td>A00003(Camera)</td><td>500</td><td>300.00</td><td>0</td></tr></table>

Figure 20-53   Composite Data Type Specification

Figure 20-54 shows the design's CCD of the *purchasing management* systems definition version 1. In the figure, actor *Supplier* has three connections with the *Purchasing_Department* component; component *Purchasing_Department* has a connection with each one of the *Quotation_Processing_GUI*, *Purchase_Order_Processing_GUI* and *Purchase_Processing_GUI* components; each one of the *Quotation_Processing_GUI*, *Purchase_Order_Processing_GUI* and *Purchase_Processing_GUI* components has a connection with the *Purchasing_Database*; component *Purchasing_Database* has a connection with the *PM_Subsystem_1* component.

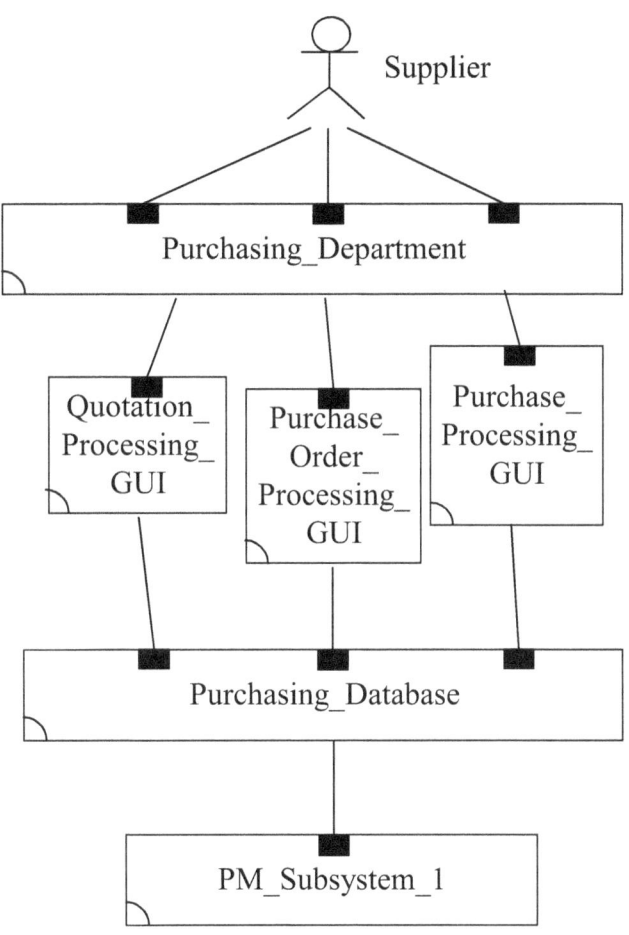

Figure 20-54   Design's CCD of the *Purchasing Management*
Systems Definition Version 1

20-1-2-3-2 Design's Systems Behavior of the Purchasing Management Systems
Definition Version 1

The entire design's systems behavior of the *purchasing management* systems
definition version 1 includes: a) *Design's SBCD* and b) *Design's IFD* of the
purchasing management systems definition version 1.

Figure 20-55 shows the design's SBCD of the *purchasing management*
systems definition version 1 in which interactions among the *Supplier* actor and the
Purchasing_Department,   *Quotation_Processing_GUI,*
Purchase_Order_Processing_GUI,   *Purchase_Processing_GUI,*
Purchasing_Database, PM_Subsystem_1 components shall draw forth the *Quotation,*
Purchase_Order and *Purchase* behaviors.

270

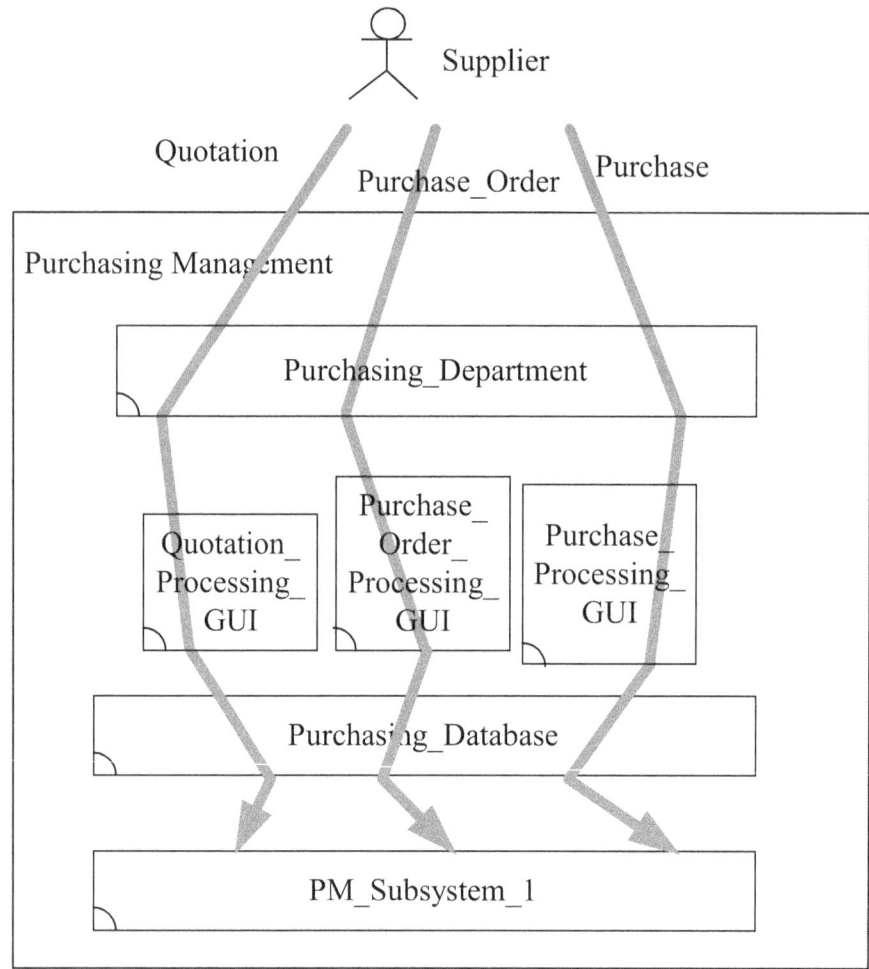

Figure 20-55   Design's SBCD of the *Purchasing Management*
Systems Definition Version 1

The overall design's behavior of the *purchasing management* systems definition version 1 includes the *Quotation, Purchase_Order and Purchase* behaviors. In other words, the *Quotation, Purchase_Order and Purchase* behaviors together provide the overall design's behavior of the *purchasing management* systems definition version 1.

Be noticed that the *Quotation, Purchase_Order and Purchase* behaviors are mutually independent of each other. They tend to be executed concurrently [Hoar85, Miln89, Miln99].

The overall design's behavior of the *purchasing management* systems definition version 1 includes three individual behaviors: *Quotation, Purchase_Order and Purchase*. Each individual behavior is represented by an execution path. We use an interaction flow diagram (IFD) to define each one of these execution paths. Figure

20-56 shows the design's IFD of the *purchasing management* systems definition version 1 *Quotation* behavior. First, actor *Supplier* interacts with the *Purchasing_Department* component through the *Quotation_Verify* operation call interaction, carrying the *Quotation_Form* input parameter. Next, component *Purchasing_Department* interacts with the *Quotation_Processing_GUI* component through the *Quotation_Processing_Button_Click* operation call interaction, carrying the *Quotation_Form* input parameter. Continuingly, component *Quotation_Processing_GUI* interacts with the *Purchasing_Database* component through the *SQL_Quotation_Insert* operation call interaction, carrying the *Quotation_Query* input parameter. Finally, component *Purchasing_Database* interacts with the *PM_Subsystem_1* component through the *Infrastructure_Resources_Share* operation call interaction.

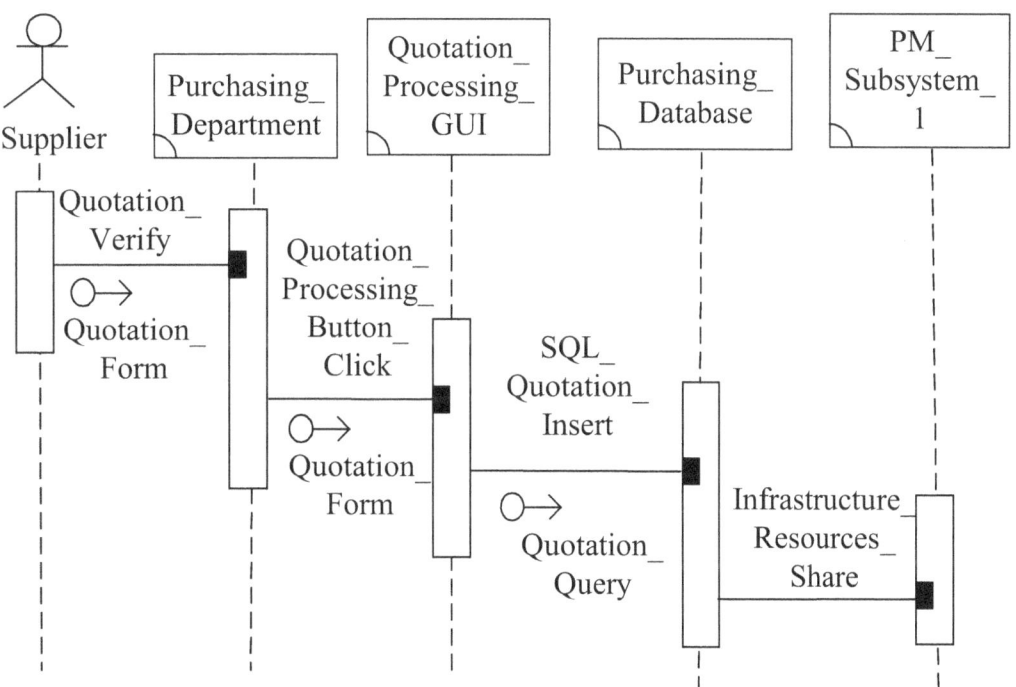

Figure 20-56   Design's IFD of the *Purchasing Management*
Systems Definition Version 1 *Quotation* Behavior

Figure 20-57 shows the design's IFD of the *purchasing management* systems definition version 1 *Purchase_Order* behavior. First, actor *Supplier* interacts with the *Purchasing_Department* component through the *Purchase_Order_Verify* operation call interaction. Next, component *Purchasing_Department* interacts with the *Purchase_Order_Processing_GUI* component through the *Purchase_Order_Processing_Button_Click* operation call interaction, carrying the *Purchase_Order_Form* input parameter. Continuingly, component

Purchase_Order_Processing_GUI interacts with the *Purchasing_Database* component through the *SQL_Purchase_Order_Insert* operation call interaction, carrying the *Purchase_Order_Query* input parameter. Continuingly, component *Purchasing_Database* interacts with the *PM_Subsystem_1* component through the *Infrastructure_Resources_Share* operation call interaction. Continuingly, component *Purchasing_Department* interacts with the *Purchase_Order_Processing_GUI* component through the *Purchase_Order_Processing_Button_Click* operation return interaction, carrying the *Purchase_Order_Report* output parameter. Finally, actor *Supplier* interacts with the *Purchasing_Department* component through the *Purchase_Order_Verify* operation return interaction, carrying the *Purchase_Order_Report* output parameter.

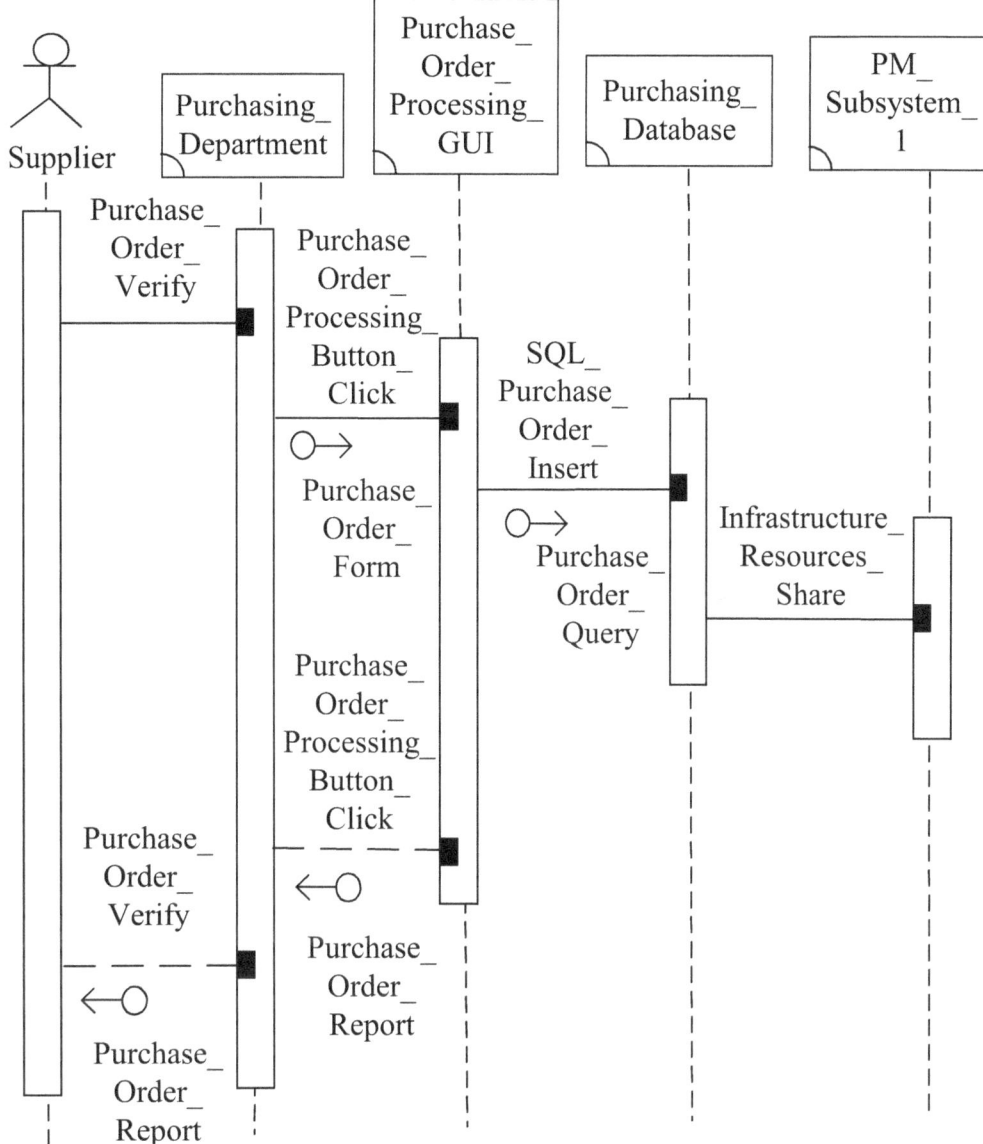

Figure 20-57   Design's IFD of the *Purchasing Management*
Systems Definition Version 1 *Purchase_Order* Behavior

Figure 20-58 shows the design's IFD of the *purchasing management* systems definition version 1 *Purchase* behavior. First, actor *Supplier* interacts with the *Purchasing_Department* component through the *Purchase_Verify* operation call interaction, carrying the *Products_In* and *Invoice* input parameters. Next, component *Purchasing_Department* interacts with the *Purchase_Processing_GUI* component through the *Purchase_Processing_Button_Click* operation call interaction, carrying the *Purchase_Form* input parameter. Continuingly, component *Purchase_Processing_GUI* interacts with the *Purchasing_Database* component through the *SQL_Purchase_Insert* operation call interaction, carrying the *Purchase_Query* input parameter. Continuingly, component *Purchasing_Database*

274

interacts   with   the   *PM_Subsystem_1*   component   through   the
Infrastructure_Resources_Share operation call interaction. Finally, actor *Supplier*
interacts with the *Purchasing_Department* component through the *Purchase_Verify*
operation return interaction, carrying the *Products_Out* output parameter.

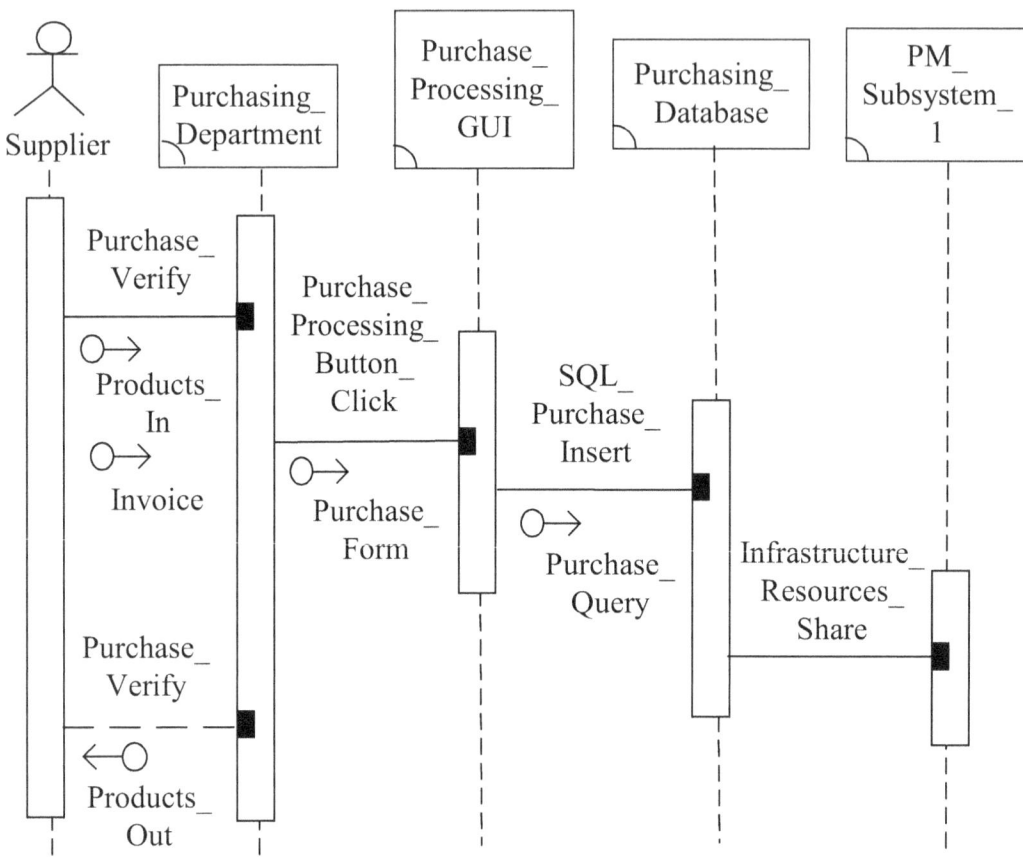

Figure 20-58   Design's IFD of the *Purchasing Management*
Systems Definition Version 1 *Purchase* Behavior

20-1-2-4 Implementation View of the Purchasing Management Systems Definition
Version 1

The implementation view of the *purchasing management* systems definition
version 1 is shown in Figure 20-59.

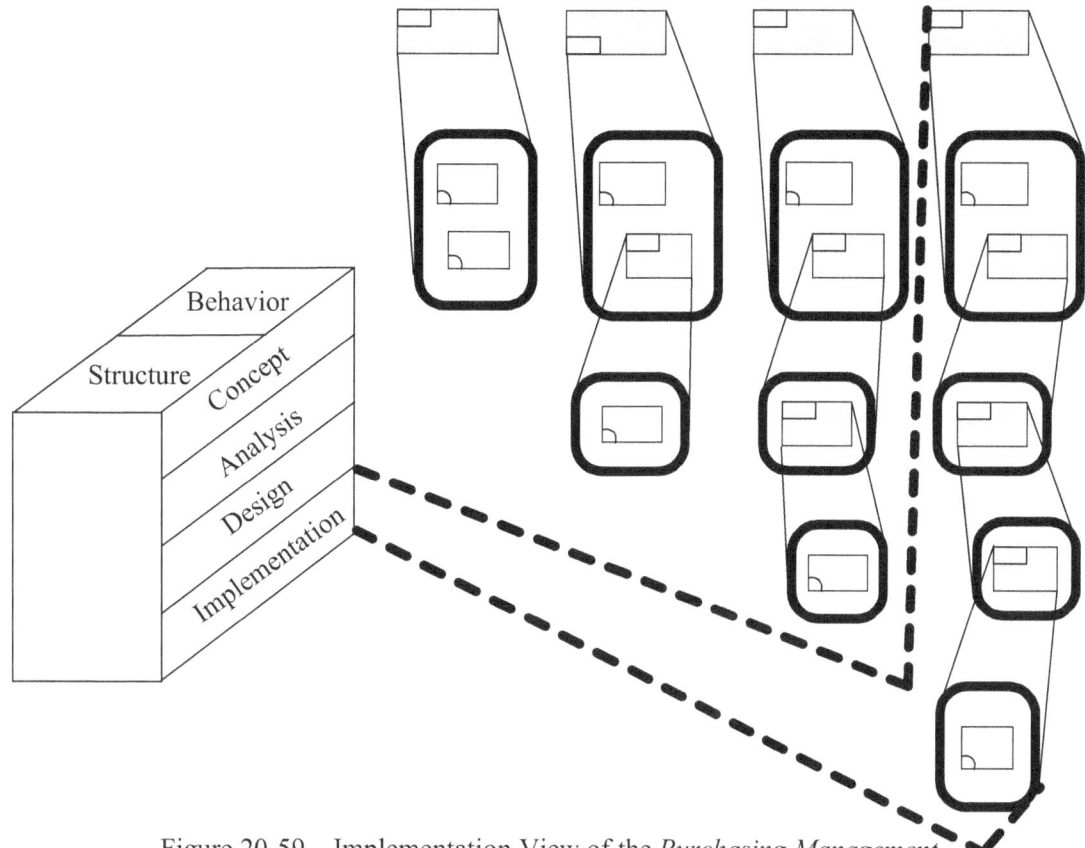

Figure 20-59   Implementation View of the *Purchasing Management*
Systems Definition Version 1

The implementation view of the *purchasing management* systems definition version 1 consists of: a) implementation's systems structure of the *purchasing management* systems definition version 1 and b) implementation's systems behavior of the *purchasing management* systems definition version 1.

20-1-2-4-1 Implementation's Systems Structure of the Purchasing Management Systems Definition Version 1

The entire implementation's systems structure of the *purchasing management* systems definition version 1 includes: a) *Implementation's AHD*, b) *Implementation's FD*, c) *Implementation's COD* and d) *Implementation's CCD* of the *purchasing management* systems definition version 1.

We first draw the implementation's AHD of the *purchasing management* systems definition version 1. As shown in Figure 20-60, *Purchasing Management* is composed of *Purchasing_Department* and *PM_Subsystem_3*; *PM_Subsystem_3* is composed of *Quotation_Processing_GUI*, *Purchase_Order_Processing_GUI*, *Purchase_Processing_GUI* and *PM_Subsystem_2*; *PM_Subsystem_2* is composed of *Purchasing_Database* and *PM_Subsystem_1*; *PM_Subsystem_1* is composed of

276

Network_Operating_System. In the figure, *Purchasing Management, PM_Subsystem_3, PM_Subsystem_2 and PM_Subsystem_1* are aggregated systems while *Purchasing_Department, Quotation_Processing_GUI, Purchase_Order_Processing_GUI, Purchase_Processing_GUI, Purchasing_Database* and *Network_Operating_System* are non-aggregated systems.

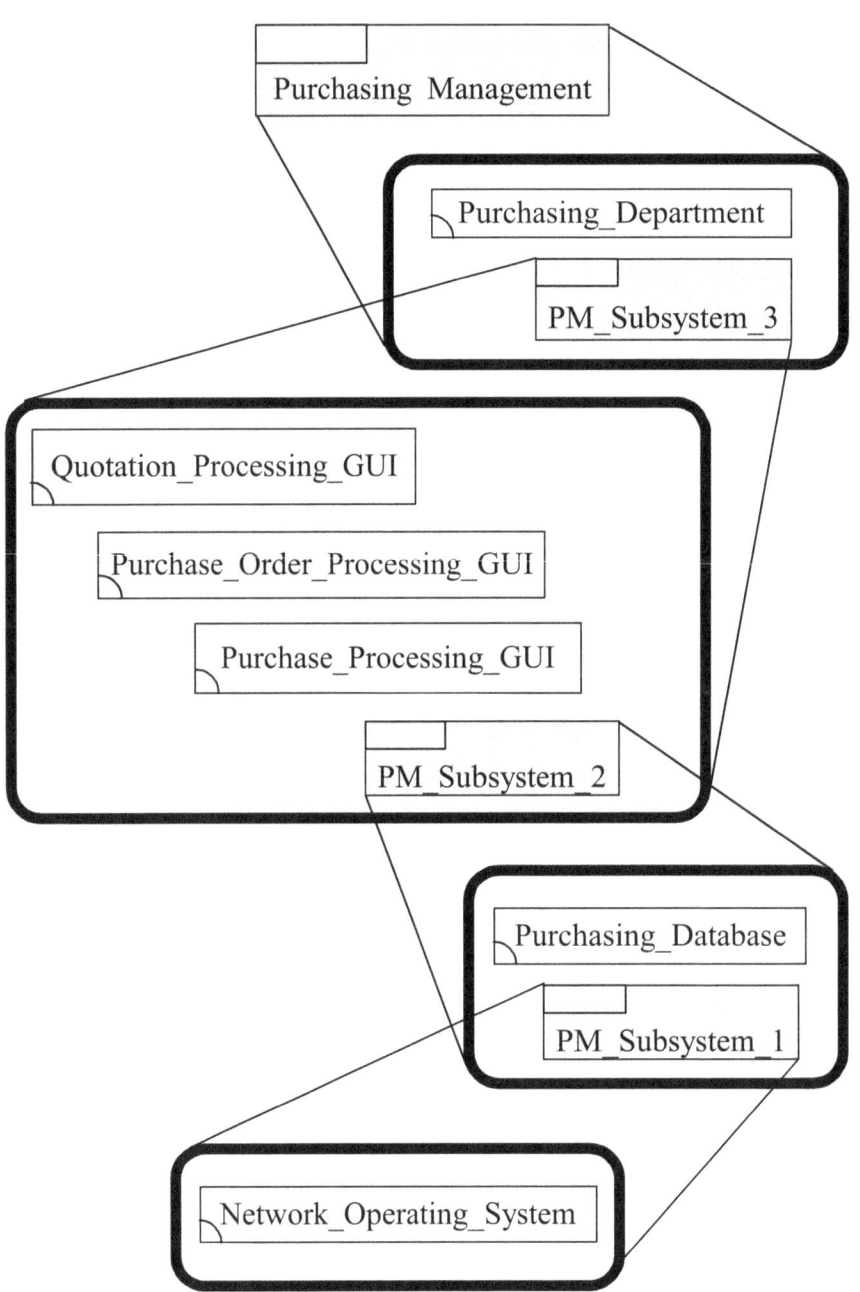

Figure 20-60   Implementation's AHD of the *Purchasing Management*
Systems Definition Version 1

Figure 20-61 shows the implementation's FD of the *purchasing management* systems definition version 1. In the figure, *Business_Layer* contains the *Purchasing_Department* component; *Application_Layer* contains the *Quotation_Processing_GUI*, *Purchase_Order_Processing_GUI* and *Purchase_Processing_GUI* components; *Data_Layer* contains the *Purchasing_Database* component; *Technology_Layer* contains the *Network_Operating_System* component.

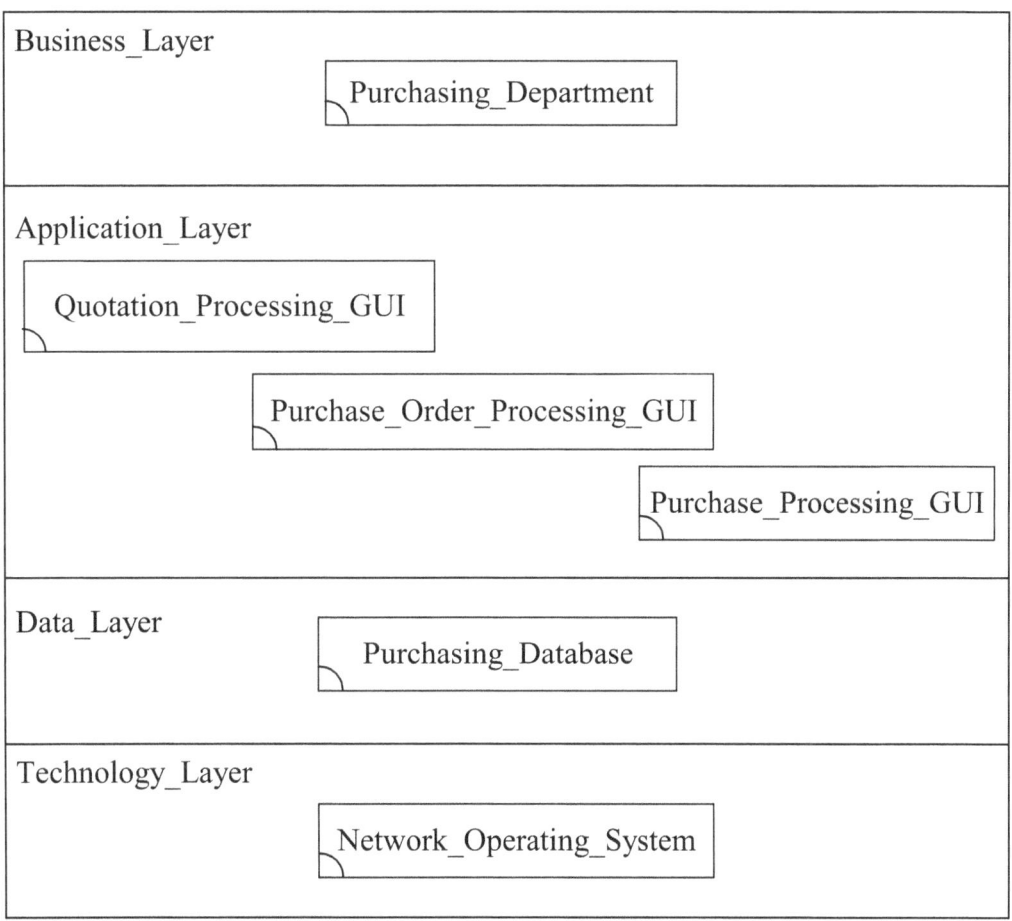

Figure 20-61   Implementation's FD of the *Purchasing Management Systems Definition Version 1*

Figure 20-62 shows the implementation's COD of the *purchasing management* systems definition version 1. In the figure, component *Purchasing_Department* has three operations: *Quotation_Verify*, *Purchase_Order_Verify* and *Purchase_Verify*; component

Quotation_Processing_GUI has one operation: *Quotation_Processing_Button_Click*; component *Purchase_Order_Processing_GUI* has one operation: *Purchase_Order_Processing_Button_Click*; component *Purchase_Processing_GUI* has one operation: *Purchase_Processing_Button_Click*; component *Purchasing_Database* has three operations: *SQL_Quotation_Insert*, *SQL_Purchase_Order_Insert* and *SQL_Purchase_Insert*; component *Network_Operating_System* has one operation: *Infrastructure_Resources_Share*.

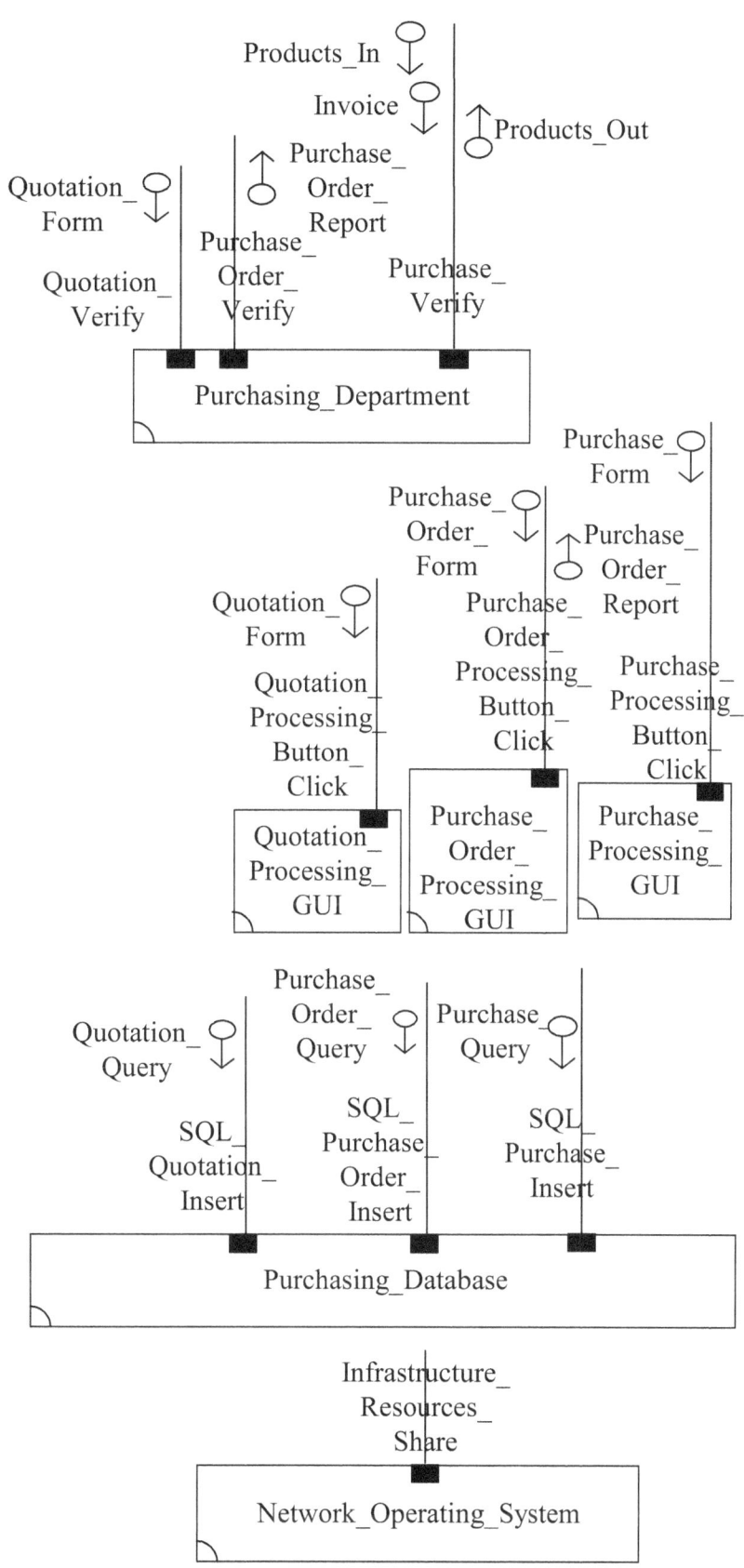

Figure 20-62   Implementation's COD of the *Purchasing Management
Systems Definition Version 1*

The operation formula of *Quotation_Verify* is *Quotation_Verify(In Quotation_Form)*. The operation formula of *Purchase_Order_Verify* is *Purchase_Order_Verify(Out Purchase_Order_Report)*. The operation formula of *Purchase_Verify* is *Purchase_Verify(In Products_In, Invoice; Out Products_Out)*. The operation formula of *Quotation_Processing_Button_Click* is *Quotation_Processing_Button_Click(In Quotation_Form)*. The operation formula of *Purchase_Order_Processing_Button_Click* is *Purchase_Order_Processing_Button_Click(In Purchase_Order_Form; Out Purchase_Order_Report)*. The operation formula of *Purchase_Processing_Button_Click* is *Purchase_Processing_Button_Click(In Purchase_Form)*. The operation formula of *SQL_Quotation_Insert* is *SQL_Quotation_Insert(In Quotation_Query)*. The operation formula of *SQL_Purchase_Order_Insert* is *SQL_Purchase_Order_Insert (In Purchase_Order_Query)*. The operation formula of *SQL_Purchase_Insert* is *SQL_Purchase_Insert (In Purchase_Query)*. The operation formula of *Infrastructure_Resources_Share* is *Infrastructure_Resources_Share*.

Figure 20-63 shows the composite data type specification of the *Quotation_Form* input parameter occurring in the *Quotation_Verify(In Quotation_Form)* and *Quotation_Processing_Button_Click(In Quotation_Form)* operation formulas.

Parameter	*Quotation_Form*

Data Type	TABLE of
	  Date : Text
	  SupplierName: Text
	  ProductNo : Text
	  Quantity : Integer
	  UnitPrice : Real
	  Total : Real
	End TABLE ;

Instances	**Quotation Form** Date: 2011/10/25 SupplierName : Johnson Corp. ProductNo Quantity UnitPrice A 0 0 0 0 1 (P e n) 3 0 0 1 0 0 . 0 0 _A00002(Mouse)____400____200.00__ _A00003(Camera)___500____300.00__ Total : 260,000.00

Figure 20-63   Composite Data Type Specification

Figure 20-64 shows the composite data type specification of the *Purchase_Order_Report* output parameter occurring in the *Purchase_Order_Verify(Out Purchase_Order_Report)* and *Purchase_Order_Processing_Button_Click(In Purchase_Order_Form; Out Purchase_Order_Report)* operation formulas.

Parameter	*Purchase_Order_Report*
Data Type	TABLE of Date : Text SupplierName: Text ProductNo : Text Quantity : Integer End TABLE ;
Instances	Date : 20111118 SupplierName : Johnson Corp. <table><tr><th>ProductNo</th><th>Quantity</th></tr><tr><td>A00001(Pen)</td><td>300</td></tr><tr><td>A00002(Mouse)</td><td>400</td></tr><tr><td>A00003(Camera)</td><td>500</td></tr></table>

Figure 20-64   Composite Data Type Specification

Figure 20-65 shows the primitive data type specification of the *Products_In* input parameter occurring in the *Purchase_Verify(In Products_In, Invoice; Out Products_Out)* operation formula.

Parameter	Data Type	Instances
Products_In	Physical Object	Pen, Mouse, Camera

Figure 20-65   Primitive Data Type Specification

Figure 20-66 shows the composite data type specification of the *Invoice* input parameter occurring in the *Purchase_Verify(In Products_In, Invoice; Out Products_Out)* operation formula.

Parameter	Invoice

Data Type	TABLE of
	  Date : Text
	  SupplierName: Text
	  ProductNo : Text
	  Quantity : Integer
	  UnitPrice : Real
	  Total : Real
	End TABLE ;

Instances	Date : 20111130 SupplierName : Johnson Corp. | ProductNo	Quantity	UnitPrice
---	---	---
A00001(Pen)	300	100.00
A00002(Mouse)	400	200.00
A00003(Camera)	500	300.00	 Total : 260,000.00

Figure 20-66   Composite Data Type Specification

Figure 20-67 shows the primitive data type specification of the *Products_Out* output parameter occurring in the *Purchase_Verify(In Products_In, Invoice; Out Products_Out)* operation formula.

Parameter	Data Type	Instances
Products_Out	Physical Object	Pen, Mouse, Camera

Figure 20-67   Primitive Data Type Specification

Figure 20-68 shows the composite data type specification of the *Purchase_Order_Form* input parameter occurring in the *Purchase_Order_Processing_Button_Click(In Purchase_Order_Form; Out Purchase_Order_Report)* operation formula.

Parameter	*Purchase_Order_Form*

Data Type	TABLE of
	  Date : Text
	  SupplierName: Text
	  ProductNo : Text
	  Quantity : Integer
	End TABLE ;

Instances	**Purchase Order Form** Date: 2011/11/18 SupplierName : Johnson Corp. ProductNo Quantity __A00001(Pen) ____300____ __A00002(Mouse)____400___ __A00003(Camera)___500___

Figure 20-68　　　Composite Data Type Specification

Figure 20-69 shows the composite data type specification of the *Purchase_Form* input parameter occurring in the *Purchase_Processing_Button_Click(In Purchase_Form)* operation formula

Parameter	*Purchase_Form*

Data Type	TABLE of
	  Date : Text
	  SupplierName: Text
	  ProductNo : Text
	  Quantity : Integer
	  UnitPrice : Real
	  ReturnQuantity : Integer
	  Total : Real
	End TABLE ;

Instances	**Purchase Form** Date: 2011/12/12 SupplierName : Johnson Corp. ProductNo Quantity UnitPrice ReturnQuantity A00001(Pen) 300 100.00 0 A00002(Mouse) 390 200.00 10 A00003(Camera)500 300.00 0 Total : 258,000.00

Figure 20-69   Composite Data Type Specification

Figure 20-70 shows the composite data type specification of the *Quotation_Query* input parameter occurring in the *SQL_Quotation_Insert(In Quotation_Query)* operation formula.

286

Parameter	*Quotation_Query*

Data Type	TABLE of
	  Date : Text
	  SupplierName: Text
	  ProductNo : Text
	  Quantity : Integer
	  UnitPrice : Real
	  Total : Real
	End TABLE ;

Instances	<table><tr><td>Date</td><td>SupplierName</td><td>Total</td></tr><tr><td>20111025</td><td>Johnson Corp.</td><td>260,000.00</td></tr></table> <table><tr><td>ProductNo</td><td>Quantity</td><td>UnitPrice</td></tr><tr><td>A00001(Pen)</td><td>300</td><td>100.00</td></tr><tr><td>A00002(Mouse)</td><td>400</td><td>200.00</td></tr><tr><td>A00003(Camera)</td><td>500</td><td>300.00</td></tr></table>

Figure 20-70   Composite Data Type Specification

Figure 20-71 shows the composite data type specification of the *Purchase_Order_Query* input parameter occurring in the *SQL_Purchase_Order_Insert(In Purchase_Order_Query)* operation formula.

Parameter	*Purchase_Order_Query*

Data Type	TABLE of
	  Date : Text
	  SupplierName: Text
	  ProductNo : Text
	  Quantity : Integer
	End TABLE ;

Instances	

Date	SupplierName
20111118	Johnson Corp.

ProductNo	Quantity
A00001(Pen)	300
A00002(Mouse)	400
A00003(Camera)	500

Figure 20-71   Composite Data Type Specification

Figure 20-72 shows the composite data type specification of the *Purchase_Query* input parameter occurring in the *SQL_Purchase_Insert (In Purchase_Query)* operation formula.

Parameter	*Purchase_Query*

Data Type	TABLE of
	  Date : Text
	  SupplierName: Text
	  ProductNo : Text
	  Quantity : Integer
	  UnitPrice : Real
	  ReturnQuantity : Integer
	  Total : Real
	End TABLE ;

Instances	

Date	SupplierName	Total
20111212	Johnson Corp.	258,000.00

ProductNo	Quantity	UnitPrice	ReturnQuantity
A00001(Pen)	300	100.00	0
A00002(Mouse)	390	200.00	10
A00003(Camera)	500	300.00	0

Figure 20-72   Composite Data Type Specification

Figure 20-73 shows the implementation's CCD of the *purchasing management* systems definition version 1. In the figure, actor *Supplier* has three connections with the *Purchasing_Department* component; component *Purchasing_Department* has a connection with each one of the *Quotation_Processing_GUI*, *Purchase_Order_Processing_GUI* and *Purchase_Processing_GUI* components; each one of the *Quotation_Processing_GUI*, *Purchase_Order_Processing_GUI* and *Purchase_Processing_GUI* components has a connection with the *Purchasing_Database*; component *Purchasing_Database* has a connection with the *Network_Operating_System* component.

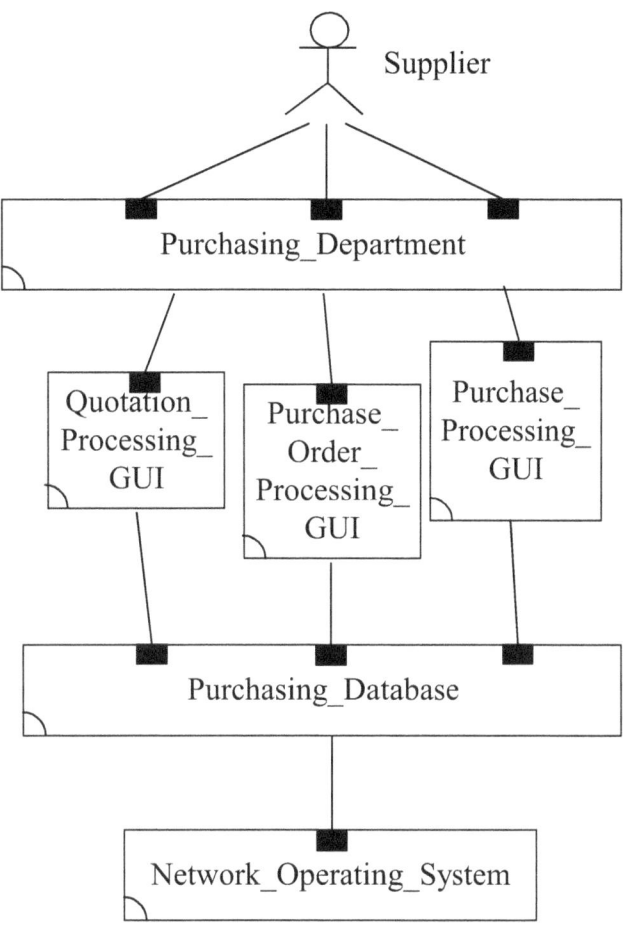

Figure 20-73   Implementation's CCD of the *Purchasing Management Systems Definition Version 1*

20-1-2-4-2 Implementation's Systems Behavior of the Purchasing Management Systems Definition Version 1

The entire implementation's systems behavior of the *purchasing management* systems definition version 1 includes: a) *Implementation's SBCD* and b) *Implementation's IFD* of the *purchasing management* systems definition version 1.

Figure 20-74 shows the implementation's SBCD of the *purchasing management* systems definition version 1 in which interactions among the *Supplier* actor and the *Purchasing_Department*, *Quotation_Processing_GUI*, *Purchase_Order_Processing_GUI*, *Purchase_Processing_GUI*, *Purchasing_Database*, *Network_Operating_System* components shall draw forth the *Quotation*, *Purchase_Order* and *Purchase* behaviors.

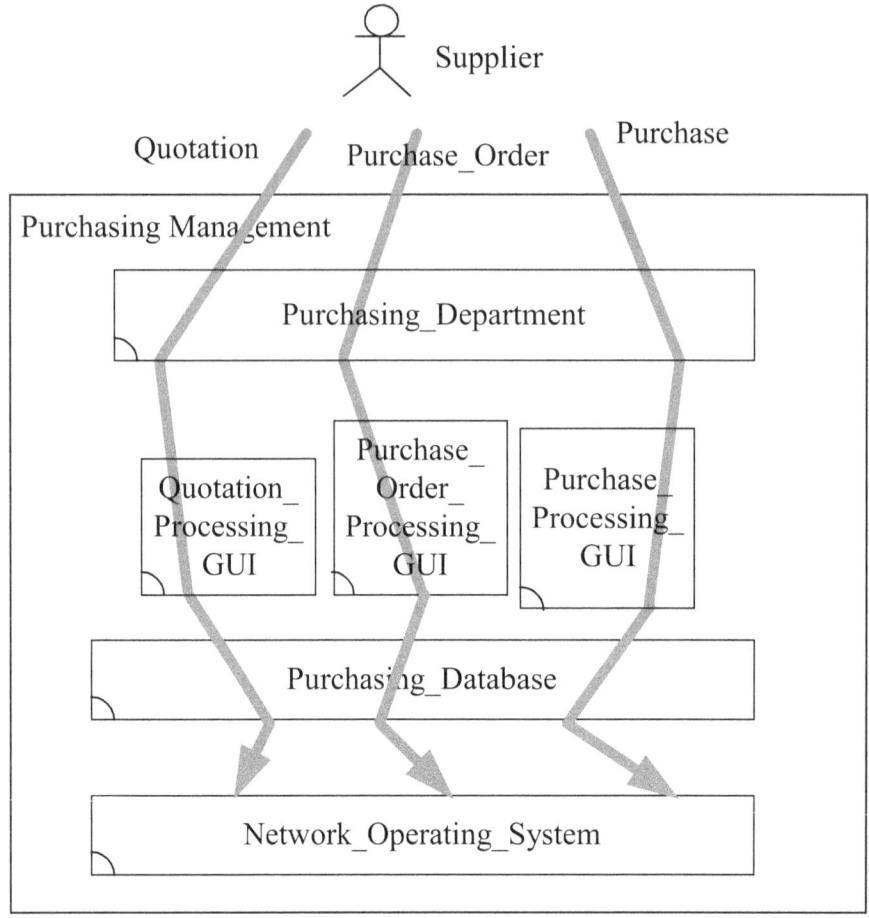

Figure 20-74   Implementation's SBCD of the *Purchasing Management*
Systems Definition Version 1

The overall implementation's behavior of the *purchasing management* systems definition version 1 includes the *Quotation, Purchase_Order and Purchase* behaviors. In other words, the *Quotation, Purchase_Order and Purchase* behaviors together provide the overall implementation's behavior of the *purchasing management* systems definition version 1.

Be noticed that the *Quotation, Purchase_Order and Purchase* behaviors are mutually independent of each other. They tend to be executed concurrently [Hoar85, Miln89, Miln99].

The overall implementation's behavior of the *purchasing management* systems definition version 1 includes three individual behaviors: *Quotation, Purchase_Order and Purchase*. Each individual behavior is represented by an execution path. We use an interaction flow diagram (IFD) to define each one of these execution paths. Figure 20-75 shows the implementation's IFD of the *purchasing management* systems definition version 1 *Quotation* behavior. First, actor *Supplier* interacts with the *Purchasing_Department* component through the *Quotation_Verify* operation call interaction, carrying the *Quotation_Form* input parameter. Next, component *Purchasing_Department* interacts with the *Quotation_Processing_GUI* component through the *Quotation_Processing_Button_Click* operation call interaction, carrying the *Quotation_Form* input parameter. Continuingly, component *Quotation_Processing_GUI* interacts with the *Purchasing_Database* component through the *SQL_Quotation_Insert* operation call interaction, carrying the *Quotation_Query* input parameter. Finally, component *Purchasing_Database* interacts with the *Network_Operating_System* component through the *Infrastructure_Resources_Share* operation call interaction.

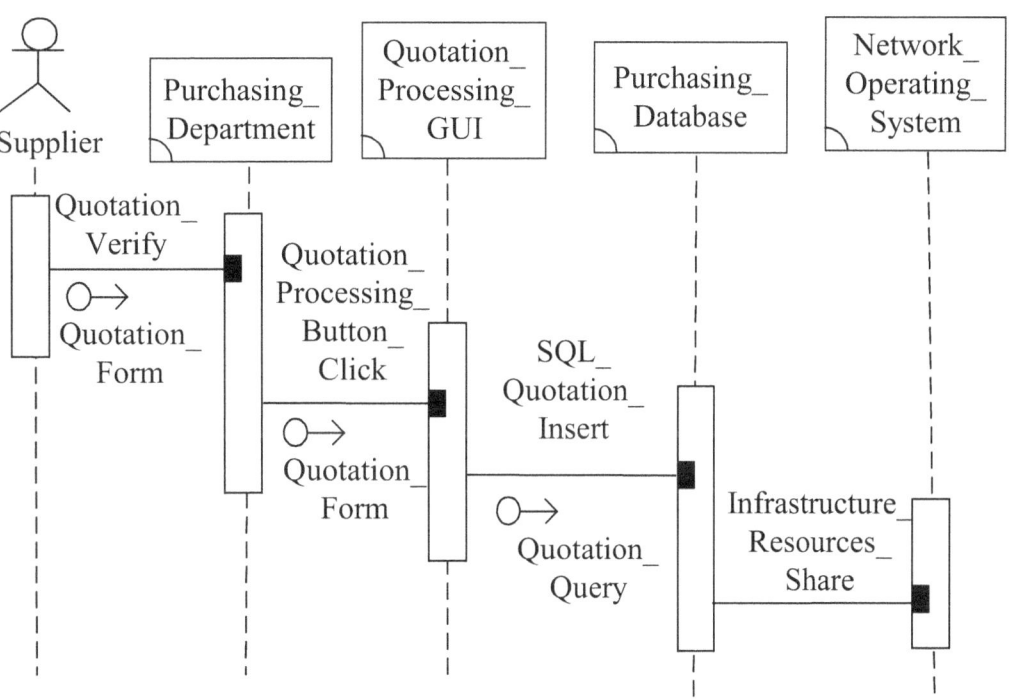

Figure 20-75   Implementation's IFD of the *Purchasing Management* Systems Definition Version 1 *Quotation* Behavior

Figure 20-76 shows the implementation's IFD of the *purchasing management* systems definition version 1 *Purchase_Order* behavior. First, actor *Supplier* interacts with the *Purchasing_Department* component through the *Purchase_Order_Verify* operation call interaction. Next, component *Purchasing_Department* interacts with the *Purchase_Order_Processing_GUI* component through the *Purchase_Order_Processing_Button_Click* operation call interaction, carrying the *Purchase_Order_Form* input parameter. Continuingly, component *Purchase_Order_Processing_GUI* interacts with the *Purchasing_Database* component through the *SQL_Purchase_Order_Insert* operation call interaction,

292

carrying the *Purchase_Order_Query* input parameter. Continuingly, component *Purchasing_Database* interacts with the *Network_Operating_System* component through the *Infrastructure_Resources_Share* operation call interaction. Continuingly, component *Purchasing_Department* interacts with the *Purchase_Order_Processing_GUI* component through the *Purchase_Order_Processing_Button_Click* operation return interaction, carrying the *Purchase_Order_Report* output parameter. Finally, actor *Supplier* interacts with the *Purchasing_Department* component through the *Purchase_Order_Verify* operation return interaction, carrying the *Purchase_Order_Report* output parameter.

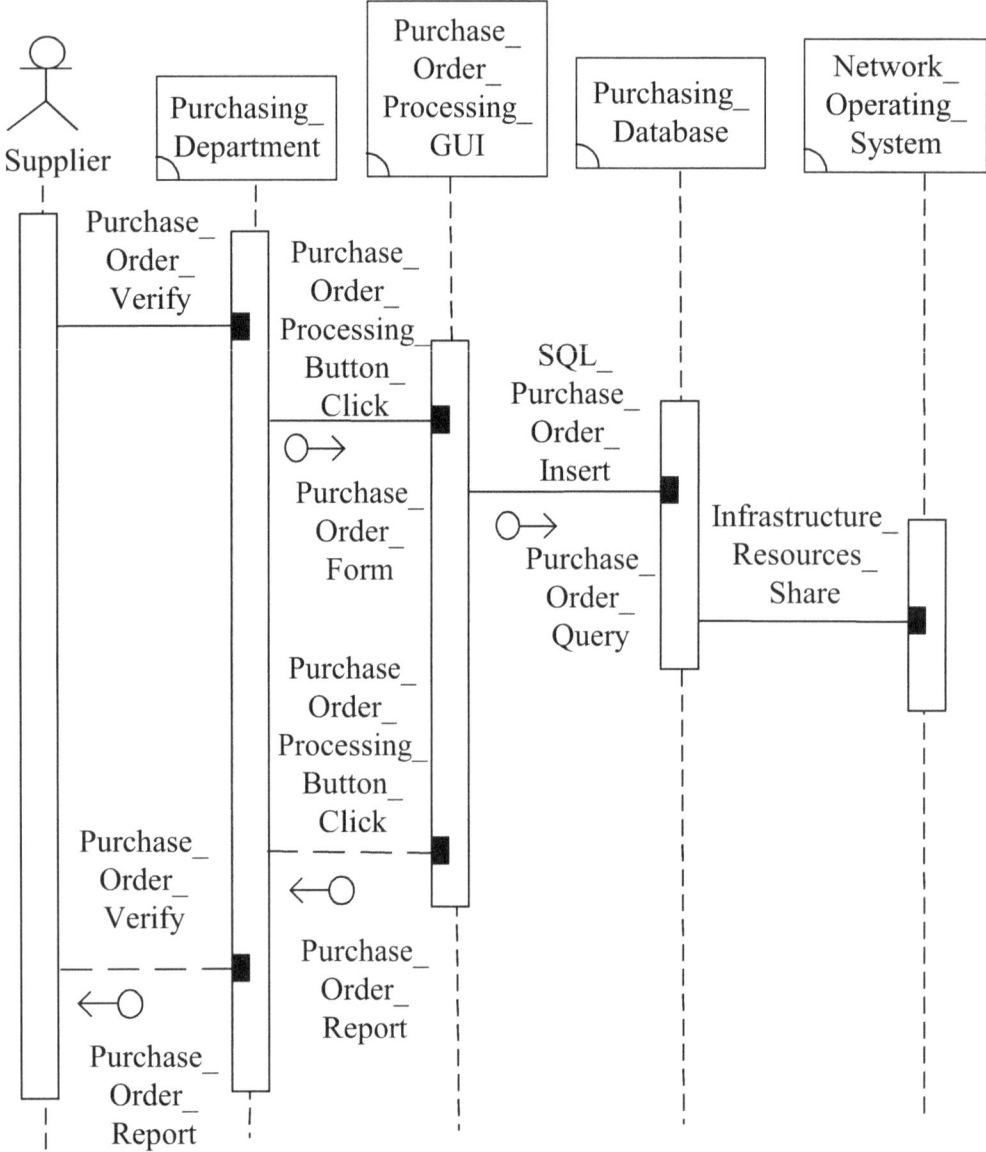

Figure 20-76   Implementation's IFD of the *Purchasing Management* Systems Definition Version 1 *Purchase_Order* Behavior

Figure 20-77 shows the implementation's IFD of the *purchasing management* systems definition version 1 *Purchase* behavior. First, actor *Supplier* interacts with the *Purchasing_Department* component through the *Purchase_Verify* operation call interaction, carrying the *Products_In* and *Invoice* input parameters. Next, component *Purchasing_Department* interacts with the *Purchase_Processing_GUI* component through the *Purchase_Processing_Button_Click* operation call interaction, carrying the *Purchase_Form* input parameter. Continuingly, component *Purchase_Processing_GUI* interacts with the *Purchasing_Database* component through the *SQL_Purchase_Insert* operation call interaction, carrying the *Purchase_Query* input parameter. Continuingly, component *Purchasing_Database* interacts with the *PM_Subsystem_1* component through the *Infrastructure_Resources_Share* operation call interaction. Finally, actor *Supplier* interacts with the *Purchasing_Department* component through the *Purchase_Verify* operation return interaction, carrying the *Products_Out* output parameter.

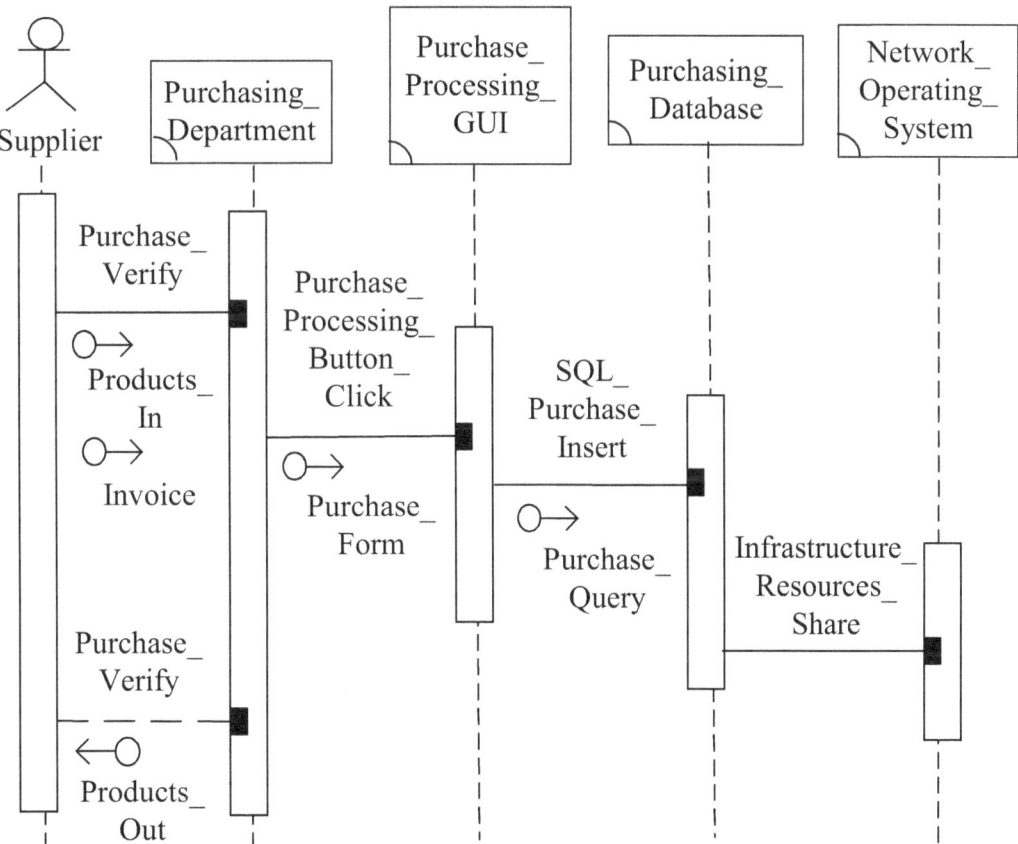

Figure 20-77　Implementation's IFD of the *Purchasing Management* Systems Definition Version 1 *Purchase* Behavior

20-2 Purchasing Management Systems Definition Strategy/Version 2

The *purchasing management* systems motivation model shown in Figure 20-78, being a higher-order system, has the *purchasing management* systems definition strategy 2 as its input and the *purchasing management* systems definition version 2 as its output.

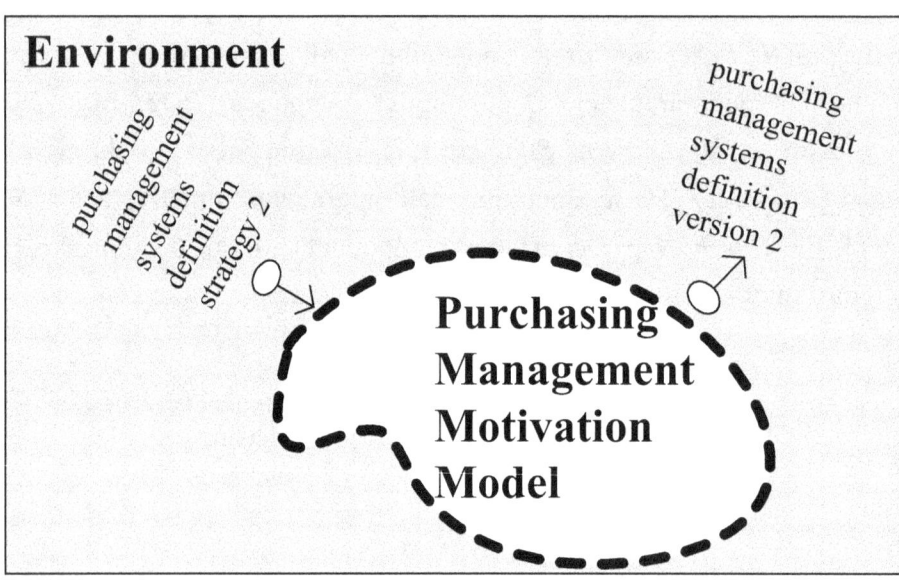

Figure 20-78   *Purchasing Management* Motivation Model is a Higher-Order System

The *purchasing management* systems definition strategy 2 mapped to the *purchasing management* systems definition version 2 can be represented as an ordered pair, as shown in Figure 20-79.

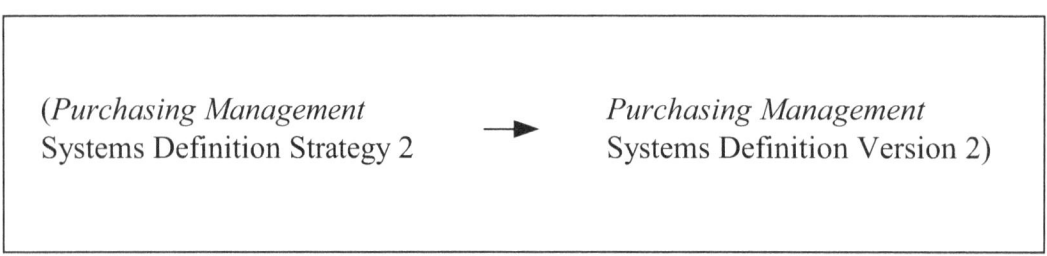

Figure 20-79   An Ordered Pair

20-2-1 Purchasing Management Systems Definition Strategy 2

Here, we use (a) goal drivers, (b) goal assumptions, (c) goal constraints and (d) SWOT analysis, to illustrate the strategic means of the "*purchasing management systems definition strategy 2*".

Goal drivers are up from the policy considerations, the goal driver is kind of why we want to have this *purchasing management* systems definition version 2. The goal drivers of *purchasing management* systems definition version 2 are: currently, a purchasing management system with only the *Quotation, Purchase_Order and Purchase* behaviors is losing its appeal to managers; a purchasing management system with the *Purchase_Requisition, Quotation, Purchase_Order and Purchase* behaviors shall attract more managers.

Goal assumptions are taking into account of those assumptions that have a positive impact on the *purchasing management* systems definition version 2. We assume that if the cost of installing the purchasing management system is affordable, then every manager will have a great desire to install one. This is the major goal assumption of this strategy.

Goal constraints are up from the policy considerations, the goal constraints are related to those restrictions which have a negative impact on the *purchasing management* systems definition version 2. If the company is short of fund to invest in this new installation, then this would become the goal constraint of this strategy.

SWOT analysis is to analyze the internal strengths, weaknesses, opportunities and threats, and so for executing this *purchasing management* systems definition strategy 2. Being a big company, it should be trivial for the company to install this *purchasing management* system with the *Purchase_Requisition, Quotation, Purchase_Order and Purchase* behaviors. This is the internal strength of this company. However, kind of bulky makes the company may not react fast enough to carry out this new installation. This is the internal weakness of this company.

20-2-2 Purchasing Management Systems Definition Version 2

Using the SBC multi-level (hierarchical) view, an architect goes through: a) concept view, b) analysis view, c) design view and d) implementation view for the *purchasing management* systems definition version 2 as shown in Figure 20-80.

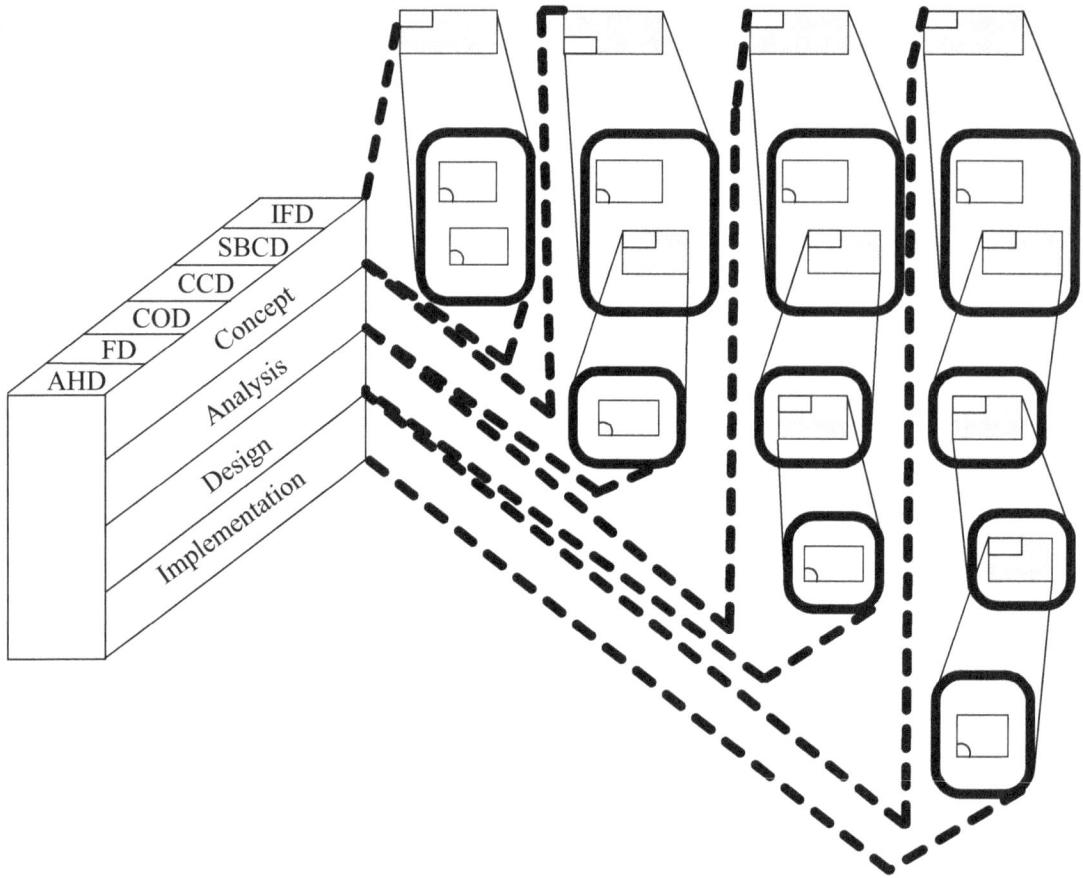

Figure 20-80   *Purchasing Management* Systems Definition Version 2

20-2-2-1 Concept View of the Purchasing Management Systems Definition Version 2

The concept view of the *purchasing management* systems definition version 2 is shown in Figure 20-81.

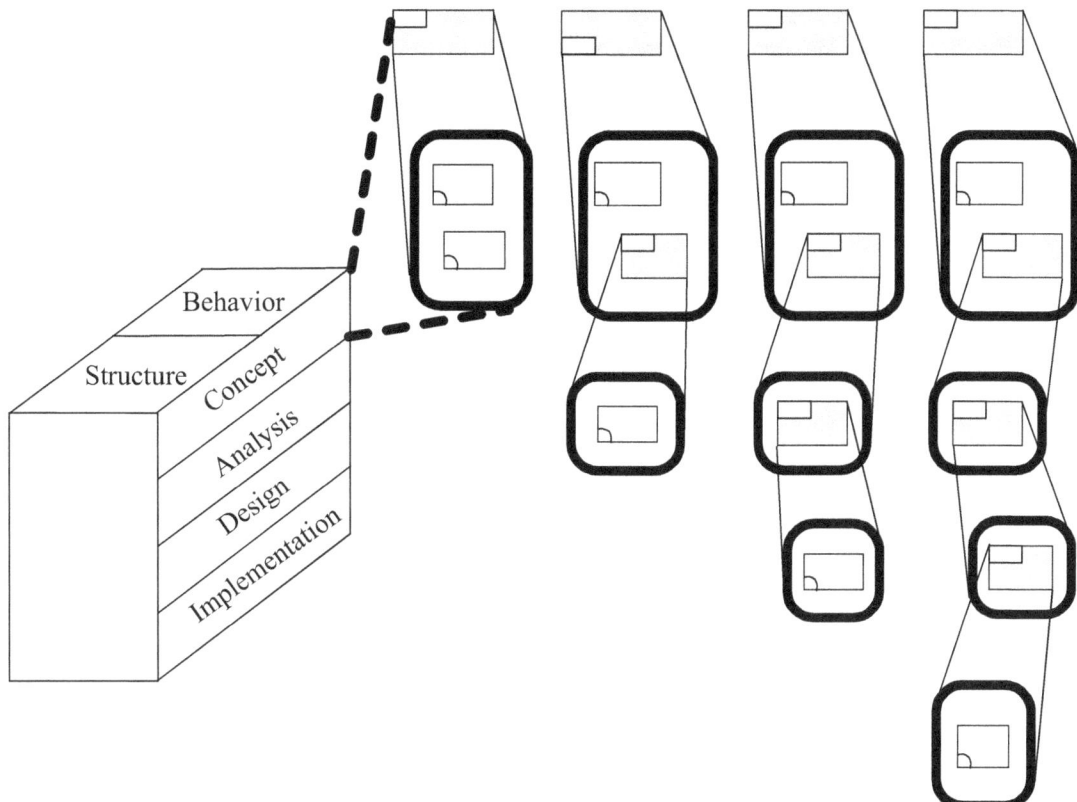

Figure 20-81   Concept View of the *Purchasing Management* Systems Definition Version 2

The concept view of the *purchasing management* systems definition version 2 consists of: a) concept's systems structure of the *purchasing management* systems definition version 2 and b) concept's systems behavior of the *purchasing management* systems definition version 2.

20-2-2-1-1 Concept's Systems Structure of the Purchasing Management Systems Definition Version 2

The entire concept's systems structure of the *purchasing management* systems definition version 2 includes: a) *Concept's AHD*, b) *Concept's FD*, c) *Concept's COD* and d) *Concept's CCD* of the *purchasing management* systems definition version 2.

We first draw the concept's AHD of the *purchasing management* systems definition version 2. As shown in Figure 20-82, *Purchasing Management* is composed of *Purchase_Order_Coordinator*, *Purchasing_Department* and *PM_Subsystem_3*. In the figure, *Purchasing Management* is an aggregated system while *Purchase_Order_Coordinator*, *Purchasing_Department* and *PM_Subsystem_3* are non-aggregated systems.

298

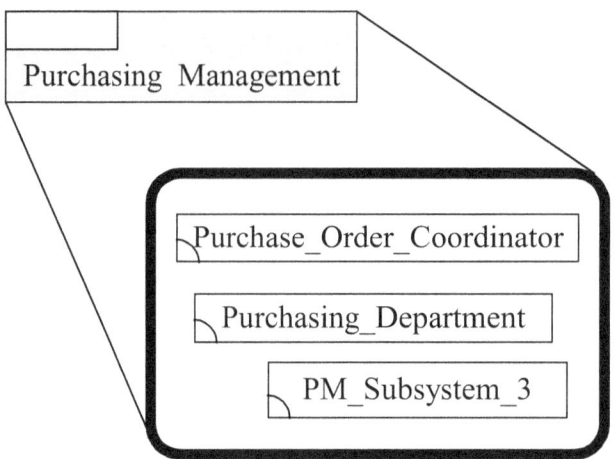

Figure 20-82　Concept's AHD of the *Purchasing Management*
Systems Definition Version 2

Figure 20-83 shows the concept's FD of the *purchasing management* systems definition version 2. In the figure, *Business_Layer* contains the *Purchase_Order_Coordinator*, *Purchasing_Department* and *PM_Subsystem_3* components.

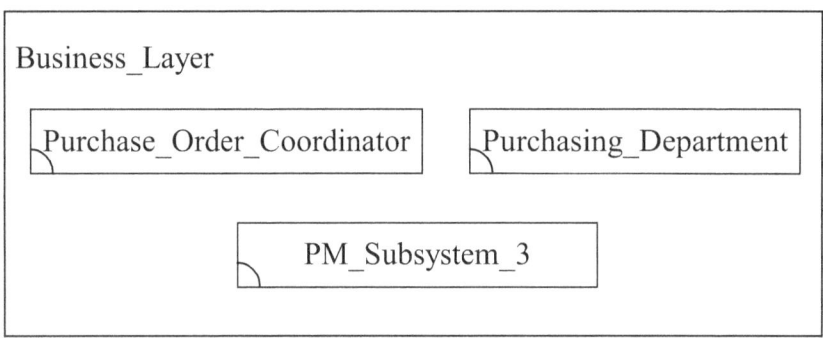

Figure 20-83　Concept's FD of the *Purchasing Management*
Systems Definition Version 2

Figure 20-84 shows the concept's COD of the *purchasing management* systems definition version 2. In the figure, component *Purchasing_Department* has three operations: *Quotation_Verify*, *Purchase_Order_Verify* and *Purchase_Verify*; component *PM_Subsystem_3* has three operations: *Quotation_Processing_Button_Click*, *Purchase_Order_Processing_Button_Click and Purchase_Processing_Button_Click*.

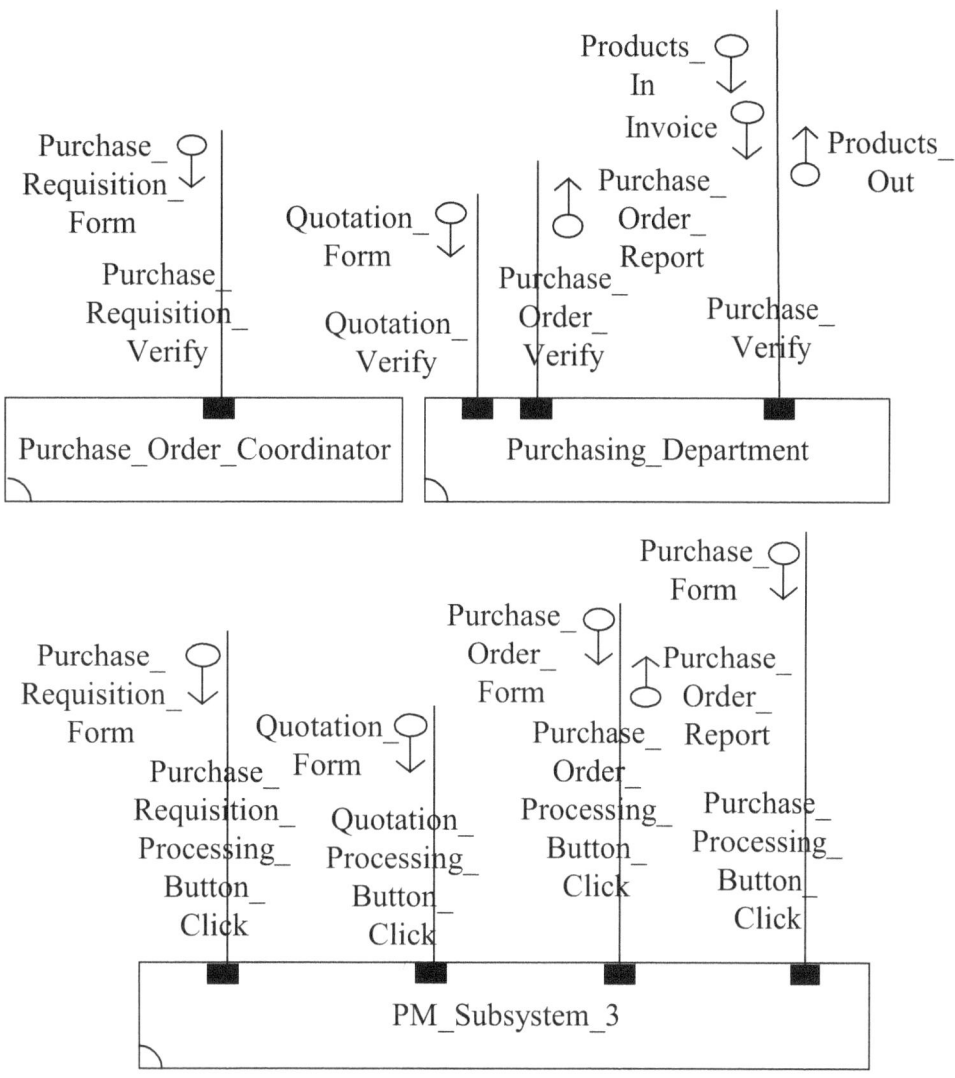

Figure 20-84   Concept's COD of the *Purchasing Management*
Systems Definition Version 2

The operation formula of *Purchase_Requisition_Verify* is *Purchase_Requisition_Verify(In Purchase_Requisition_Form)*. The operation formula of *Quotation_Verify* is *Quotation_Verify(In Quotation_Form)*. The operation formula of *Purchase_Order_Verify* is *Purchase_Order_Verify(Out Purchase_Order_Report)*. The operation formula of *Purchase_Verify* is *Purchase_Verify(In Products_In, Invoice; Out Products_Out)*. The operation formula of *Purchase_Requisition_Processing_Button_Click* is *Purchase_Requisition_Processing_Button_Click(In Purchase_Requisition_Form)*. The operation formula of *Quotation_Processing_Button_Click* is *Quotation_Processing_Button_Click(In Quotation_Form)*. The operation formula of *Purchase_Order_Processing_Button_Click* is

Purchase_Order_Processing_Button_Click(In   Purchase_Order_Form;   Out Purchase_Order_Report).   The   operation   formula   of *Purchase_Processing_Button_Click   is   Purchase_Processing_Button_Click(In Purchase_Form).*

Figure 20-85 shows the composite data type specification of the *Purchase_Requisition_Form*   input   parameter   occurring   in   the *Purchase_Requisition_Verify(In   Purchase_Requisition_Form)   and Purchase_Requisition_Processing_Button_Click(In   Purchase_Requisition_Form)* operation formulas.

Parameter	*Purchase_Requisition_Form*

Data Type	TABLE of
	  Date : Text
	  OD : Text
	  ProductNo : Text
	  Quantity : Integer
	End TABLE ;

Instances	**Purchase Requisition Form** Date: 2011/10/17 Originating_Department : Sales Dept. ProductNo Quantity __A00001(Pen)_____300_____ __A00002(Mouse)_____400_____ __A00003(Camera)_____500_____

Figure 20-85   Composite Data Type Specification

Figure 20-86 shows the composite data type specification of the *Quotation_Form* input parameter occurring in the *Quotation_Verify(In Quotation_Form)* and *Quotation_Processing_Button_Click(In   Quotation_Form)* operation formulas.

Parameter	*Quotation_Form*

Data Type	TABLE of
	  Date : Text
	  SupplierName: Text
	  ProductNo : Text
	  Quantity : Integer
	  UnitPrice : Real
	  Total : Real
	End TABLE ;

Instances	**Quotation Form** Date: 2011/10/25 SupplierName : Johnson Corp. ProductNo Quantity UnitPrice A00001(Pen) 300 100.00 _A00002(Mouse)____400____200.00__ _A00003(Camera)___500____300.00__ Total : 260,000.00

Figure 20-86　　Composite Data Type Specification

Figure 20-87 shows the composite data type specification of the *Purchase_Order_Report* output parameter occurring in the *Purchase_Order_Verify(Out Purchase_Order_Report)* and *Purchase_Order_Processing_Button_Click(In Purchase_Order_Form; Out Purchase_Order_Report)* operation formulas.

Parameter	*Purchase_Order_Report*

Data Type	TABLE of
	  Date : Text
	  SupplierName: Text
	  ProductNo : Text
	  Quantity : Integer
	End TABLE ;

Instances	Date : 20111118 SupplierName : Johnson Corp. | ProductNo	Quantity | |---|---| | A00001(Pen) | 300 | | A00002(Mouse) | 400 | | A00003(Camera) | 500 |

Figure 20-87   Composite Data Type Specification

Figure 20-88 shows the primitive data type specification of the *Products_In* input parameter occurring in the *Purchase_Verify(In Products_In, Invoice; Out Products_Out)* operation formula.

Parameter	Data Type	Instances
Products_In	Physical Object	Pen, Mouse, Camera

Figure 20-88   Primitive Data Type Specification

Figure 20-89 shows the composite data type specification of the *Invoice* input parameter occurring in the *Purchase_Verify(In Products_In, Invoice; Out Products_Out)* operation formula.

Parameter	*Invoice*
Data Type	TABLE of Date : Text SupplierName: Text ProductNo : Text Quantity : Integer UnitPrice : Real Total : Real End TABLE ;
Instances	Date : 20111130 SupplierName :　Johnson Corp. | ProductNo	Quantity	UnitPrice
---	---	---
A00001(Pen)	300	100.00
A00002(Mouse)	400	200.00
A00003(Camera)	500	300.00	 Total : 260,000.00

Figure 20-89　　Composite Data Type Specification

Figure 20-90 shows the primitive data type specification of the *Products_Out* output parameter occurring in the *Purchase_Verify(In Products_In, Invoice; Out Products_Out)* operation formula.

Parameter	Data Type	Instances
Products_Out	Physical Object	Pen, Mouse, Camera

Figure 20-90　　Primitive Data Type Specification

Figure 20-91 shows the composite data type specification of the *Purchase_Order_Form* input parameter occurring in the *Purchase_Order_Processing_Button_Click(In Purchase_Order_Form; Out Purchase_Order_Report)* operation formula.

Parameter	*Purchase_Order_Form*

Data Type	TABLE of
	  Date : Text
	  SupplierName: Text
	  ProductNo : Text
	  Quantity : Integer
	End TABLE ;

Instances	**Purchase Order Form** Date: 2011/11/18 SupplierName : Johnson Corp. ProductNo Quantity __A00001(Pen) ____300 ___ __A00002(Mouse)____400___ __A00003(Camera)___500___

Figure 20-91   Composite Data Type Specification

Figure 20-92 shows the composite data type specification of the *Purchase_Form* input parameter occurring in the *Purchase_Processing_Button_Click(In Purchase_Form)* operation formula.

Parameter	*Purchase_Form*

Data Type	TABLE of
	   Date : Text
	   SupplierName: Text
	   ProductNo : Text
	   Quantity : Integer
	   UnitPrice : Real
	   ReturnQuantity : Integer
	   Total : Real
	End TABLE ;

Instances	**Purchase Form** Date: 2011/12/12 SupplierName : Johnson Corp. ProductNo Quantity UnitPrice ReturnQuantity A00001(Pen) 300 100.00 0 A00002(Mouse) 390 200.00 10 A00003(Camera)500 300.00 0 Total : 258,000.00

Figure 20-92   Composite Data Type Specification

Figure 20-93 shows the concept's CCD of the *purchasing management* systems definition version 2. In the figure, actor *Originating_Department* has one connection with the *Purchase_Order_Coordinator* component; actor *Supplier* has three connections with the *Purchasing_Department* component; component *Purchase_Order_Coordinator* has one connection with the *PM_Subsystem_3* component; component *Purchasing_Department* has three connections with the *PM_Subsystem_3* component.

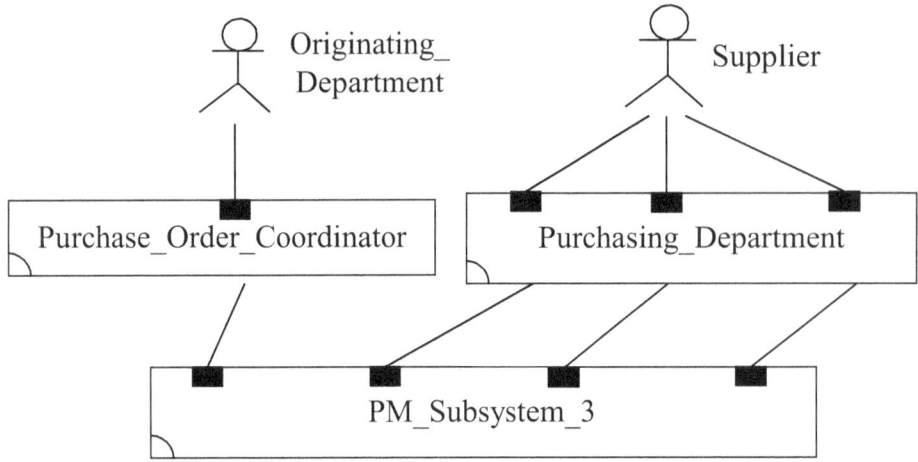

Figure 20-93   Concept's CCD of the *Purchasing Management*
Systems Definition Version 2

20-2-2-1-2 Concept's Systems Behavior of the Purchasing Management Systems Definition Version 2

The entire concept's systems behavior of the *purchasing management* systems definition version 2 includes: a) *Concept's SBCD* and b) *Concept's IFD* of the *purchasing management* systems definition version 2.

Figure 20-94 shows the concept's SBCD of the *purchasing management* systems definition version 2 in which interactions among the *Originating_Department*, *Supplier* actors and the *Purchase_Order_Coordinator*, *Purchasing_Department*, *PM_Subsystem_3* components shall draw forth the *Purchase_Requisition*, *Quotation*, *Purchase_Order and Purchase* behaviors.

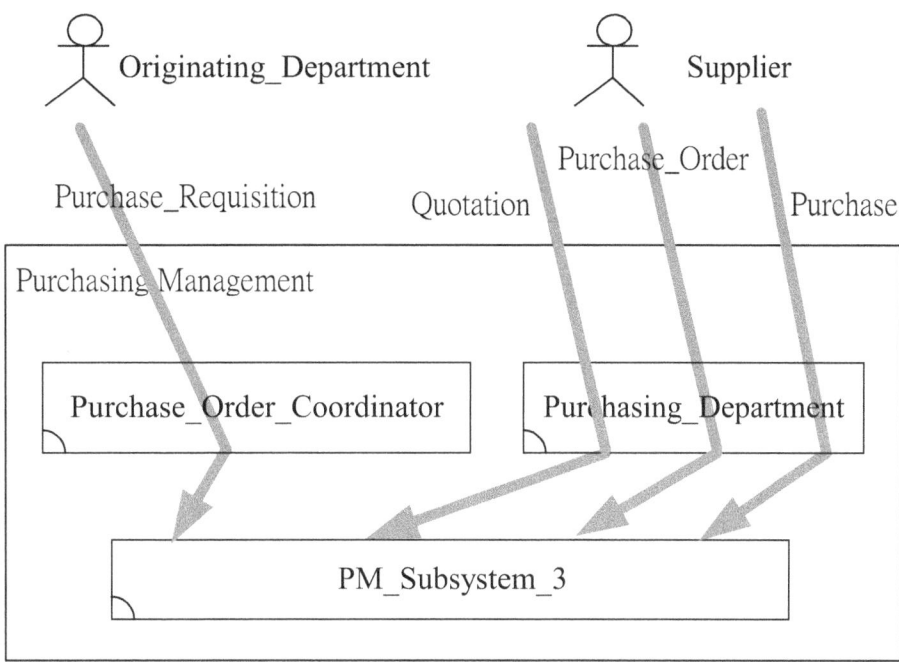

Figure 20-94   Concept's SBCD of the *Purchasing Management*
Systems Definition Version 2

The overall concept's behavior of the *purchasing management* systems definition version 2 includes the *Purchase_Requisition, Quotation, Purchase_Order and Purchase* behaviors. In other words, the *Purchase_Requisition, Quotation, Purchase_Order and Purchase* behaviors together provide the overall concept's behavior of the *purchasing management* systems definition version 2.

Be noticed that the *Purchase_Requisition, Quotation, Purchase_Order and Purchase* behaviors are mutually independent of each other. They tend to be executed concurrently [Hoar85, Miln89, Miln99].

The overall concept's behavior of the *purchasing management* systems definition version 2 includes four individual behaviors: *Purchase_Requisition, Quotation, Purchase_Order and Purchase*. Each individual behavior is represented by an execution path. We use an interaction flow diagram (IFD) to define each one of these execution paths. Figure 20-95 shows the concept's IFD of the *purchasing management* systems definition version 2 *Purchase_Requisition* behavior. First, actor *Originating_Department* interacts with the *Purchase_Order_Coordinator* component through the *Purchase_Requisition_Verify* operation call interaction, carrying the *Purchase_Requisition_Form* input parameter. Finally, component *Purchase_Order_Coordinator* interacts with the *PM_Subsystem_3* component through the *Purchase_Requisition_Processing_Button_Click* operation call interaction, carrying the *Purchase_Requisition_Form* input parameter.

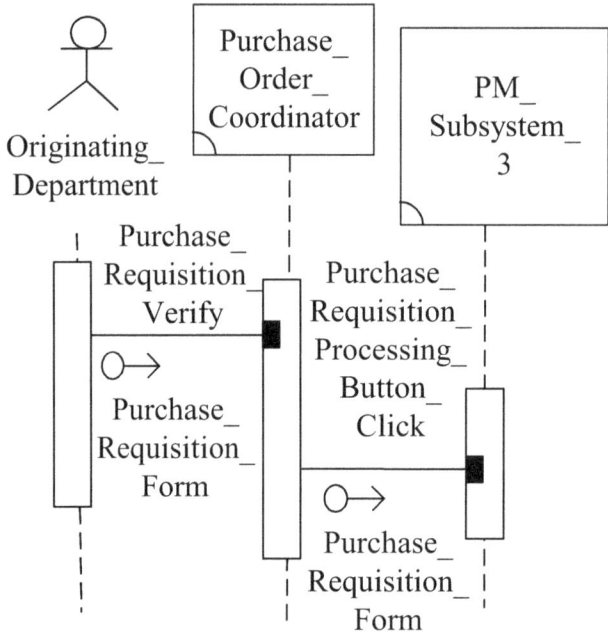

Figure 20-95   Concept's IFD of the *Purchasing Management Systems Definition Version 2 Purchase_Requisition* Behavior

Figure 20-96 shows the concept's IFD of the *purchasing management* systems definition version 2 *Quotation* behavior. First, actor *Supplier* interacts with the *Purchasing_Department* component through the *Quotation_Verify* operation call interaction, carrying the *Quotation_Form* input parameter. Finally, component *Purchasing_Department* interacts with the *PM_Subsystem_3* component through the *Quotation_Processing_Button_Click* operation call interaction, carrying the *Quotation_Form* input parameter.

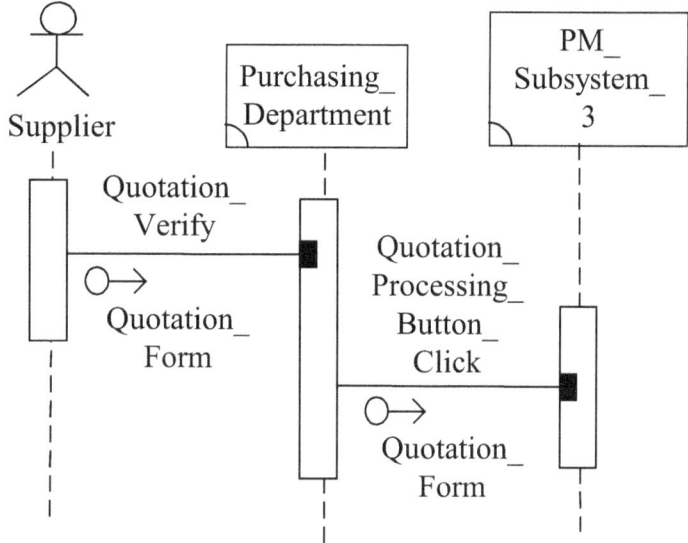

Figure 20-96   Concept's IFD of the *Purchasing Management*
Systems Definition Version 2 *Quotation* Behavior

Figure 20-97 shows the concept's IFD of the *purchasing management* systems definition version 2 *Purchase_Order* behavior. First, actor *Supplier* interacts with the *Purchasing_Department* component through the *Purchase_Order_Verify* operation call interaction. Next, component *Purchasing_Department* interacts with the *PM_Subsystem_3* component through the *Purchase_Order_Processing_Button_Click* operation call interaction, carrying the *Purchase_Order_Form* input parameter and *Purchase_Order_Report* output parameter. Finally, actor *Supplier* interacts with the *Purchasing_Department* component through the *Purchase_Order_Verify* operation return interaction, carrying the *Purchase_Order_Report* output parameter.

310

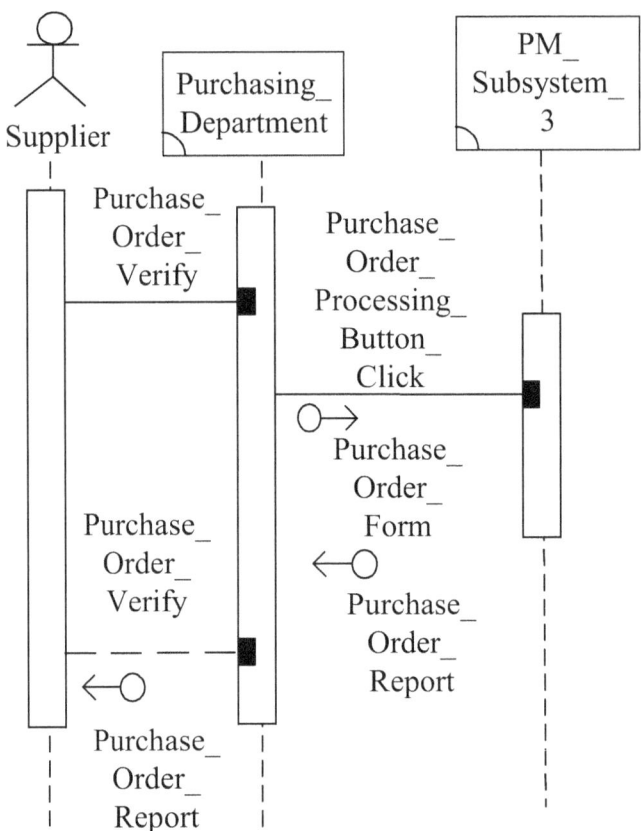

Figure 20-97   Concept's IFD of the *Purchasing Management*
Systems Definition Version 2 *Purchase_Order* Behavior

Figure 20-98 shows the concept's IFD of the *purchasing management* systems definition version 2 *Purchase* behavior. First, actor *Supplier* interacts with the *Purchasing_Department* component through the *Purchase_Verify* operation call interaction, carrying the *Products_In* and *Invoice* input parameters. Next, component *Purchasing_Department* interacts with the *PM_Subsystem_3* component through the *Purchase_Processing_Button_Click* operation call interaction, carrying the *Purchase_Form* input parameter. Finally, actor *Supplier* interacts with the *Purchasing_Department* component through the *Purchase_Verify* operation return interaction, carrying the *Products_Out* output parameter.

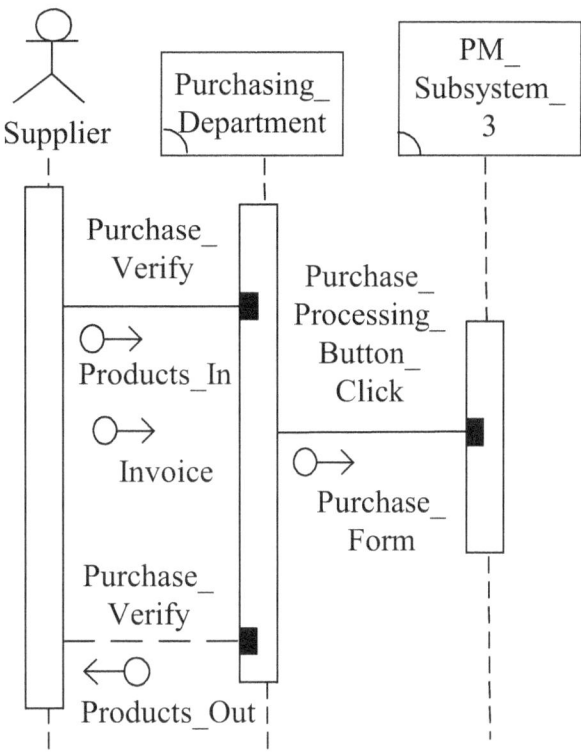

Figure 20-98   Concept's IFD of the *Purchasing Management*
Systems Definition Version 2 *Purchase* Behavior

20-2-2-2 Analysis View of the Purchasing Management Systems Definition Version 2

The analysis view of the *purchasing management* systems definition version 2 is shown in Figure 20-99.

312

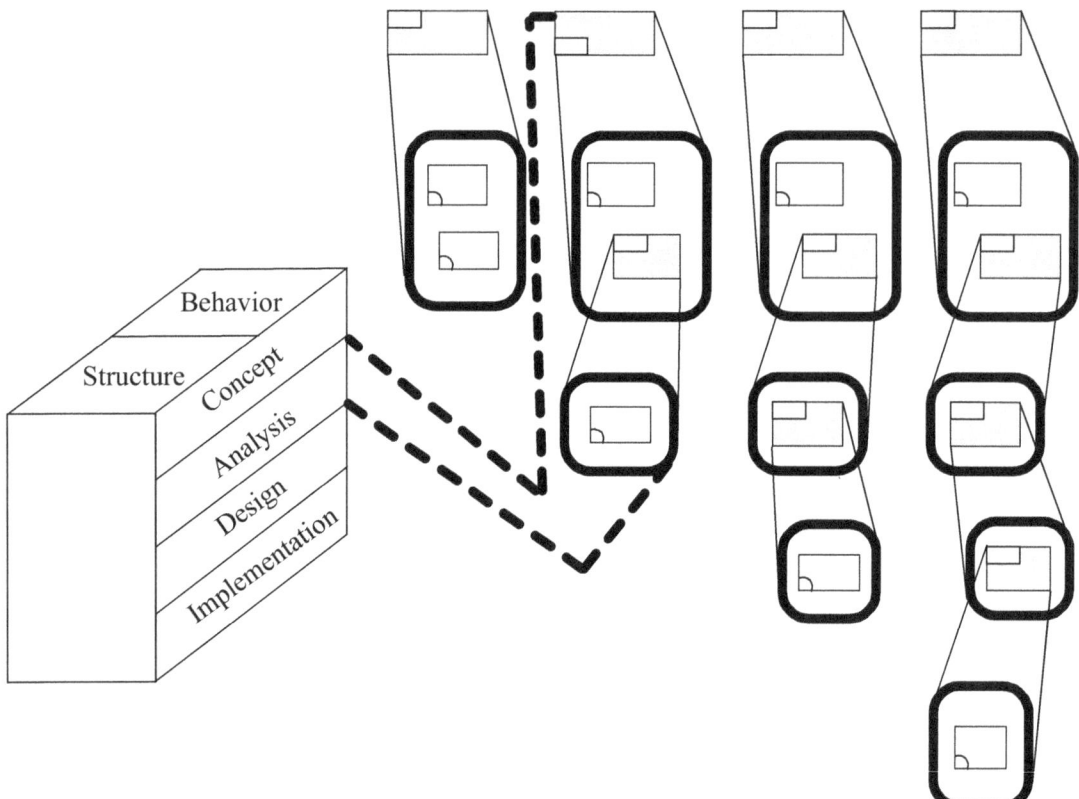

Figure 20-99   Analysis View of the *Purchasing Management* Systems Definition Version 2

The analysis view of the *purchasing management* systems definition version 2 consists of: a) analysis' systems structure of the *purchasing management* systems definition version 2 and b) analysis' systems behavior of the *purchasing management* systems definition version 2.

20-2-2-2-1 Analysis' Systems Structure of the Purchasing Management Systems Definition Version 2

The entire analysis' systems structure of the *purchasing management* systems definition version 2 includes: a) *Analysis' AHD*, b) *Analysis' FD*, c) *Analysis' COD* and d) *Analysis' CCD* of the *purchasing management* systems definition version 2.

We first draw the analysis' AHD of the *purchasing management* systems definition version 2. As shown in Figure 20-100, *Purchasing Management* is composed of *Purchase_Order_Coordinator*, *Purchasing_Department* and *PM_Subsystem_3*; *PM_Subsystem_3* is composed of *Purchase_Requisition_Processing_GUI*, *Quotation_Processing_GUI*, *Purchase_Order_Processing_GUI*, *Purchase_Processing_GUI* and *PM_Subsystem_2*. In the figure, *Purchasing Management* and *PM_Subsystem_3* are aggregated systems while *Purchase_Order_Coordinator*, *Purchasing_Department*,

Purchase_Requisition_Processing_GUI,   *Quotation_Processing_GUI,* *Purchase_Order_Processing_GUI, Purchase_Processing_GUI* and *PM_Subsystem_2* are non-aggregated systems.

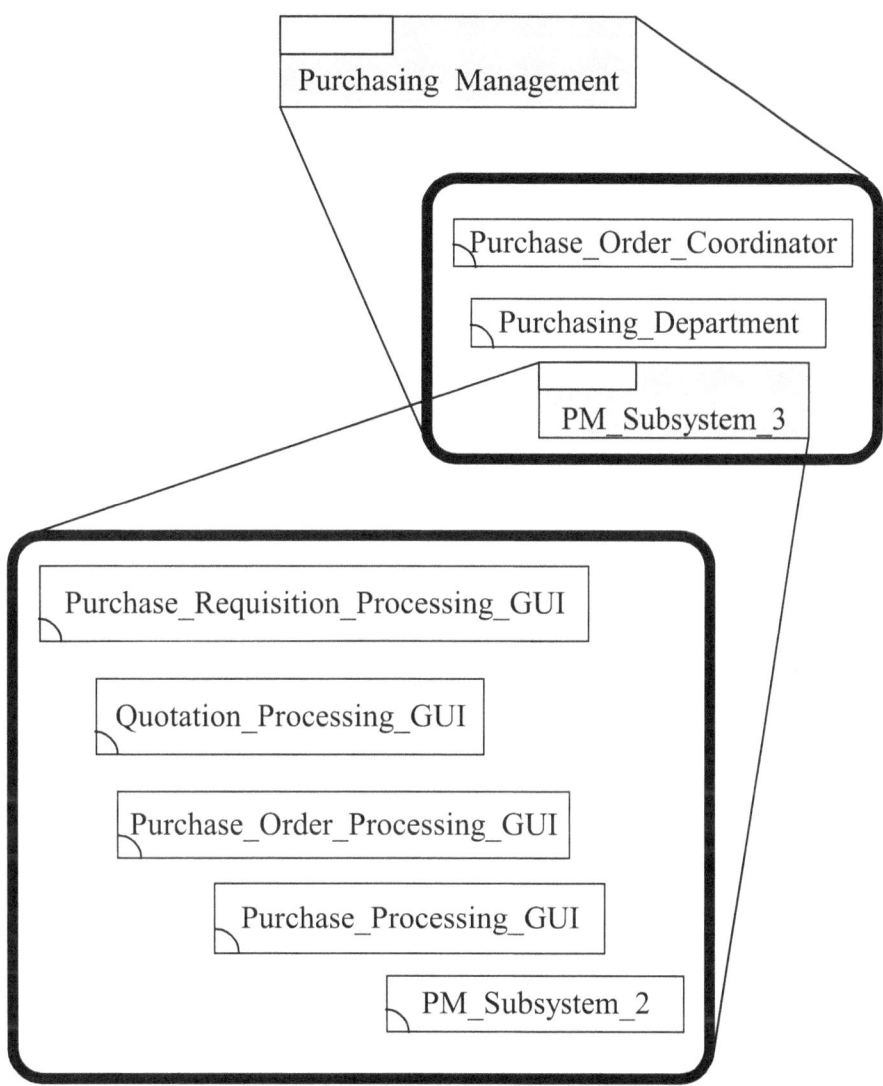

Figure 20-100   Analysis' AHD of the *Purchasing Management*
Systems Definition Version 2

Figure 20-101 shows the analysis' FD of the *purchasing management* systems definition version 2. In the figure, *Business_Layer* contains the *Purchase_Order_Coordinator* and *Purchasing_Department* components; *Application_Layer* contains the *Purchase_Requisition_Processing_GUI,* *Quotation_Processing_GUI,* *Purchase_Order_Processing_GUI,* *Purchase_Processing_GUI* and *PM_Subsystem_2* components.

314

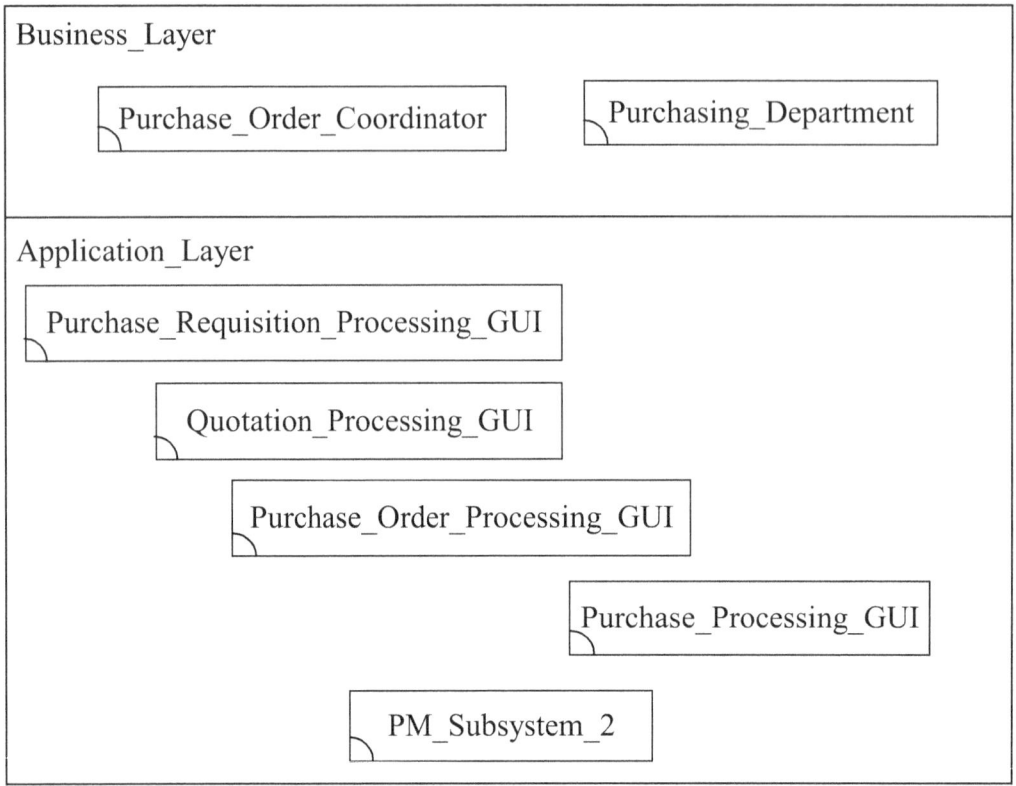

Figure 20-101   Analysis' FD of the *Purchasing Management*
Systems Definition Version 2

Figure 20-102 shows the analysis' COD of the *purchasing management* systems definition version 2. In the figure, component *Purchase_Order_Coordinator* has one operation: *Purchase_Requisition_Verify*; component *Purchasing_Department* has three operations: *Quotation_Verify*, *Purchase_Order_Verify* and *Purchase_Verify*; component *Purchase_Requisition_Processing_GUI* has one operation: *Purchase_Requisition_Processing_Button_Click*; component *Quotation_Processing_GUI* has one operation: *Quotation_Processing_Button_Click*; component *Purchase_Order_Processing_GUI* has one operation: *Purchase_Order_Processing_Button_Click*; component *Purchase_Processing_GUI* has one operation: *Purchase_Processing_Button_Click*; component *PM_Subsystem_2* has four operations: *SQL_Purchase_Requisition_Insert*, *SQL_Quotation_Insert*, *SQL_Purchase_Order_Insert* and *SQL_Purchase_Insert*.

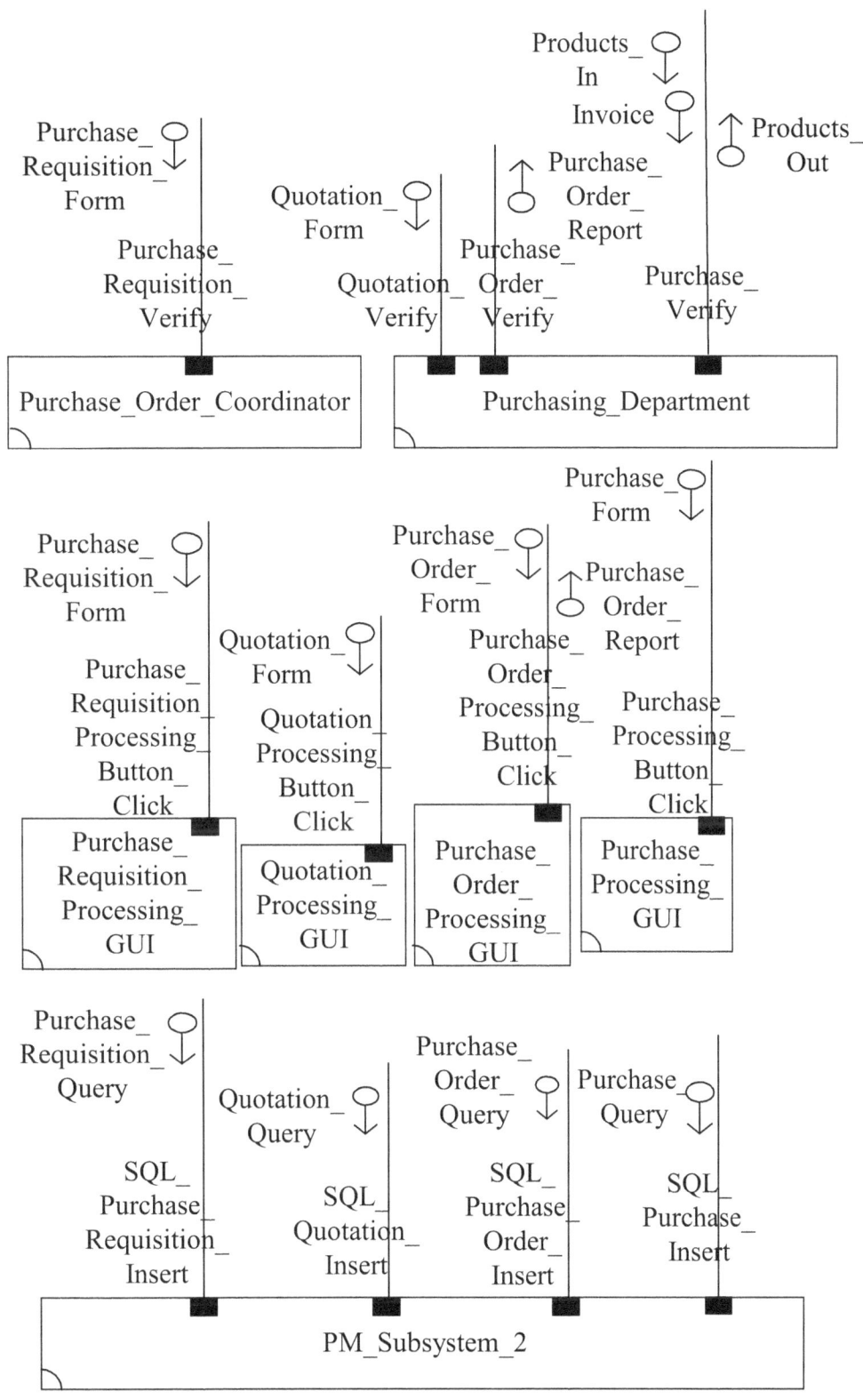

Figure 20-102   Analysis' COD of the *Purchasing Management*
Systems Definition Version 2

The operation formula of *Purchase_Requisition_Verify* is *Purchase_Requisition_Verify(In Purchase_Requisition_Form)*. The operation formula of *Quotation_Verify* is *Quotation_Verify(In Quotation_Form)*. The operation formula of *Purchase_Order_Verify* is *Purchase_Order_Verify(Out Purchase_Order_Report)*. The operation formula of *Purchase_Verify* is *Purchase_Verify(In Products_In, Invoice; Out Products_Out)*. The operation formula of *Purchase_Requisition_Processing_Button_Click* is *Purchase_Requisition_Processing_Button_Click(In Purchase_Requisition_Form)*. The operation formula of *Quotation_Processing_Button_Click* is *Quotation_Processing_Button_Click(In Quotation_Form)*. The operation formula of *Purchase_Order_Processing_Button_Click* is *Purchase_Order_Processing_Button_Click(In Purchase_Order_Form; Out Purchase_Order_Report)*. The operation formula of *Purchase_Processing_Button_Click* is *Purchase_Processing_Button_Click(In Purchase_Form)*. The operation formula of *SQL_Purchase_Requisition_Insert* is *SQL_Purchase_Requisition_Insert(In Purchase_Requisition_Query)*. The operation formula of *SQL_Quotation_Insert* is *SQL_Quotation_Insert(In Quotation_Query)*. The operation formula of *SQL_Purchase_Order_Insert* is *SQL_Purchase_Order_Insert (In Purchase_Order_Query)*. The operation formula of *SQL_Purchase_Insert* is *SQL_Purchase_Insert (In Purchase_Query)*.

Figure 20-103 shows the composite data type specification of the *Purchase_Requisition_Form* input parameter occurring in the *Purchase_Requisition_Verify(In Purchase_Requisition_Form)* and *Purchase_Requisition_Processing_Button_Click(In Purchase_Requisition_Form)* operation formulas.

Parameter	*Purchase_Requisition_Form*

Data Type	TABLE of
	  Date : Text
	  OD : Text
	  ProductNo : Text
	  Quantity : Integer
	End TABLE ;

Instances	**Purchase Requisition Form** Date: 2011/10/17 Originating_Department : Sales Dept. ProductNo Quantity __A00001(Pen)_____300_____ __A00002(Mouse)_____400_____ __A00003(Camera)_____500_____

Figure 20-103   Composite Data Type Specification

Figure 20-104 shows the composite data type specification of the *Quotation_Form* input parameter occurring in the *Quotation_Verify(In Quotation_Form)* and *Quotation_Processing_Button_Click(In Quotation_Form)* operation formulas.

Parameter	*Quotation_Form*

Data Type	TABLE of
	  Date : Text
	  SupplierName: Text
	  ProductNo : Text
	  Quantity : Integer
	  UnitPrice : Real
	  Total : Real
	End TABLE ;

Instances	**Quotation Form** Date: 2011/10/25 SupplierName : Johnson Corp. ProductNo Quantity UnitPrice A 0 0 0 0 1 (P e n) 3 0 0 1 0 0 . 0 0 _A00002(Mouse)____400____200.00__ _A00003(Camera)___500____300.00__ Total : 260,000.00

Figure 20-104   Composite Data Type Specification

Figure 20-105 shows the composite data type specification of the *Purchase_Order_Report* output parameter occurring in the *Purchase_Order_Verify(Out Purchase_Order_Report)* and *Purchase_Order_Processing_Button_Click(In Purchase_Order_Form; Out Purchase_Order_Report)* operation formulas.

Parameter	*Purchase_Order_Report*
Data Type	TABLE of Date : Text SupplierName: Text ProductNo : Text Quantity : Integer End TABLE ;
Instances	Date : 20111118 SupplierName : Johnson Corp. <table><tr><th>ProductNo</th><th>Quantity</th></tr><tr><td>A00001(Pen)</td><td>300</td></tr><tr><td>A00002(Mouse)</td><td>400</td></tr><tr><td>A00003(Camera)</td><td>500</td></tr></table>

Figure 20-105   Composite Data Type Specification

Figure 20-106 shows the primitive data type specification of the *Products_In* input parameter occurring in the *Purchase_Verify(In Products_In, Invoice; Out Products_Out)* operation formula.

Parameter	Data Type	Instances
Products_In	Physical Object	Pen, Mouse, Camera

Figure 20-106   Primitive Data Type Specification

Figure 20-107 shows the composite data type specification of the *Invoice* input parameter occurring in the *Purchase_Verify(In Products_In, Invoice; Out Products_Out)* operation formula.

Parameter	*Invoice*
Data Type	TABLE of 　Date : Text 　SupplierName: Text 　ProductNo : Text 　Quantity : Integer 　UnitPrice : Real 　Total : Real End TABLE ;
Instances	Date : 20111130 SupplierName : Johnson Corp. <table><tr><th>ProductNo</th><th>Quantity</th><th>UnitPrice</th></tr><tr><td>A00001(Pen)</td><td>300</td><td>100.00</td></tr><tr><td>A00002(Mouse)</td><td>400</td><td>200.00</td></tr><tr><td>A00003(Camera)</td><td>500</td><td>300.00</td></tr></table> Total : 260,000.00

Figure 20-107   Composite Data Type Specification

Figure 20-108 shows the primitive data type specification of the *Products_Out* output parameter occurring in the *Purchase_Verify(In Products_In, Invoice; Out Products_Out)* operation formula.

Parameter	Data Type	Instances
Products_Out	Physical Object	Pen, Mouse, Camera

Figure 20-108   Primitive Data Type Specification

Figure 20-109 shows the composite data type specification of the *Purchase_Order_Form* input parameter occurring in the *Purchase_Order_Processing_Button_Click(In Purchase_Order_Form; Out Purchase_Order_Report)* operation formula.

Parameter	*Purchase_Order_Form*

Data Type	TABLE of
	  Date : Text
	  SupplierName: Text
	  ProductNo : Text
	  Quantity : Integer
	End TABLE ;

Instances	**Purchase Order Form** Date: 2011/11/18 SupplierName : Johnson Corp. ProductNo Quantity __A00001(Pen) ____300___ __A00002(Mouse)___400__ __A00003(Camera)___500___

Figure 20-109   Composite Data Type Specification

Figure 20-110 shows the composite data type specification of the *Purchase_Form* input parameter occurring in the *Purchase_Processing_Button_Click(In Purchase_Form)* operation formula.

Parameter	*Purchase_Form*

Data Type	TABLE of
	  Date : Text
	  SupplierName: Text
	  ProductNo : Text
	  Quantity : Integer
	  UnitPrice : Real
	  ReturnQuantity : Integer
	  Total : Real
	End TABLE ;

Instances	**Purchase Form** Date: 2011/12/12 SupplierName : Johnson Corp. ProductNo Quantity UnitPrice ReturnQuantity A00001(Pen) 300 100.00 0 A00002(Mouse) 390 200.00 10 A00003(Camera)500 300.00 0 Total : 258,000.00

Figure 20-110   Composite Data Type Specification

Figure 20-111 shows the composite data type specification of the *Purchase_Requisition_Query* input parameter occurring in the *SQL_Purchase_Requisition_Insert(In Purchase_Requisition_Query)* operation formula.

Parameter	*Purchase_Requisition_Query*

Data Type	TABLE of
	  Date : Text
	  OD : Text
	  ProductNo : Text
	  Quantity : Integer
	End TABLE ;

Instances	

Date	Originating_Department :
20111017	Sales Dept.

ProductNo	Quantity
A00001(Pen)	300
A00002(Mouse)	400
A00003(Camera)	500

Figure 20-111   Composite Data Type Specification

Figure 20-112 shows the composite data type specification of the *Quotation_Query* input parameter occurring in the *SQL_Quotation_Insert(In Quotation_Query)* operation formula.

Parameter	*Quotation_Query*

Data Type	TABLE of
	  Date : Text
	  SupplierName: Text
	  ProductNo : Text
	  Quantity : Integer
	  UnitPrice : Real
	  Total : Real
	End TABLE ;

Instances	

Date	SupplierName	Total
20111025	Johnson Corp.	260,000.00

ProductNo	Quantity	UnitPrice
A00001(Pen)	300	100.00
A00002(Mouse)	400	200.00
A00003(Camera)	500	300.00

Figure 20-112   Composite Data Type Specification

Figure 20-113 shows the composite data type specification of the *Purchase_Order_Query* input parameter occurring in the *SQL_Purchase_Order_Insert(In Purchase_Order_Query)* operation formula.

Parameter	*Purchase_Order_Query*

Data Type	TABLE of
	  Date : Text
	  SupplierName: Text
	  ProductNo : Text
	  Quantity : Integer
	End TABLE ;

Instances	

Date	SupplierName
20111118	Johnson Corp.

ProductNo	Quantity
A00001(Pen)	300
A00002(Mouse)	400
A00003(Camera)	500

Figure 20-113   Composite Data Type Specification

Figure 20-114 shows the composite data type specification of the *Purchase_Query* input parameter occurring in the *SQL_Purchase_Insert (In Purchase_Query)* operation formula.

Parameter	*Purchase_Query*

Data Type	TABLE of
	  Date : Text
	  SupplierName: Text
	  ProductNo : Text
	  Quantity : Integer
	  UnitPrice : Real
	  ReturnQuantity : Integer
	  Total : Real
	End TABLE ;

Instances	<table><tr><td>Date</td><td>SupplierName</td><td>Total</td></tr><tr><td>20111212</td><td>Johnson Corp.</td><td>258,000.00</td></tr></table> <table><tr><td>ProductNo</td><td>Quantity</td><td>UnitPrice</td><td>ReturnQuantity</td></tr><tr><td>A00001(Pen)</td><td>300</td><td>100.00</td><td>0</td></tr><tr><td>A00002(Mouse)</td><td>390</td><td>200.00</td><td>10</td></tr><tr><td>A00003(Camera)</td><td>500</td><td>300.00</td><td>0</td></tr></table>

Figure 20-114   Composite Data Type Specification

Figure 20-115 shows the analysis' CCD of the *purchasing management* systems definition version 2. In the figure, actor *Originating_Department* has a connection with the *Purchase_Order_Coordinator* component; actor *Supplier* has three connections with the *Purchasing_Department* component; component *Purchase_Order_Coordinator* has a connection with the *Purchase_Requisition_Processing_GUI* component; component *Purchasing_Department* has a connection with each one of the *Quotation_Processing_GUI*, *Purchase_Order_Processing_GUI* and *Purchase_Processing_GUI* components; each one of the *Purchase_Requisition_Processing_GUI*, *Quotation_Processing_GUI*, *Purchase_Order_Processing_GUI* and *Purchase_Processing_GUI* components has a connection with the *PM_Subsystem_2* component.

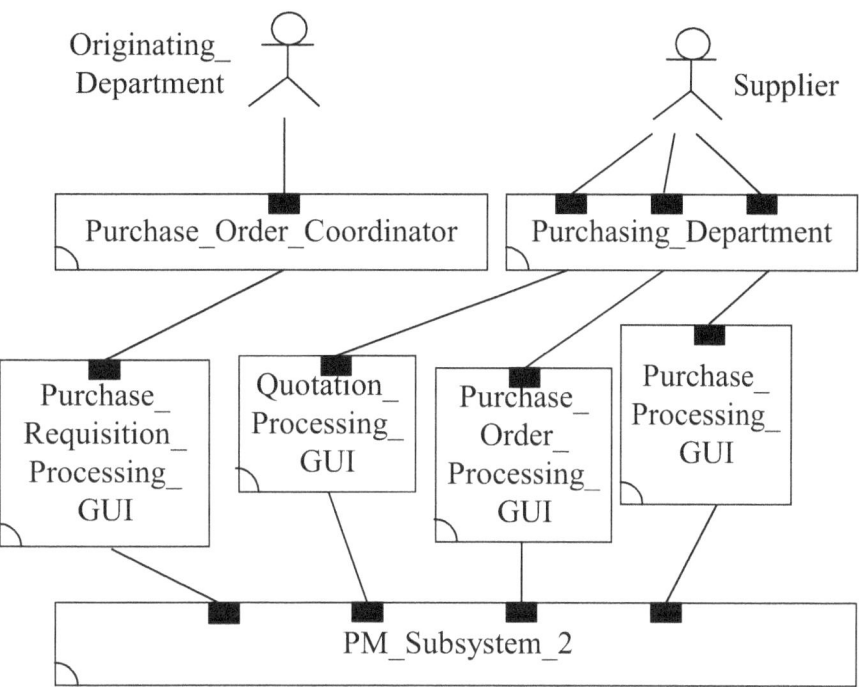

Figure 20-115   Analysis' CCD of the *Purchasing Management*
Systems Definition Version 2

20-2-2-2-2 Analysis' Systems Behavior of the Purchasing Management Systems Definition Version 2

The entire analysis' systems behavior of the *purchasing management* systems definition version 2 includes: a) *Analysis' SBCD* and b) *Analysis' IFD* of the *purchasing management* systems definition version 2.

Figure 20-116 shows the analysis' SBCD of the *purchasing management* systems definition version 2 in which interactions among the *Originating_Department*, *Supplier* actors and the *Purchase_Order_Coordinator*, *Purchasing_Department*, *Purchase_Requisition_Processing_GUI*,   *Quotation_Processing_GUI*, *Purchase_Order_Processing_GUI*, *Purchase_Processing_GUI*, *PM_Subsystem_2* components shall draw forth the *Purchase_Requisition*, *Quotation*, *Purchase_Order* and *Purchase* behaviors.

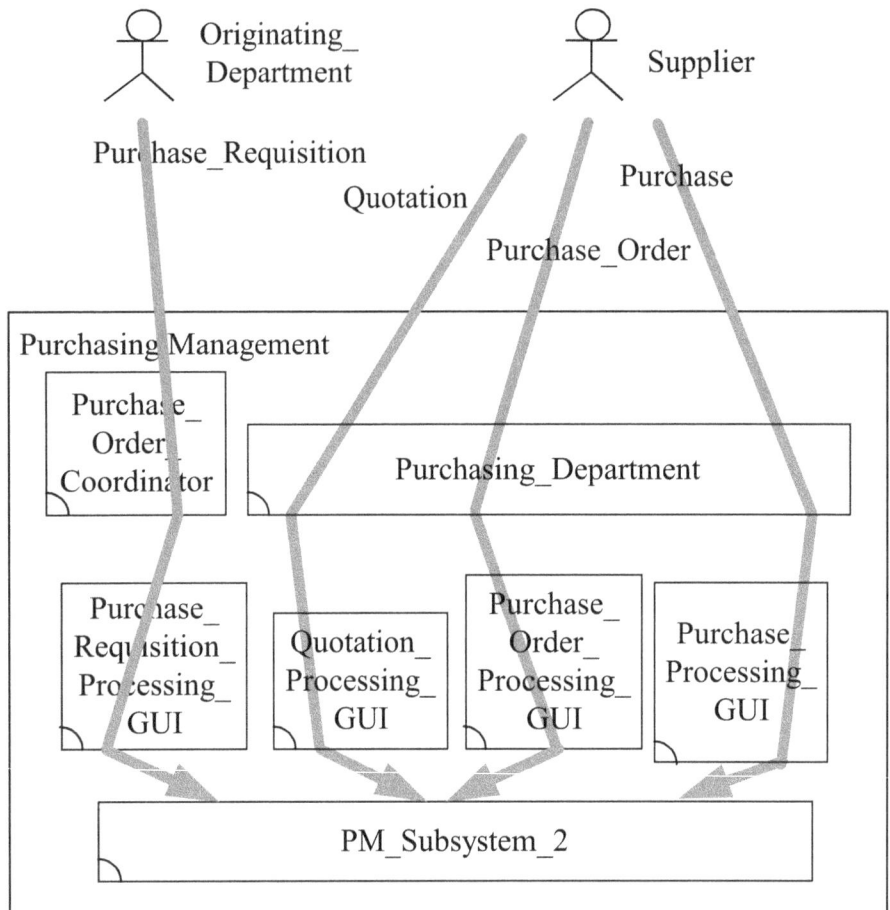

Figure 20-116   Analysis' SBCD of the *Purchasing Management*
Systems Definition Version 2

The overall analysis' behavior of the *purchasing management* systems definition version 2 includes the *Purchase_Requisition, Quotation, Purchase_Order and Purchase* behaviors. In other words, the *Purchase_Requisition, Quotation, Purchase_Order and Purchase* behaviors together provide the overall analysis' behavior of the *purchasing management* systems definition version 2.

Be noticed that the *Purchase_Requisition, Quotation, Purchase_Order and Purchase* behaviors are mutually independent of each other. They tend to be executed concurrently [Hoar85, Miln89, Miln99].

The overall analysis' behavior of the *purchasing management* systems definition version 2 includes four individual behaviors: *Purchase_Requisition, Quotation, Purchase_Order and Purchase*. Each individual behavior is represented by an execution path. We use an interaction flow diagram (IFD) to define each one of these execution paths. Figure 20-117 shows the analysis' IFD of the *purchasing management* systems definition version 2 *Purchase_Requisition* behavior. First, actor *Originating_Department* interacts with the *Purchase_Order_Coordinator* component through the *Purchase_Requisition_Verify* operation call interaction, carrying the *Purchase_Requisition_Form* input parameter. Next, component *Purchase_Order_Coordinator* interacts with the *Purchase_Requisition_Processing_GUI* component through the *Purchase_Requisition_Processing_Button_Click* operation call interaction, carrying the *Purchase_Requisition_Form* input parameter. Finally, component *Purchase_Requisition_Processing_GUI* interacts with the *PM_Subsystem_2* component through the *SQL_Purchase_Requisition_Insert* operation call interaction, carrying the *Purchase_Requisition_Query* input parameter.

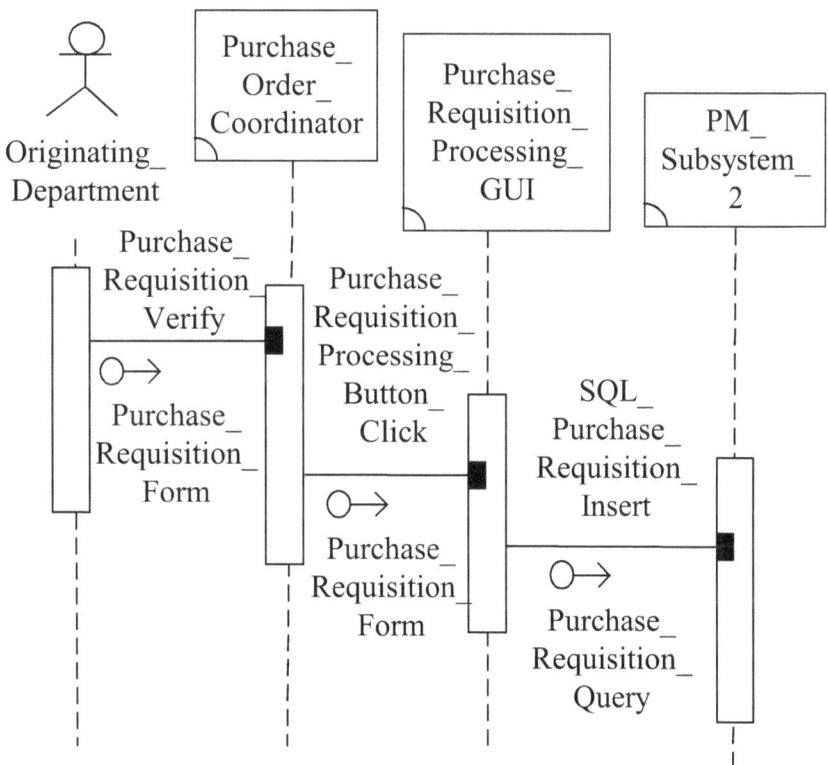

Figure 20-117   Analysis' IFD of the *Purchasing Management* Systems Definition Version 2 *Purchase_Requisition* Behavior

Figure 20-118 shows the analysis' IFD of the *purchasing management* systems definition version 2 *Quotation* behavior. First, actor *Supplier* interacts with the *Purchasing_Department* component through the *Quotation_Verify* operation call interaction, carrying the *Quotation_Form* input parameter. Next, component *Purchasing_Department* interacts with the *Quotation_Processing_GUI* component through the *Quotation_Processing_Button_Click* operation call interaction, carrying the *Quotation_Form* input parameter. Finally, component *Quotation_Processing_GUI* interacts with the *PM_Subsystem_2* component through the *SQL_Quotation_Insert* operation call interaction, carrying the *Quotation_Query* input parameter.

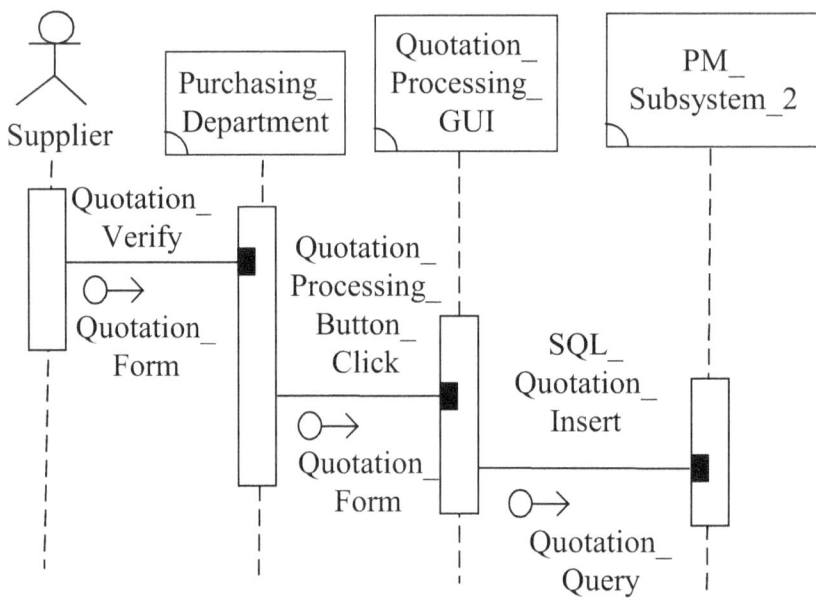

Figure 20-118   Analysis' IFD of the *Purchasing Management Systems Definition Version 2 Quotation* Behavior

Figure 20-119 shows the analysis' IFD of the *purchasing management systems definition version 2 Purchase_Order* behavior. First, actor *Supplier* interacts with the *Purchasing_Department* component through the *Purchase_Order_Verify* operation call interaction. Next, component *Purchasing_Department* interacts with the *Purchase_Order_Processing_GUI* component through the *Purchase_Order_Processing_Button_Click* operation call interaction, carrying the *Purchase_Order_Form* input parameter. Continuingly, component *Purchase_Order_Processing_GUI* interacts with the *PM_Subsystem_2* component through the *SQL_Purchase_Order_Insert* operation call interaction, carrying the *Purchase_Order_Query* input parameter. Continuingly, component *Purchasing_Department* interacts with the *Purchase_Order_Processing_GUI* component through the *Purchase_Order_Processing_Button_Click* operation return interaction, carrying the *Purchase_Order_Report* output parameter. Finally, actor *Supplier* interacts with the *Purchasing_Department* component through the *Purchase_Order_Verify* operation return interaction, carrying the *Purchase_Order_Report* output parameter.

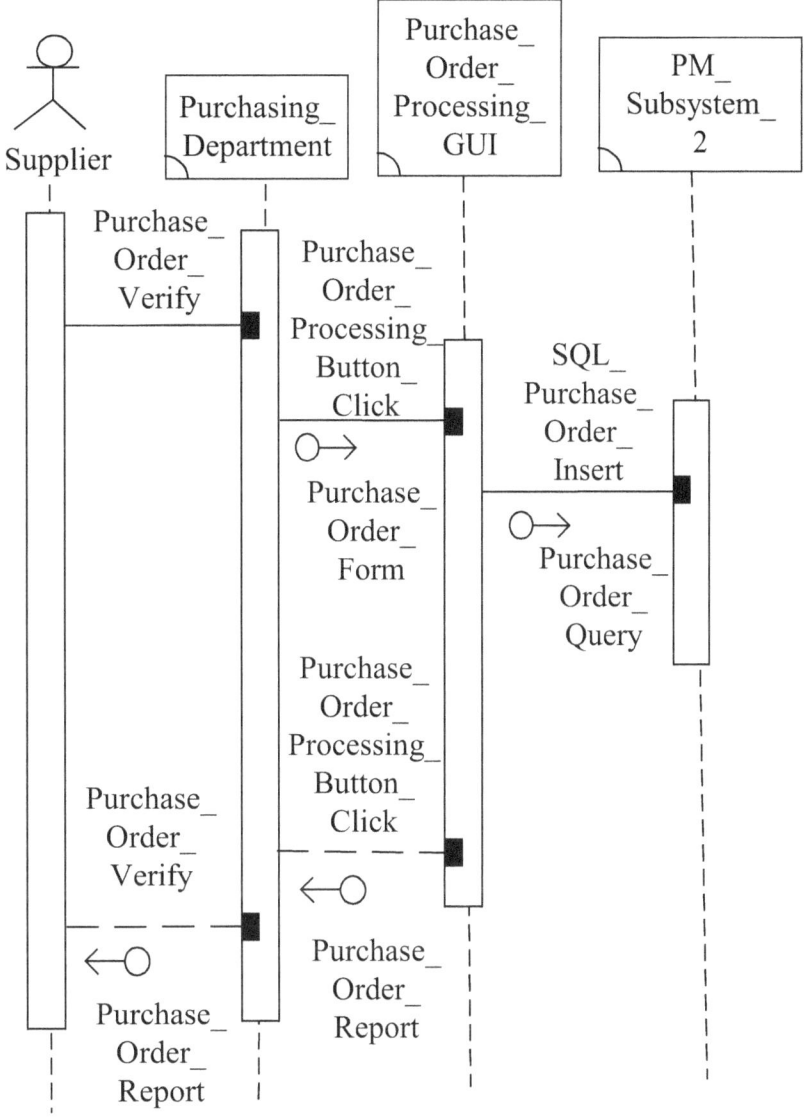

Figure 20-119   Analysis' IFD of the *Purchasing Management*
Systems Definition Version 2 *Purchase_Order* Behavior

Figure 20-120 shows the analysis' IFD of the *purchasing management* systems definition version 2 *Purchase* behavior. First, actor *Supplier* interacts with the *Purchasing_Department* component through the *Purchase_Verify* operation call interaction, carrying the *Products_In* and *Invoice* input parameters. Next, component *Purchasing_Department* interacts with the *Purchase_Processing_GUI* component through the *Purchase_Processing_Button_Click* operation call interaction, carrying the *Purchase_Form* input parameter. Continuingly, component *Purchase_Processing_GUI* interacts with the *PM_Subsystem_2* component through the *SQL_Purchase_Insert* operation call interaction, carrying the *Purchase_Query* input parameter. Finally, actor *Supplier* interacts with the *Purchasing_Department* component through the *Purchase_Verify* operation return interaction, carrying the *Products_Out* output parameter.

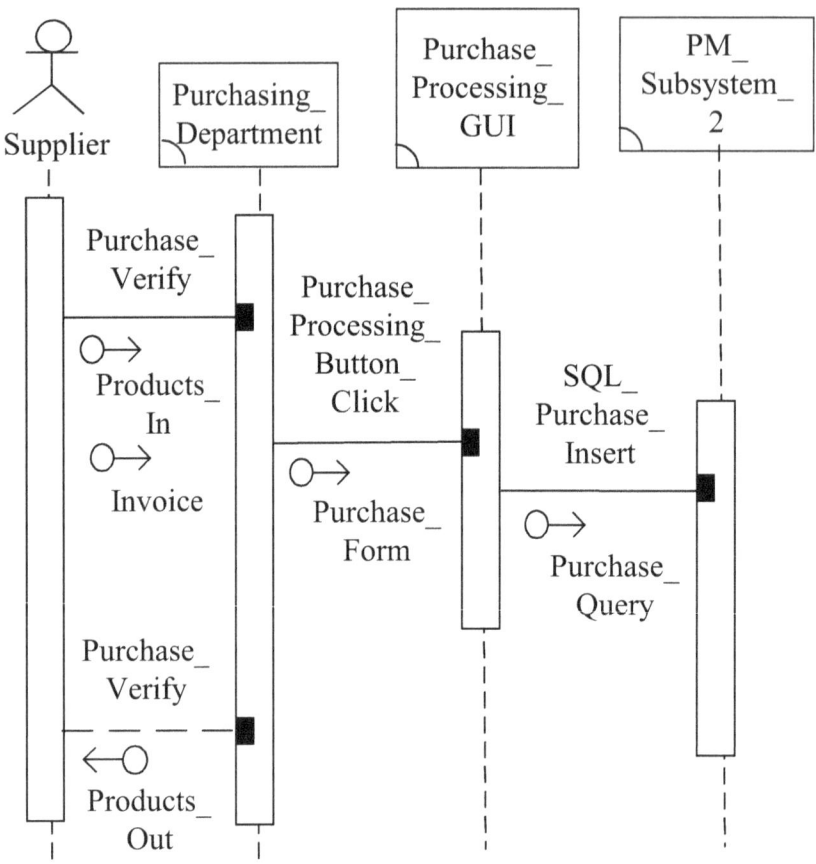

Figure 20-120   Analysis' IFD of the *Purchasing Management* Systems Definition Version 2 *Purchase* Behavior

20-2-2-3 Design View of the Purchasing Management Systems Definition Version 2

The design view of the *purchasing management* systems definition version 2 is shown in Figure 20-121.

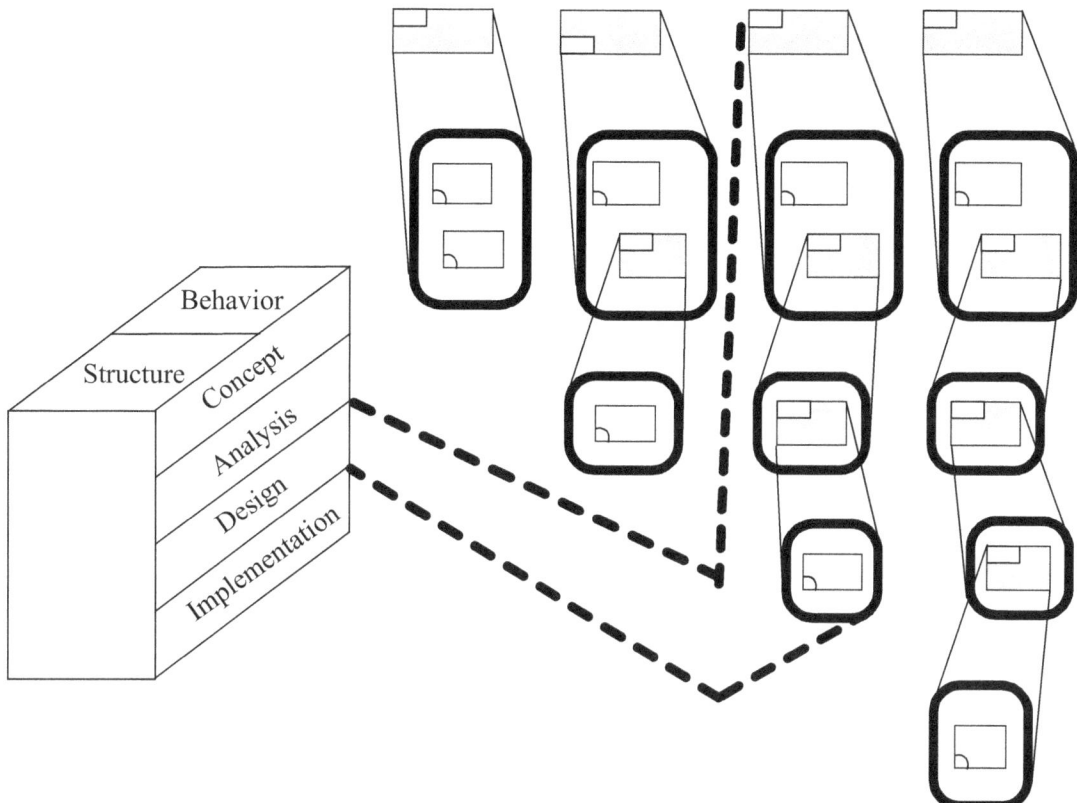

Figure 20-121   Design View of the *Purchasing Management* Systems Definition Version 2

The design view of the *purchasing management* systems definition version 2 consists of: a) design's systems structure of the *purchasing management* systems definition version 2 and b) design's systems behavior of the *purchasing management* systems definition version 2.

20-2-2-3-1 Design's Systems Structure of the Purchasing Management Systems Definition Version 2

The entire design's systems structure of the *purchasing management* systems definition version 2 includes: a) *Design's AHD*, b) *Design's FD*, c) *Design's COD* and d) *Design's CCD* of the *purchasing management* systems definition version 2.

We first draw the design's AHD of the *purchasing management* systems definition version 2. As shown in Figure 20-122, *Purchasing Management* is composed of *Purchase_Order_Coordinator*, *Purchasing_Department* and *PM_Subsystem_3*; *PM_Subsystem_3* is composed of *Purchase_Requisition_Processing_GUI*, *Quotation_Processing_GUI*, *Purchase_Order_Processing_GUI*, *Purchase_Processing_GUI* and *PM_Subsystem_2*; *PM_Subsystem_2* is composed of *Purchasing_Database* and *PM_Subsystem_1*. In the figure, *Purchasing Management, PM_Subsystem_3 and PM_Subsystem_2* are aggregated systems while *Purchase_Order_Coordinator, Purchasing_Department,*

334

Purchase_Requisition_Processing_GUI,   *Quotation_Processing_GUI,*

Purchase_Order_Processing_GUI,   *Purchase_Processing_GUI,*

Purchasing_Database and *PM_Subsystem_1* are non-aggregated systems.

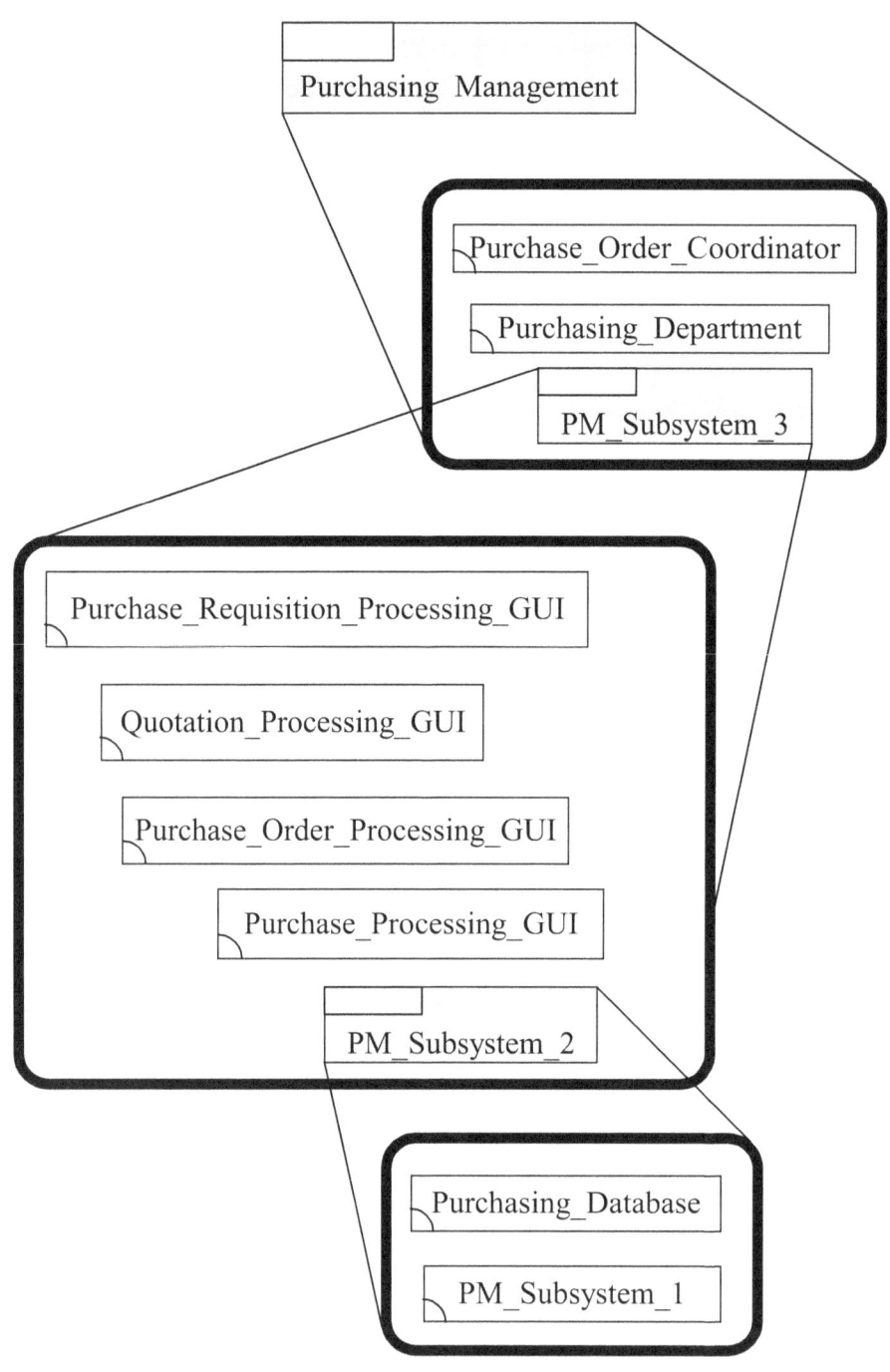

Figure 20-122   Design's AHD of the *Purchasing Management*
Systems Definition Version 2

Figure 20-123 shows the design's FD of the *purchasing management* systems definition version 2. In the figure, *Business_Layer* contains the *Purchase_Order_Coordinator* and *Purchasing_Department* components; *Application_Layer* contains the *Purchase_Requisition_Processing_GUI*, *Quotation_Processing_GUI*, *Purchase_Order_Processing_GUI* and *Purchase_Processing_GUI* components; *Data_Layer* contains the *Purchasing_Database* and *PM_Subsystem_1* components.

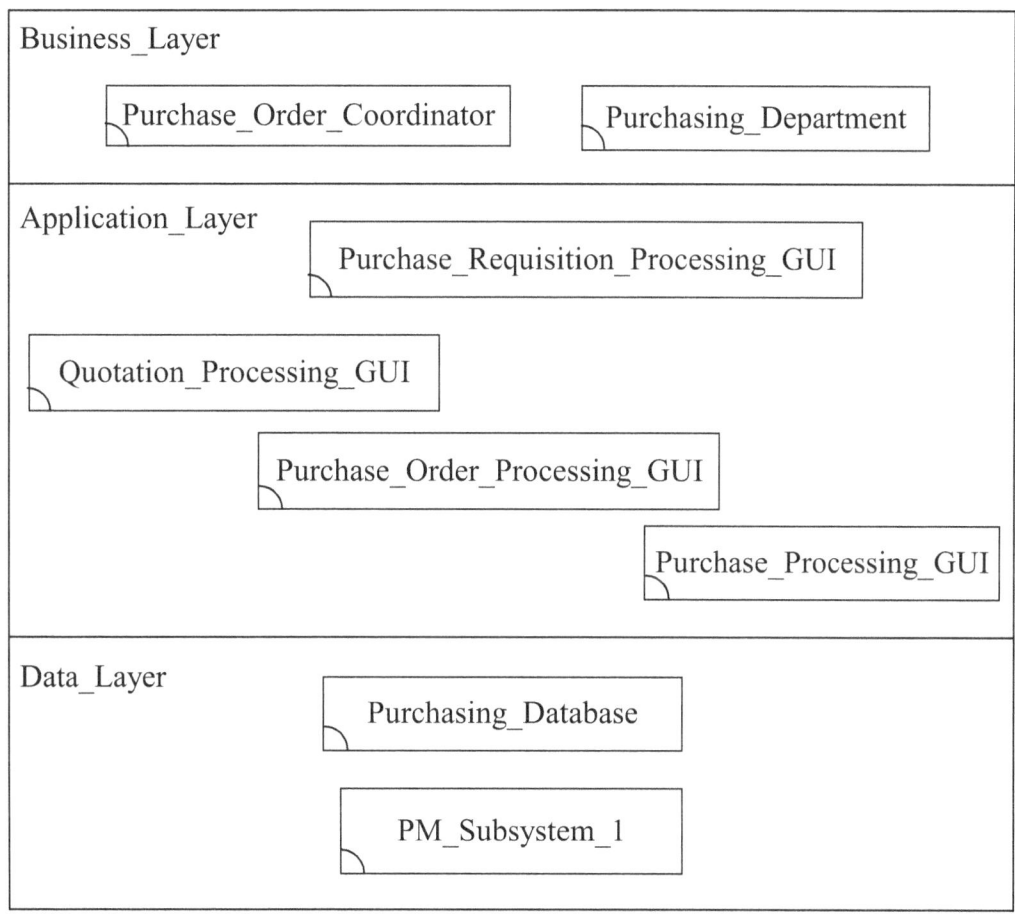

Figure 20-123   Design's FD of the *Purchasing Management*
Systems Definition Version 2

Figure 20-124 shows the design's COD of the *purchasing management* systems definition version 2. In the figure, component *Purchase_Order_Coordinator* has one operation: *Purchase_Requisition_Verify*; component *Purchasing_Department* has three operations: *Quotation_Verify*, *Purchase_Order_Verify* and *Purchase_Verify*; component *Purchase_Requisition_Processing_GUI* has one operation: *Purchase_Requisition_Processing_Button_Click*;   component

Quotation_Processing_GUI has one operation: *Quotation_Processing_Button_Click*; component *Purchase_Order_Processing_GUI* has one operation: *Purchase_Order_Processing_Button_Click*; component *Purchase_Processing_GUI* has one operation: *Purchase_Processing_Button_Click*; component *Purchasing_Database* has four operations: *SQL_Purchase_Requisition_Insert*, *SQL_Quotation_Insert*, *SQL_Purchase_Order_Insert* and *SQL_Purchase_Insert*; component *PM_Subsystem_1* has one operation: *Infrastructure_Resources_Share*.

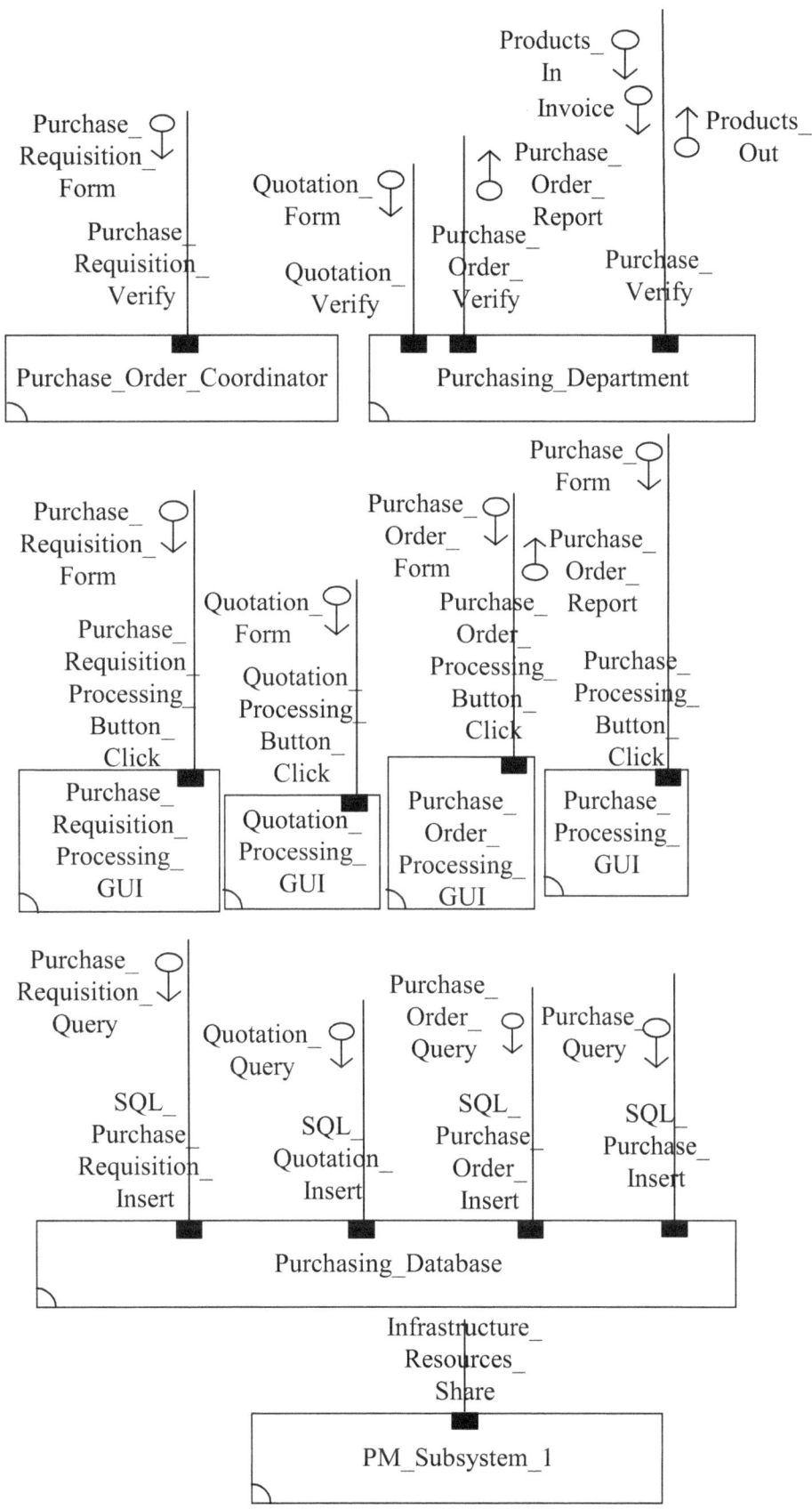

Figure 20-124   Design's COD of the *Purchasing Management Systems Definition Version 2*

The operation formula of *Purchase_Requisition_Verify* is *Purchase_Requisition_Verify(In Purchase_Requisition_Form)*. The operation formula of *Quotation_Verify* is *Quotation_Verify(In Quotation_Form)*. The operation formula of *Purchase_Order_Verify* is *Purchase_Order_Verify(Out Purchase_Order_Report)*. The operation formula of *Purchase_Verify* is *Purchase_Verify(In Products_In, Invoice; Out Products_Out)*. The operation formula of *Purchase_Requisition_Processing_Button_Click* is *Purchase_Requisition_Processing_Button_Click(In Purchase_Requisition_Form)*. The operation formula of *Quotation_Processing_Button_Click* is *Quotation_Processing_Button_Click(In Quotation_Form)*. The operation formula of *Purchase_Order_Processing_Button_Click* is *Purchase_Order_Processing_Button_Click(In Purchase_Order_Form; Out Purchase_Order_Report)*. The operation formula of *Purchase_Processing_Button_Click* is *Purchase_Processing_Button_Click(In Purchase_Form)*. The operation formula of *SQL_Purchase_Requisition_Insert* is *SQL_Purchase_Requisition_Insert(In Purchase_Requisition_Query)*. The operation formula of *SQL_Quotation_Insert* is *SQL_Quotation_Insert(In Quotation_Query)*. The operation formula of *SQL_Purchase_Order_Insert* is *SQL_Purchase_Order_Insert (In Purchase_Order_Query)*. The operation formula of *SQL_Purchase_Insert* is *SQL_Purchase_Insert (In Purchase_Query)*. The operation formula of *Infrastructure_Resources_Share* is *Infrastructure_Resources_Share*.

Figure 20-125 shows the composite data type specification of the *Purchase_Requisition_Form* input parameter occurring in the *Purchase_Requisition_Verify(In Purchase_Requisition_Form)* and *Purchase_Requisition_Processing_Button_Click(In Purchase_Requisition_Form)* operation formulas.

Parameter	*Purchase_Requisition_Form*

Data Type	TABLE of
	   Date : Text
	   OD : Text
	   ProductNo : Text
	   Quantity : Integer
	 End TABLE ;

Instances	**Purchase Requisition Form** Date: 2011/10/17 Originating_Department : Sales Dept. ProductNo Quantity __A00001(Pen)_____300_____ __A00002(Mouse)_____400_____ __A00003(Camera)_____500_____

Figure 20-125   Composite Data Type Specification

Figure 20-126 shows the composite data type specification of the *Quotation_Form* input parameter occurring in the *Quotation_Verify(In Quotation_Form)* and *Quotation_Processing_Button_Click(In Quotation_Form)* operation formulas.

Parameter	*Quotation_Form*

Data Type	TABLE of
	  Date : Text
	  SupplierName: Text
	  ProductNo : Text
	  Quantity : Integer
	  UnitPrice : Real
	  Total : Real
	End TABLE ;

Instances	**Quotation Form** Date: 2011/10/25 SupplierName : Johnson Corp. ProductNo Quantity UnitPrice A00001(Pen) 300 100.00 _A00002(Mouse)____400____200.00__ _A00003(Camera)___500____300.00__ Total : 260,000.00

Figure 20-126   Composite Data Type Specification

Figure 20-127 shows the composite data type specification of the *Purchase_Order_Report* output parameter occurring in the *Purchase_Order_Verify(Out Purchase_Order_Report)* and *Purchase_Order_Processing_Button_Click(In Purchase_Order_Form; Out Purchase_Order_Report)* operation formulas.

341

Parameter	*Purchase_Order_Report*

Data Type	TABLE of
	  Date : Text
	  SupplierName: Text
	  ProductNo : Text
	  Quantity : Integer
	End TABLE ;

Instances	Date : 20111118 SupplierName : Johnson Corp. <table><tr><th>ProductNo</th><th>Quantity</th></tr><tr><td>A00001(Pen)</td><td>300</td></tr><tr><td>A00002(Mouse)</td><td>400</td></tr><tr><td>A00003(Camera)</td><td>500</td></tr></table>

Figure 20-127   Composite Data Type Specification

Figure 20-128 shows the primitive data type specification of the *Products_In* input parameter occurring in the *Purchase_Verify(In Products_In, Invoice; Out Products_Out)* operation formula.

Parameter	Data Type	Instances
Products_In	Physical Object	Pen, Mouse, Camera

Figure 20-128   Primitive Data Type Specification

Figure 20-129 shows the composite data type specification of the *Invoice* input parameter occurring in the *Purchase_Verify(In Products_In, Invoice; Out Products_Out)* operation formula.

Parameter	*Invoice*
Data Type	TABLE of Date : Text SupplierName: Text ProductNo : Text Quantity : Integer UnitPrice : Real Total : Real End TABLE ;
Instances	Date : 20111130 SupplierName : Johnson Corp. ProductNo / Quantity / UnitPrice: A00001(Pen) — 300 — 100.00 A00002(Mouse) — 400 — 200.00 A00003(Camera) — 500 — 300.00 Total : 260,000.00

Figure 20-129   Composite Data Type Specification

Figure 20-130 shows the primitive data type specification of the *Products_Out* output parameter occurring in the *Purchase_Verify(In Products_In, Invoice; Out Products_Out)* operation formula.

Parameter	Data Type	Instances
Products_Out	Physical Object	Pen, Mouse, Camera

Figure 20-130   Primitive Data Type Specification

Figure 20-131 shows the composite data type specification of the *Purchase_Order_Form* input parameter occurring in the *Purchase_Order_Processing_Button_Click(In Purchase_Order_Form; Out Purchase_Order_Report)* operation formula.

Parameter	*Purchase_Order_Form*
Data Type	TABLE of Date : Text SupplierName: Text ProductNo : Text Quantity : Integer End TABLE ;
Instances	**Purchase Order Form** Date: 2011/11/18 SupplierName : Johnson Corp. ProductNo Quantity __A00001(Pen) ____300 ___ __A00002(Mouse)____400___ __A00003(Camera)___500___

Figure 20-131   Composite Data Type Specification

Figure 20-132 shows the composite data type specification of the *Purchase_Form* input parameter occurring in the *Purchase_Processing_Button_Click(In Purchase_Form)* operation formula

Parameter	*Purchase_Form*

Data Type	TABLE of
	  Date : Text
	  SupplierName: Text
	  ProductNo : Text
	  Quantity : Integer
	  UnitPrice : Real
	  ReturnQuantity : Integer
	  Total : Real
	End TABLE ;

Instances	**Purchase Form** Date: 2011/12/12 SupplierName : Johnson Corp. ProductNo Quantity UnitPrice ReturnQuantity A00001(Pen) 300 100.00 0 A00002(Mouse) 390 200.00 10 A00003(Camera)500 300.00 0 Total : 258,000.00

Figure 20-132   Composite Data Type Specification

Figure 20-133 shows the composite data type specification of the *Purchase_Requisition_Query* input parameter occurring in the *SQL_Purchase_Requisition_Insert(In Purchase_Requisition_Query)* operation formula.

Parameter	*Purchase_Requisition_Query*
Data Type	TABLE of Date : Text OD : Text ProductNo : Text Quantity : Integer End TABLE ;
Instances	<table><tr><td>Date</td><td colspan="2">Originating_Department :</td></tr><tr><td>20111017</td><td colspan="2">Sales Dept.</td></tr><tr><td></td><td>ProductNo</td><td>Quantity</td></tr><tr><td></td><td>A00001(Pen)</td><td>300</td></tr><tr><td></td><td>A00002(Mouse)</td><td>400</td></tr><tr><td></td><td>A00003(Camera)</td><td>500</td></tr></table>

Figure 20-133   Composite Data Type Specification

Figure 20-134 shows the composite data type specification of the *Quotation_Query* input parameter occurring in the *SQL_Quotation_Insert(In Quotation_Query)* operation formula.

Parameter	*Quotation_Query*
Data Type	TABLE of Date : Text SupplierName: Text ProductNo : Text Quantity : Integer UnitPrice : Real Total : Real End TABLE ;
Instances	<table><tr><th>Date</th><th>SupplierName</th><th>Total</th></tr><tr><td>20111025</td><td>Johnson Corp.</td><td>260,000.00</td></tr></table> <table><tr><th>ProductNo</th><th>Quantity</th><th>UnitPrice</th></tr><tr><td>A00001(Pen)</td><td>300</td><td>100.00</td></tr><tr><td>A00002(Mouse)</td><td>400</td><td>200.00</td></tr><tr><td>A00003(Camera)</td><td>500</td><td>300.00</td></tr></table>

Figure 20-134   Composite Data Type Specification

Figure 20-135 shows the composite data type specification of the *Purchase_Order_Query* input parameter occurring in the *SQL_Purchase_Order_Insert(In Purchase_Order_Query)* operation formula.

Parameter	*Purchase_Order_Query*
Data Type	TABLE of 　Date : Text 　SupplierName: Text 　ProductNo : Text 　Quantity : Integer End TABLE ;
Instances	

Date	SupplierName
20111118	Johnson Corp.

ProductNo	Quantity
A00001(Pen)	300
A00002(Mouse)	400
A00003(Camera)	500

Figure 20-135   Composite Data Type Specification

Figure 20-136 shows the composite data type specification of the *Purchase_Query* input parameter occurring in the *SQL_Purchase_Insert (In Purchase_Query)* operation formula.

Parameter	*Purchase_Query*

Data Type	TABLE of
	  Date : Text
	  SupplierName: Text
	  ProductNo : Text
	  Quantity : Integer
	  UnitPrice : Real
	  ReturnQuantity : Integer
	  Total : Real
	End TABLE ;

Instances	(see table below)

Date	SupplierName	Total
20111212	Johnson Corp.	258,000.00

ProductNo	Quantity	UnitPrice	ReturnQuantity
A00001(Pen)	300	100.00	0
A00002(Mouse)	390	200.00	10
A00003(Camera)	500	300.00	0

Figure 20-136   Composite Data Type Specification

Figure 20-137 shows the design's CCD of the *purchasing management* systems definition version 2. In the figure, actor *Originating_Department* has a connection with the *Purchase_Order_Coordinator* component; actor *Supplier* has three connections with the *Purchasing_Department* component; component *Purchase_Order_Coordinator* has a connection with the *Purchase_Requisition_Processing_GUI* component; component *Purchasing_Department* has a connection with each one of the *Quotation_Processing_GUI*, *Purchase_Order_Processing_GUI* and *Purchase_Processing_GUI* components; each one of the *Purchase_Requisition_Processing_GUI*, *Quotation_Processing_GUI*, *Purchase_Order_Processing_GUI* and *Purchase_Processing_GUI* components has a connection with the *Purchasing_Database*; component *Purchasing_Database* has a connection with the *PM_Subsystem_1* component.

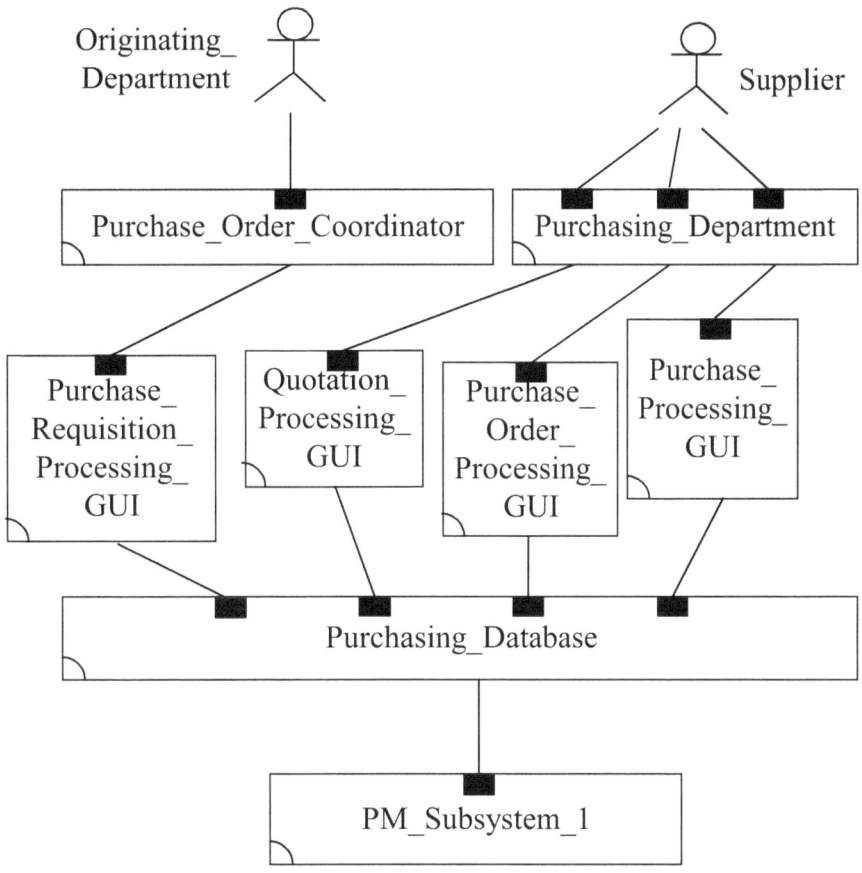

Figure 20-137   Design's CCD of the *Purchasing Management Systems Definition Version 2*

20-2-2-3-2 Design's Systems Behavior of the Purchasing Management Systems Definition Version 2

The entire design's systems behavior of the *purchasing management* systems definition version 2 includes: a) *Design's SBCD* and b) *Design's IFD* of the *purchasing management* systems definition version 2.

Figure 20-138 shows the design's SBCD of the *purchasing management* systems definition version 2 in which interactions among the *Originating_Department*, *Supplier* actors and the *Purchase_Order_Coordinator*, *Purchasing_Department*, *Purchase_Requisition_Processing_GUI*, *Quotation_Processing_GUI*, *Purchase_Order_Processing_GUI*, *Purchase_Processing_GUI*, *Purchasing_Database*, *PM_Subsystem_1* components shall draw forth the *Purchase_Requisition, Quotation, Purchase_Order* and *Purchase* behaviors.

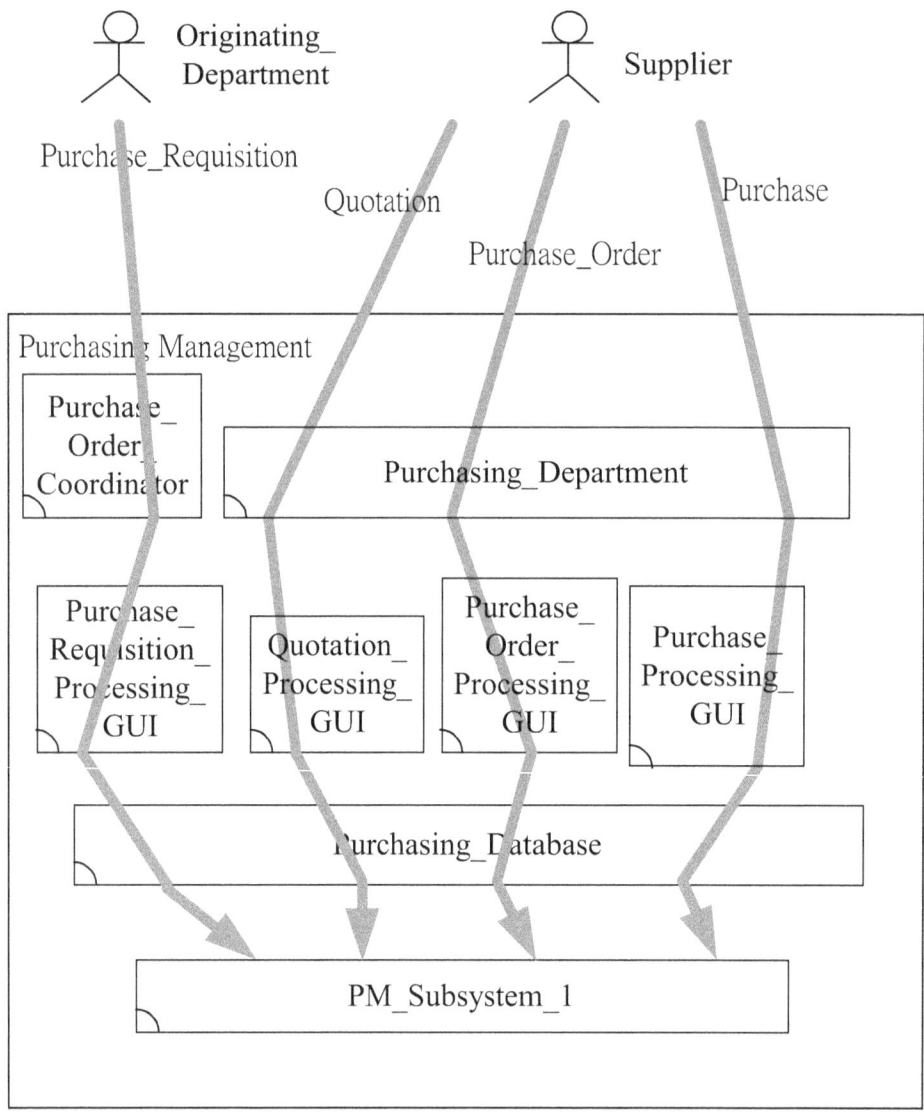

Figure 20-138   Design's SBCD of the *Purchasing Management*
Systems Definition Version 2

The overall design's behavior of the *purchasing management* systems definition version 2 includes the *Purchase_Requisition, Quotation, Purchase_Order and Purchase* behaviors. In other words, the *Purchase_Requisition, Quotation, Purchase_Order and Purchase* behaviors together provide the overall design's behavior of the *purchasing management* systems definition version 2.

Be noticed that the *Purchase_Requisition, Quotation, Purchase_Order and Purchase* behaviors are mutually independent of each other. They tend to be executed concurrently [Hoar85, Miln89, Miln99].

The overall design's behavior of the *purchasing management* systems definition version 2 includes four individual behaviors: *Purchase_Requisition,*

Quotation, Purchase_Order and Purchase. Each individual behavior is represented by an execution path. We use an interaction flow diagram (IFD) to define each one of these execution paths. Figure 20-139 shows the design's IFD of the *purchasing management* systems definition version 2 *Purchase_Requisition* behavior. First, actor *Originating_Department* interacts with the *Purchase_Order_Coordinator* component through the *Purchase_Requisition_Verify* operation call interaction, carrying the *Purchase_Requisition_Form* input parameter. Next, component *Purchase_Order_Coordinator* interacts with the *Purchase_Requisition_Processing_GUI* component through the *Purchase_Requisition_Processing_Button_Click* operation call interaction, carrying the *Purchase_Requisition_Form* input parameter. Continuingly, component *Purchase_Requisition_Processing_GUI* interacts with the *Purchasing_Database* component through the *SQL_Purchase_Requisition_Insert* operation call interaction, carrying the *Purchase_Requisition_Query* input parameter. Finally, component *Purchasing_Database* interacts with the *PM_Subsystem_1* component through the *Infrastructure_Resources_Share* operation call interaction.

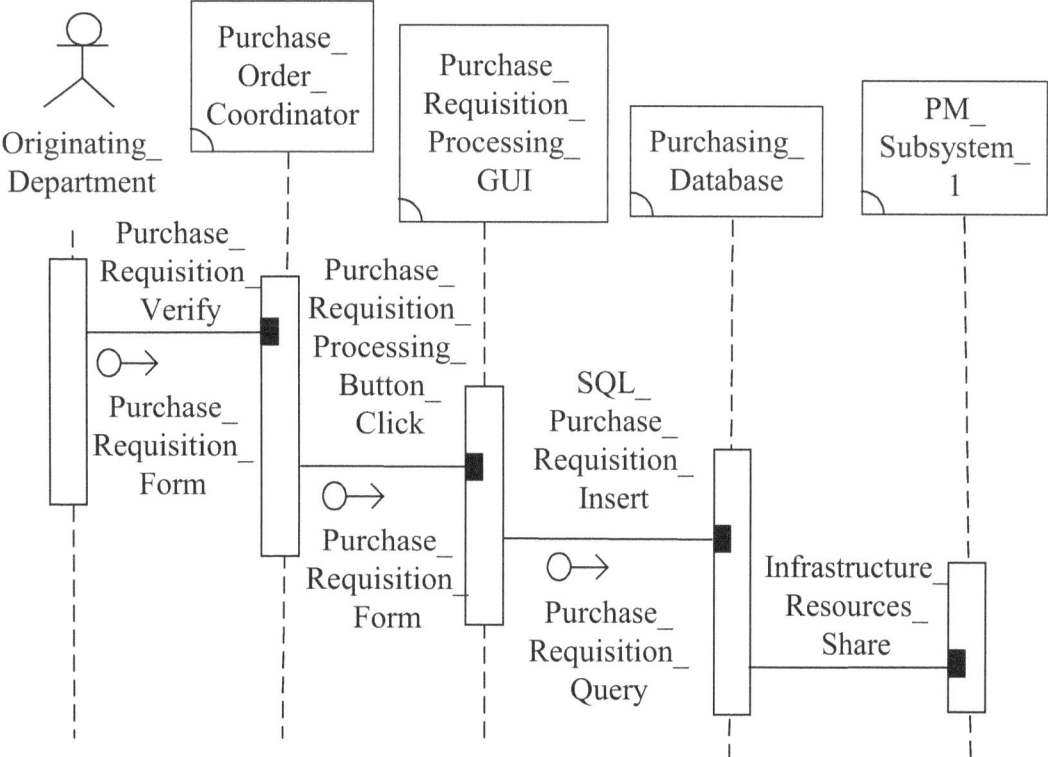

Figure 20-139   Design's IFD of the *Purchasing Management* Systems Definition Version 2 *Purchase_Requisition* Behavior

352

Figure 20-140 shows the design's IFD of the *purchasing management* systems definition version 2 *Quotation* behavior. First, actor *Supplier* interacts with the *Purchasing_Department* component through the *Quotation_Verify* operation call interaction, carrying the *Quotation_Form* input parameter. Next, component *Purchasing_Department* interacts with the *Quotation_Processing_GUI* component through the *Quotation_Processing_Button_Click* operation call interaction, carrying the *Quotation_Form* input parameter. Continuingly, component *Quotation_Processing_GUI* interacts with the *Purchasing_Database* component through the *SQL_Quotation_Insert* operation call interaction, carrying the *Quotation_Query* input parameter. Finally, component *Purchasing_Database* interacts with the *PM_Subsystem_1* component through the *Infrastructure_Resources_Share* operation call interaction.

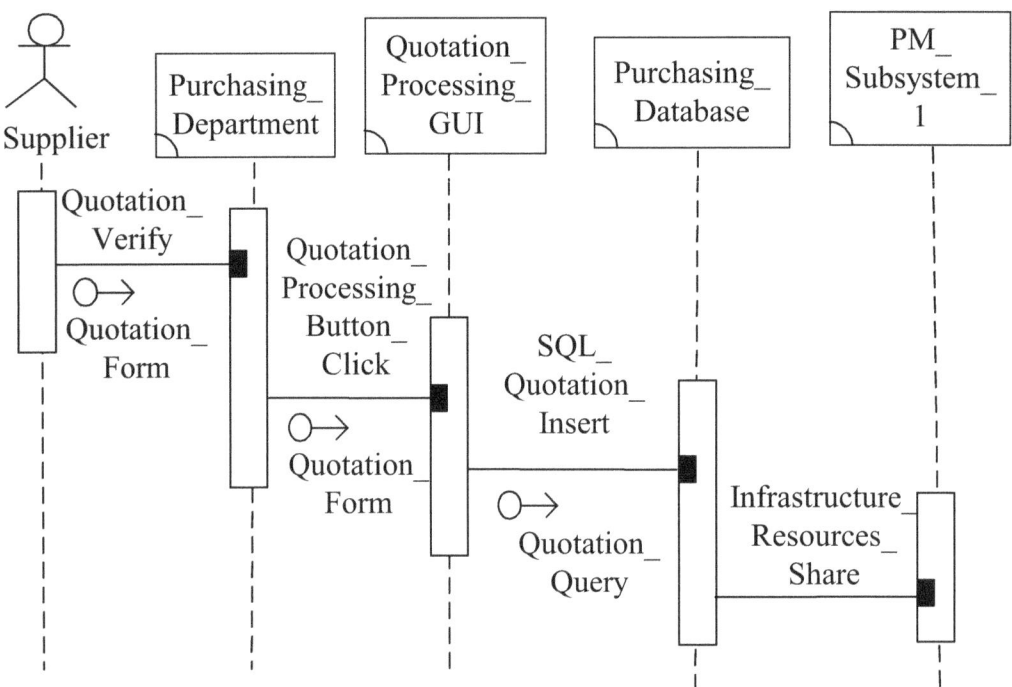

Figure 20-140   Design's IFD of the *Purchasing Management* Systems Definition Version 2 *Quotation* Behavior

Figure 20-141 shows the design's IFD of the *purchasing management* systems definition version 2 *Purchase_Order* behavior. First, actor *Supplier* interacts with the *Purchasing_Department* component through the *Purchase_Order_Verify* operation call interaction. Next, component *Purchasing_Department* interacts with the *Purchase_Order_Processing_GUI* component through the *Purchase_Order_Processing_Button_Click* operation call interaction, carrying the *Purchase_Order_Form* input parameter. Continuingly, component

Purchase_Order_Processing_GUI interacts with the *Purchasing_Database* component through the *SQL_Purchase_Order_Insert* operation call interaction, carrying the *Purchase_Order_Query* input parameter. Continuingly, component *Purchasing_Database* interacts with the *PM_Subsystem_1* component through the *Infrastructure_Resources_Share* operation call interaction. Continuingly, component *Purchasing_Department* interacts with the *Purchase_Order_Processing_GUI* component through the *Purchase_Order_Processing_Button_Click* operation return interaction, carrying the *Purchase_Order_Report* output parameter. Finally, actor *Supplier* interacts with the *Purchasing_Department* component through the *Purchase_Order_Verify* operation return interaction, carrying the *Purchase_Order_Report* output parameter.

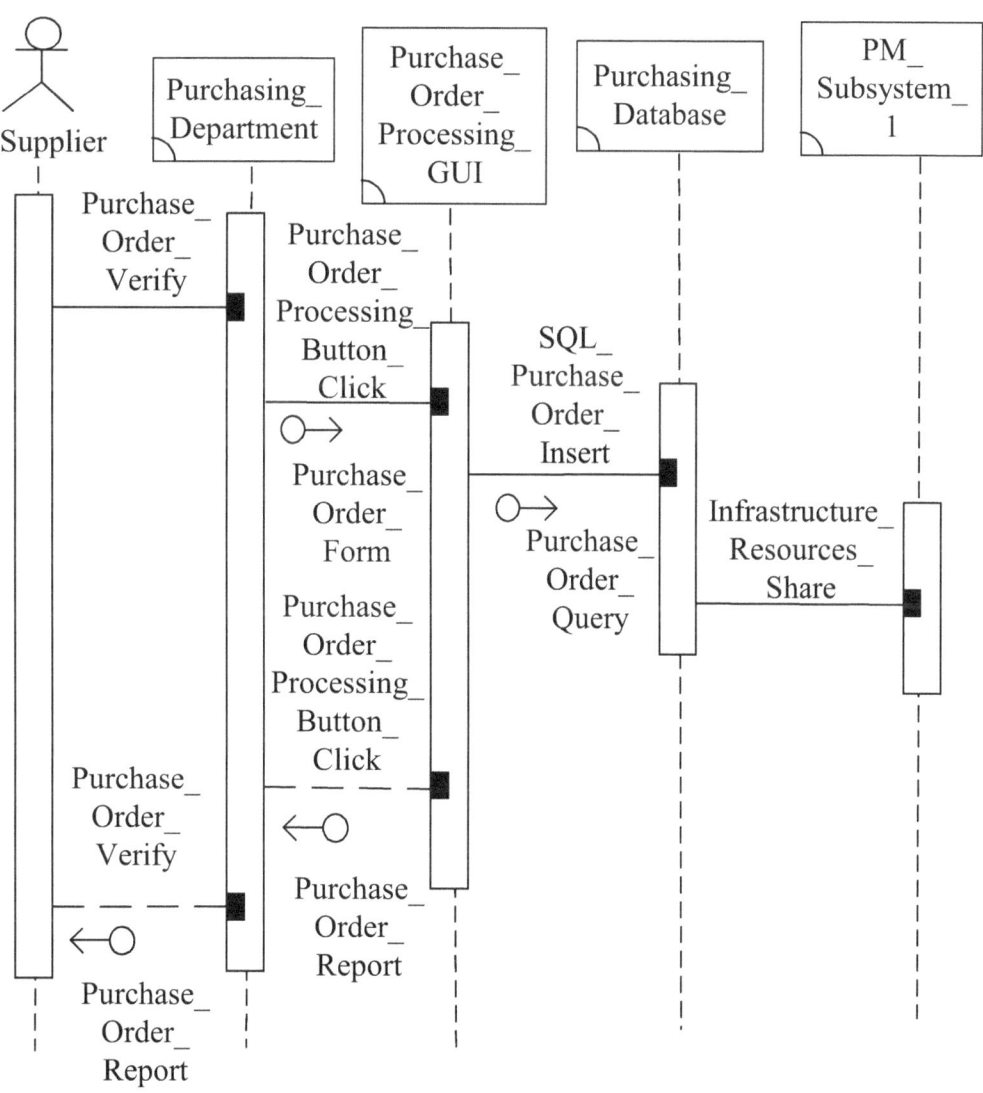

Figure 20-141   Design's IFD of the *Purchasing Management Systems Definition Version 2 Purchase_Order* Behavior

354

Figure 20-142 shows the design's IFD of the *purchasing management* systems definition version 2 *Purchase* behavior. First, actor *Supplier* interacts with the *Purchasing_Department* component through the *Purchase_Verify* operation call interaction, carrying the *Products_In* and *Invoice* input parameters. Next, component *Purchasing_Department* interacts with the *Purchase_Processing_GUI* component through the *Purchase_Processing_Button_Click* operation call interaction, carrying the *Purchase_Form* input parameter. Continuingly, component *Purchase_Processing_GUI* interacts with the *Purchasing_Database* component through the *SQL_Purchase_Insert* operation call interaction, carrying the *Purchase_Query* input parameter. Continuingly, component *Purchasing_Database* interacts with the *PM_Subsystem_1* component through the *Infrastructure_Resources_Share* operation call interaction. Finally, actor *Supplier* interacts with the *Purchasing_Department* component through the *Purchase_Verify* operation return interaction, carrying the *Products_Out* output parameter.

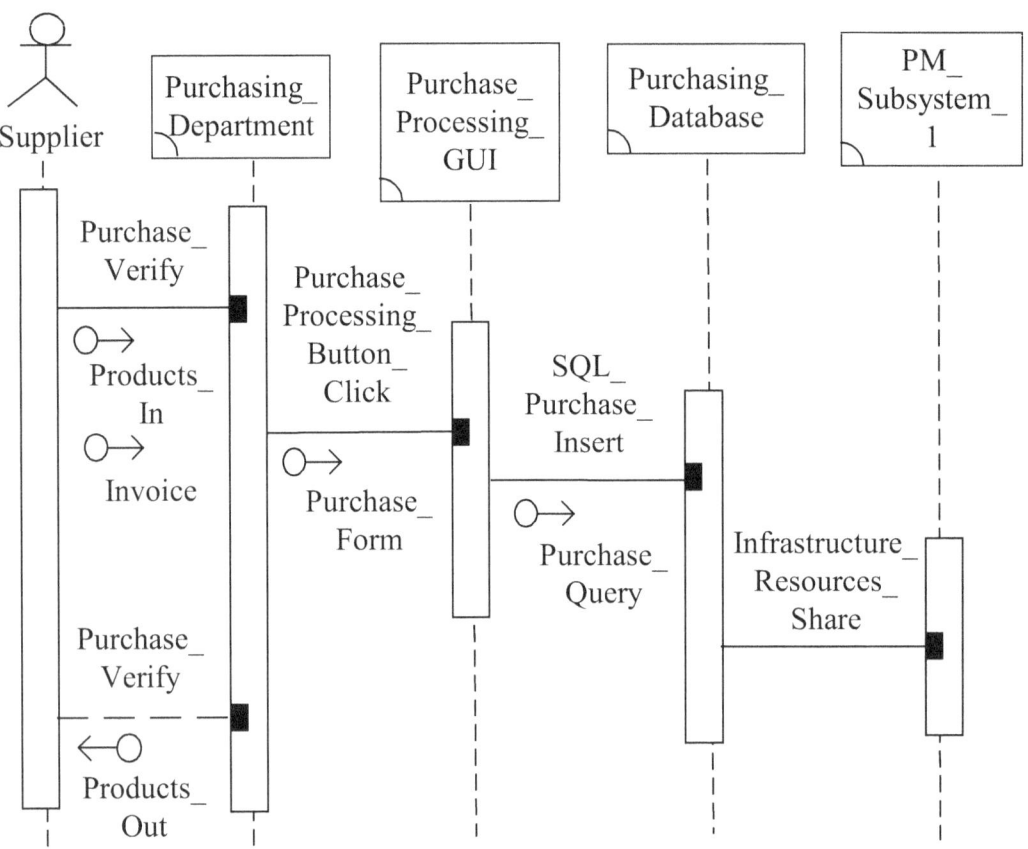

Figure 20-142   Design's IFD of the *Purchasing Management* Systems Definition Version 2 *Purchase* Behavior

20-2-2-4 Implementation View of the Purchasing Management Systems Definition Version 2

The implementation view of the *purchasing management* systems definition version 2 is shown in Figure 20-143.

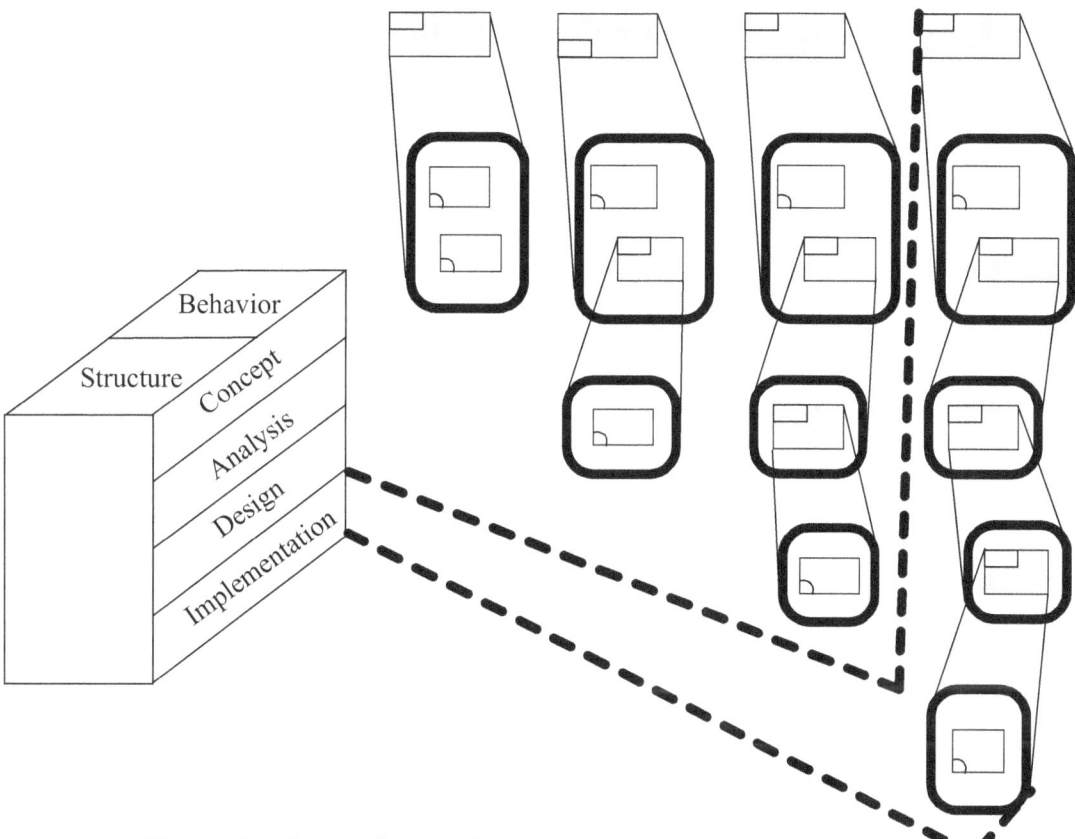

Figure 20-143　Implementation View of the *Purchasing Management*
Systems Definition Version 2

The implementation view of the *purchasing management* systems definition version 2 consists of: a) implementation's systems structure of the *purchasing management* systems definition version 2 and b) implementation's systems behavior of the *purchasing management* systems definition version 2.

20-2-2-4-1 Implementation's Systems Structure of the Purchasing Management Systems Definition Version 2

The entire implementation's systems structure of the *purchasing management* systems definition version 2 includes: a) *Implementation's AHD*, b) *Implementation's FD*, c) *Implementation's COD* and d) *Implementation's CCD* of the *purchasing management* systems definition version 2.

We first draw the implementation's AHD of the *purchasing management* systems definition version 2. As shown in Figure 20-144, *Purchasing Management* is composed of *Purchase_Order_Coordinator*, *Purchasing_Department* and *PM_Subsystem_3*; *PM_Subsystem_3* is composed of *Purchase_Requisition_Processing_GUI*, *Quotation_Processing_GUI*, *Purchase_Order_Processing_GUI*, *Purchase_Processing_GUI* and *PM_Subsystem_2*; *PM_Subsystem_2* is composed of *Purchasing_Database* and *PM_Subsystem_1*; *PM_Subsystem_1* is composed of *Network_Operating_System*. In the figure, *Purchasing Management, PM_Subsystem_3, PM_Subsystem_2 and PM_Subsystem_1* are aggregated systems while *Purchase_Order_Coordinator*, *Purchasing_Department*, *Purchase_Requisition_Processing_GUI*, *Quotation_Processing_GUI*, *Purchase_Order_Processing_GUI*, *Purchase_Processing_GUI*, *Purchasing_Database* and *Network_Operating_System* are non-aggregated systems.

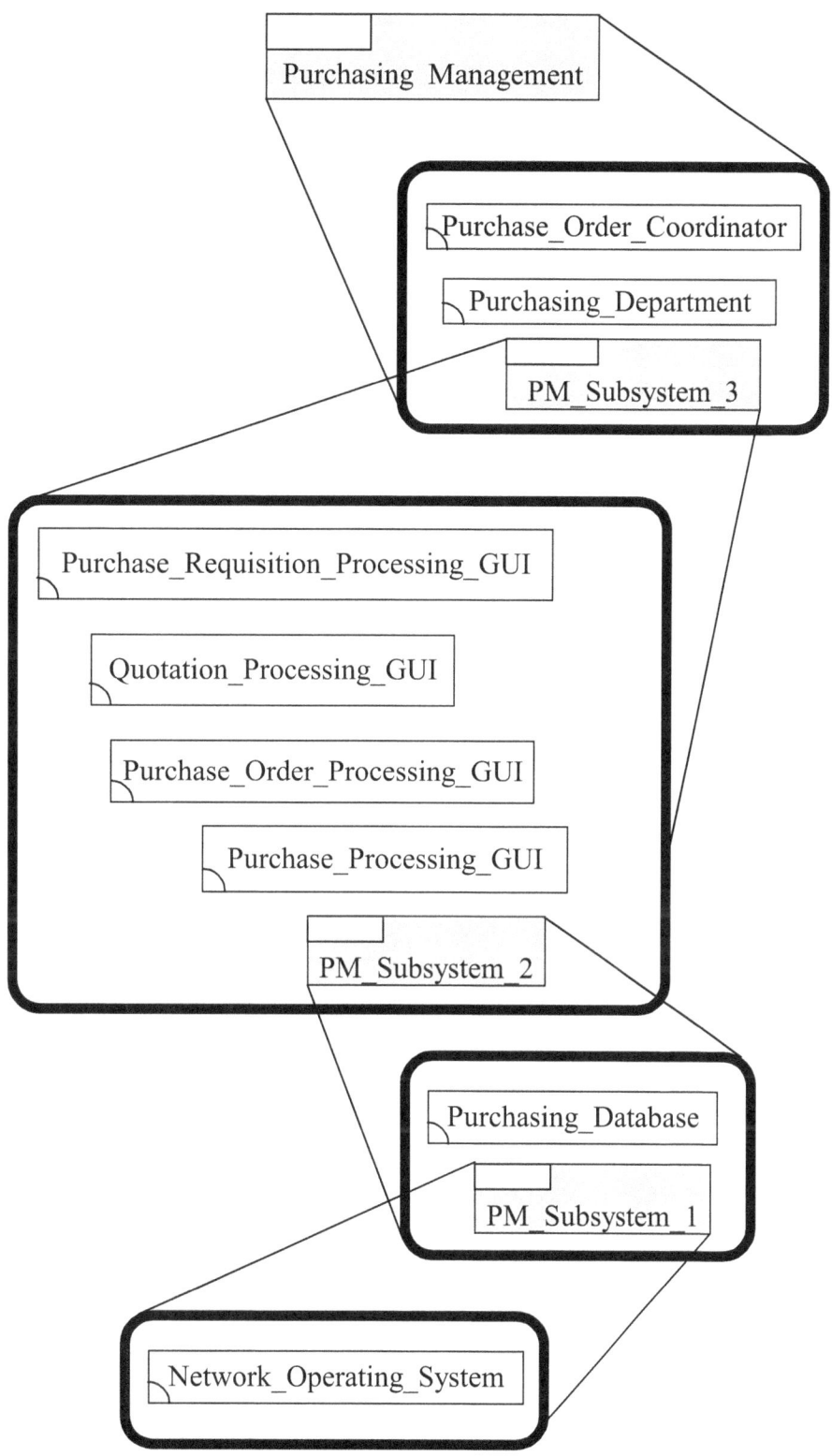

Figure 20-144   Implementation's AHD of the *Purchasing Management*
Systems Definition Version 2

Figure 20-145 shows the implementation's FD of the *purchasing management* systems definition version 2. In the figure, *Business_Layer* contains the *Purchase_Order_Coordinator* and *Purchasing_Department* components;

358

Application_Layer contains the *Purchase_Requisition_Processing_GUI*, *Quotation_Processing_GUI*, *Purchase_Order_Processing_GUI* and *Purchase_Processing_GUI* components; *Data_Layer* contains the *Purchasing_Database* component; *Technology_Layer* contains the *Network_Operating_System* component.

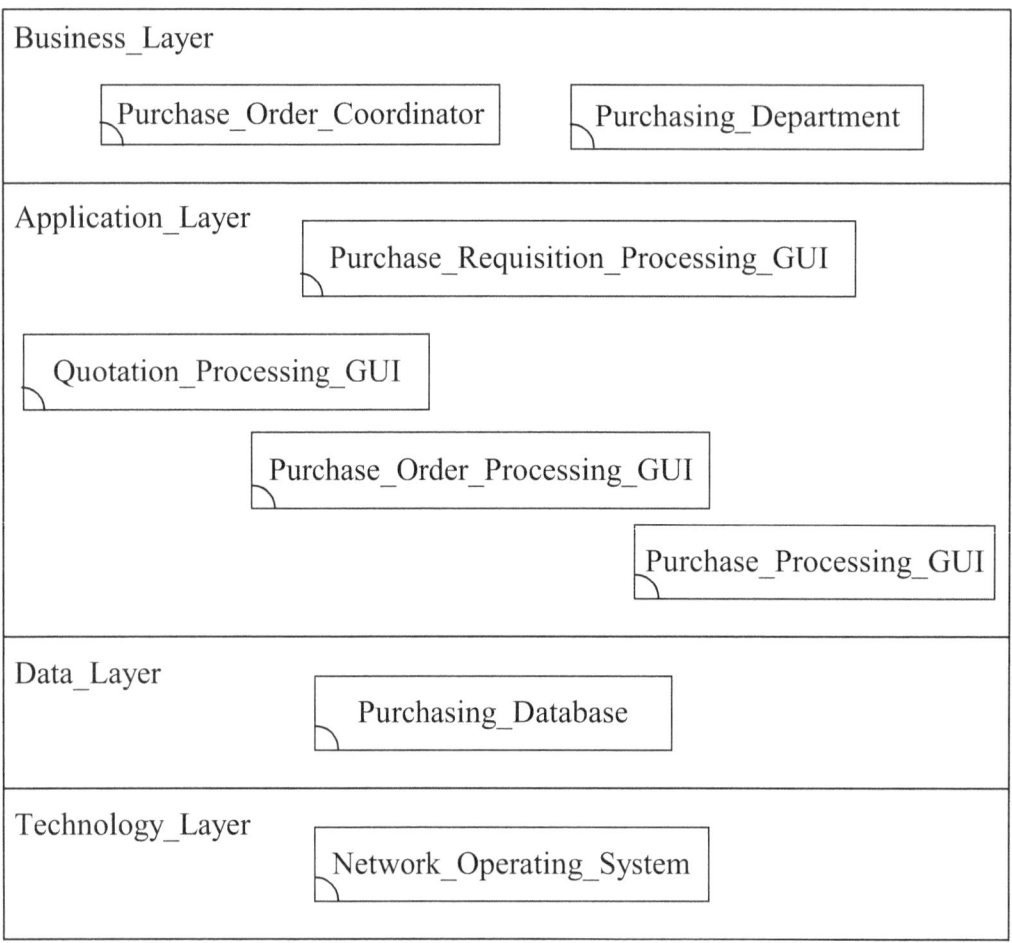

Figure 20-145   Implementation's FD of the *Purchasing Management Systems Definition Version 2*

Figure 20-146 shows the implementation's COD of the *purchasing management* systems definition version 2. In the figure, component *Purchase_Order_Coordinator* has one operation: *Purchase_Requisition_Verify*; component *Purchasing_Department* has three operations: *Quotation_Verify*, *Purchase_Order_Verify* and *Purchase_Verify*; component *Purchase_Requisition_Processing_GUI* has one operation: *Purchase_Requisition_Processing_Button_Click*; component *Quotation_Processing_GUI* has one operation: *Quotation_Processing_Button_Click*;

component *Purchase_Order_Processing_GUI* has one operation: *Purchase_Order_Processing_Button_Click*; component *Purchase_Processing_GUI* has one operation: *Purchase_Processing_Button_Click*; component *Purchasing_Database* has four operations: *SQL_Purchase_Requisition_Insert*, *SQL_Quotation_Insert*, *SQL_Purchase_Order_Insert* and *SQL_Purchase_Insert*; component *Network_Operating_System* has one operation: *Infrastructure_Resources_Share*.

360

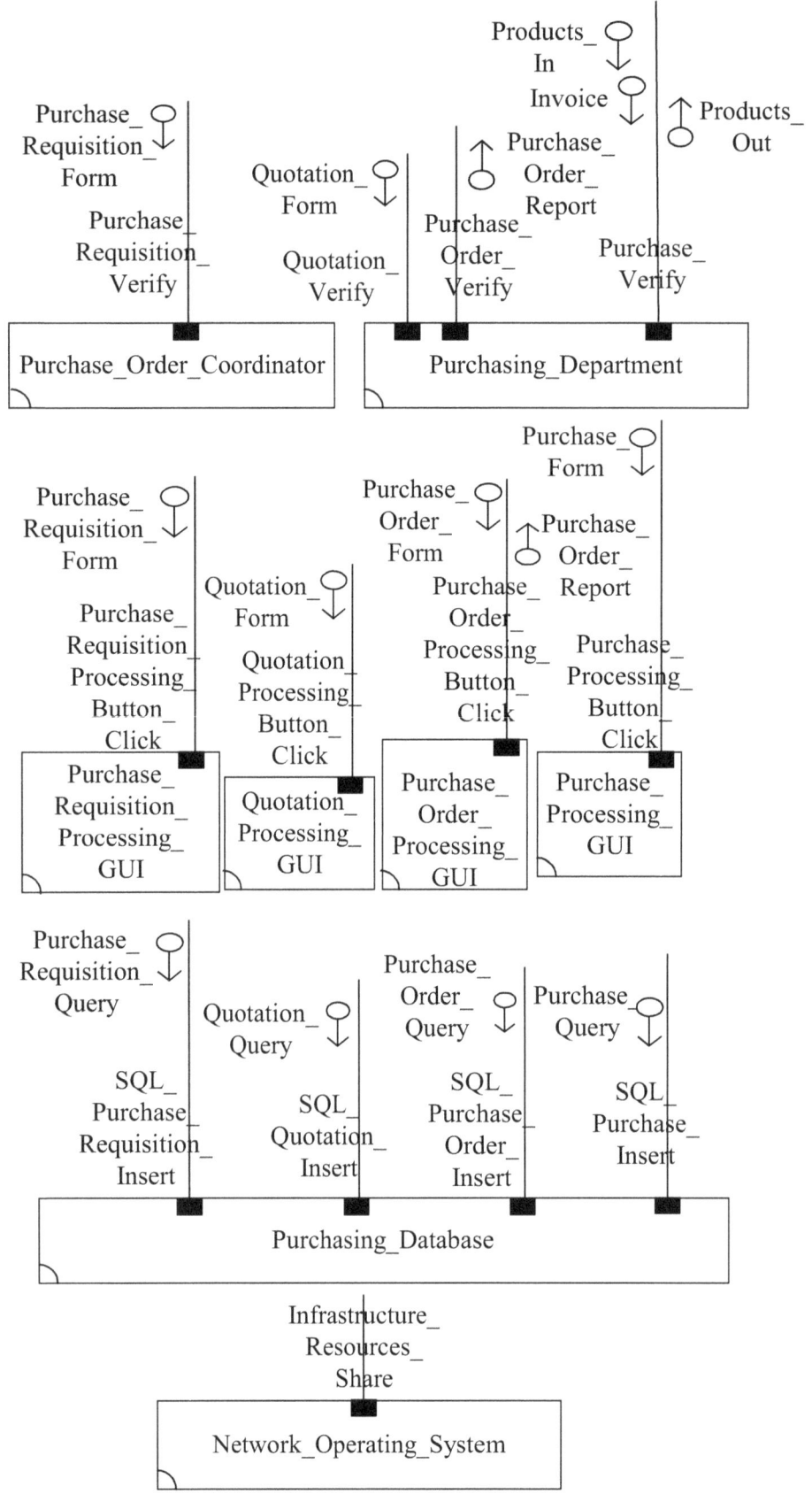

Figure 20-146   Implementation's COD of the *Purchasing Management Systems Definition Version 2*

The operation formula of *Purchase_Requisition_Verify* is *Purchase_Requisition_Verify(In Purchase_Requisition_Form)*. The operation formula of *Quotation_Verify* is *Quotation_Verify(In Quotation_Form)*. The operation formula of *Purchase_Order_Verify* is *Purchase_Order_Verify(Out Purchase_Order_Report)*. The operation formula of *Purchase_Verify* is *Purchase_Verify(In Products_In, Invoice; Out Products_Out)*. The operation formula of *Purchase_Requisition_Processing_Button_Click* is *Purchase_Requisition_Processing_Button_Click(In Purchase_Requisition_Form)*. The operation formula of *Quotation_Processing_Button_Click* is *Quotation_Processing_Button_Click(In Quotation_Form)*. The operation formula of *Purchase_Order_Processing_Button_Click* is *Purchase_Order_Processing_Button_Click(In Purchase_Order_Form; Out Purchase_Order_Report)*. The operation formula of *Purchase_Processing_Button_Click* is *Purchase_Processing_Button_Click(In Purchase_Form)*. The operation formula of *SQL_Purchase_Requisition_Insert* is *SQL_Purchase_Requisition_Insert(In Purchase_Requisition_Query)*. The operation formula of *SQL_Quotation_Insert* is *SQL_Quotation_Insert(In Quotation_Query)*. The operation formula of *SQL_Purchase_Order_Insert* is *SQL_Purchase_Order_Insert (In Purchase_Order_Query)*. The operation formula of *SQL_Purchase_Insert* is *SQL_Purchase_Insert (In Purchase_Query)*. The operation formula of *Infrastructure_Resources_Share* is *Infrastructure_Resources_Share*.

Figure 20-147 shows the composite data type specification of the *Purchase_Requisition_Form* input parameter occurring in the *Purchase_Requisition_Verify(In Purchase_Requisition_Form)* and *Purchase_Requisition_Processing_Button_Click(In Purchase_Requisition_Form)* operation formulas.

Parameter	*Purchase_Requisition_Form*

Data Type	TABLE of
	  Date : Text
	  OD : Text
	  ProductNo : Text
	  Quantity : Integer
	End TABLE ;

Instances	**Purchase Requisition Form** Date: 2011/10/17 Originating_Department : Sales Dept. ProductNo Quantity __A00001(Pen)_____300_____ __A00002(Mouse)_____400_____ __A00003(Camera)_____500_____

Figure 20-147   Composite Data Type Specification

Figure 20-148 shows the composite data type specification of the *Quotation_Form* input parameter occurring in the *Quotation_Verify(In Quotation_Form)* and *Quotation_Processing_Button_Click(In Quotation_Form)* operation formulas.

Parameter	*Quotation_Form*
Data Type	TABLE of Date : Text SupplierName: Text ProductNo : Text Quantity : Integer UnitPrice : Real Total : Real End TABLE ;
Instances	**Quotation Form** Date: 2011/10/25 SupplierName : Johnson Corp. ProductNo Quantity UnitPrice A 0 0 0 0 1 (P e n) 3 0 0 1 0 0 . 0 0 _A00002(Mouse)____400____200.00__ _A00003(Camera)___500____300.00__ Total : 260,000.00

Figure 20-148   Composite Data Type Specification

Figure 20-149 shows the composite data type specification of the *Purchase_Order_Report* output parameter occurring in the *Purchase_Order_Verify(Out Purchase_Order_Report)* and *Purchase_Order_Processing_Button_Click(In Purchase_Order_Form; Out Purchase_Order_Report)* operation formulas.

364

Parameter	*Purchase_Order_Report*
Data Type	TABLE of Date : Text SupplierName: Text ProductNo : Text Quantity : Integer End TABLE ;
Instances	Date : 20111118 SupplierName : Johnson Corp. | ProductNo	Quantity
A00001(Pen)	300
A00002(Mouse)	400
A00003(Camera)	500	

Figure 20-149   Composite Data Type Specification

Figure 20-150 shows the primitive data type specification of the *Products_In* input parameter occurring in the *Purchase_Verify(In Products_In, Invoice; Out Products_Out)* operation formula.

Parameter	Data Type	Instances
Products_In	Physical Object	Pen, Mouse, Camera

Figure 20-150   Primitive Data Type Specification

Figure 20-151 shows the composite data type specification of the *Invoice* input parameter occurring in the *Purchase_Verify(In Products_In, Invoice; Out Products_Out)* operation formula.

Parameter	*Invoice*

Data Type	TABLE of
	  Date : Text
	  SupplierName: Text
	  ProductNo : Text
	  Quantity : Integer
	  UnitPrice : Real
	  Total : Real
	End TABLE ;

Instances	Date : 20111130 SupplierName : Johnson Corp. | ProductNo	Quantity	UnitPrice
---	---	---
A00001(Pen)	300	100.00
A00002(Mouse)	400	200.00
A00003(Camera)	500	300.00	 Total : 260,000.00

Figure 20-151   Composite Data Type Specification

Figure 20-152 shows the primitive data type specification of the *Products_Out* output parameter occurring in the *Purchase_Verify(In Products_In, Invoice; Out Products_Out)* operation formula.

Parameter	Data Type	Instances
Products_Out	Physical Object	Pen, Mouse, Camera

Figure 20-152   Primitive Data Type Specification

Figure 20-153 shows the composite data type specification of the *Purchase_Order_Form* input parameter occurring in the *Purchase_Order_Processing_Button_Click(In Purchase_Order_Form; Out Purchase_Order_Report)* operation formula.

Parameter	*Purchase_Order_Form*
Data Type	TABLE of Date : Text SupplierName: Text ProductNo : Text Quantity : Integer End TABLE ;
Instances	**Purchase Order Form** Date: 2011/11/18 SupplierName : Johnson Corp. ProductNo Quantity __A00001(Pen) ____300 ___ __A00002(Mouse)____400___ __A00003(Camera)___500___

Figure 20-153   Composite Data Type Specification

Figure 20-154 shows the composite data type specification of the *Purchase_Form* input parameter occurring in the *Purchase_Processing_Button_Click(In Purchase_Form)* operation formula

Parameter	*Purchase_Form*

Data Type	TABLE of
	  Date : Text
	  SupplierName: Text
	  ProductNo : Text
	  Quantity : Integer
	  UnitPrice : Real
	  ReturnQuantity : Integer
	  Total : Real
	End TABLE ;

Instances	**Purchase Form** Date: 2011/12/12 SupplierName : Johnson Corp. ProductNo Quantity UnitPrice ReturnQuantity A00001(Pen) 300 100.00 0 A00002(Mouse) 390 200.00 10 A00003(Camera)500 300.00 0 Total : 258,000.00

Figure 20-154   Composite Data Type Specification

Figure 20-155 shows the composite data type specification of the *Quotation_Query* input parameter occurring in the *SQL_Quotation_Insert(In Quotation_Query)* operation formula.

Parameter	*Quotation_Query*

Data Type	TABLE of
	  Date : Text
	  SupplierName: Text
	  ProductNo : Text
	  Quantity : Integer
	  UnitPrice : Real
	  Total : Real
	End TABLE ;

Instances	

Date	SupplierName	Total
20111025	Johnson Corp.	260,000.00

ProductNo	Quantity	UnitPrice
A00001(Pen)	300	100.00
A00002(Mouse)	400	200.00
A00003(Camera)	500	300.00

Figure 20-155　　　Composite Data Type Specification

Figure 20-156 shows the composite data type specification of the *Purchase_Requisition_Query* input parameter occurring in the *SQL_Purchase_Requisition_Insert(In Purchase_Requisition_Query)* operation formula.

Parameter	*Purchase_Requisition_Query*

Data Type	TABLE of
	  Date : Text
	  OD : Text
	  ProductNo : Text
	  Quantity : Integer
	End TABLE ;

Instances	<table><tr><td>Date</td><td colspan="2">Originating_Department :</td></tr><tr><td>20111017</td><td colspan="2">Sales Dept.</td></tr><tr><td>ProductNo</td><td>Quantity</td></tr><tr><td>A00001(Pen)</td><td>300</td></tr><tr><td>A00002(Mouse)</td><td>400</td></tr><tr><td>A00003(Camera)</td><td>500</td></tr></table>

Figure 20-156   Composite Data Type Specification

Figure 20-157 shows the composite data type specification of the *Purchase_Order_Query* input parameter occurring in the *SQL_Purchase_Order_Insert(In Purchase_Order_Query)* operation formula.

Parameter	*Purchase_Order_Query*

Data Type	TABLE of
	  Date : Text
	  SupplierName: Text
	  ProductNo : Text
	  Quantity : Integer
	End TABLE ;

Instances	

Date	SupplierName
20111118	Johnson Corp.

ProductNo	Quantity
A00001(Pen)	300
A00002(Mouse)	400
A00003(Camera)	500

Figure 20-157   Composite Data Type Specification

Figure 20-158 shows the composite data type specification of the *Purchase_Query* input parameter occurring in the *SQL_Purchase_Insert (In Purchase_Query)* operation formula.

Parameter	*Purchase_Query*
Data Type	TABLE of 　Date : Text 　SupplierName: Text 　ProductNo : Text 　Quantity : Integer 　UnitPrice : Real 　ReturnQuantity : Integer 　Total : Real End TABLE ;
Instances	

Date	SupplierName	Total
20111212	Johnson Corp.	258,000.00

ProductNo	Quantity	UnitPrice	ReturnQuantity
A00001(Pen)	300	100.00	0
A00002(Mouse)	390	200.00	10
A00003(Camera)	500	300.00	0

Figure 20-158　　Composite Data Type Specification

Figure 20-159 shows the implementation's CCD of the *purchasing management* systems definition version 2. In the figure, actor *Originating_Department* has a connection with the *Purchase_Order_Coordinator* component; actor *Supplier* has three connections with the *Purchasing_Department* component; component *Purchase_Order_Coordinator* has a connection with the *Purchase_Requisition_Processing_GUI* component; component *Purchasing_Department* has a connection with each one of the *Quotation_Processing_GUI*, *Purchase_Order_Processing_GUI* and *Purchase_Processing_GUI* components; each one of the *Purchase_Requisition_Processing_GUI*, *Quotation_Processing_GUI*, *Purchase_Order_Processing_GUI* and *Purchase_Processing_GUI* components has a connection with the *Purchasing_Database*; component *Purchasing_Database* has a connection with the *Network_Operating_System* component.

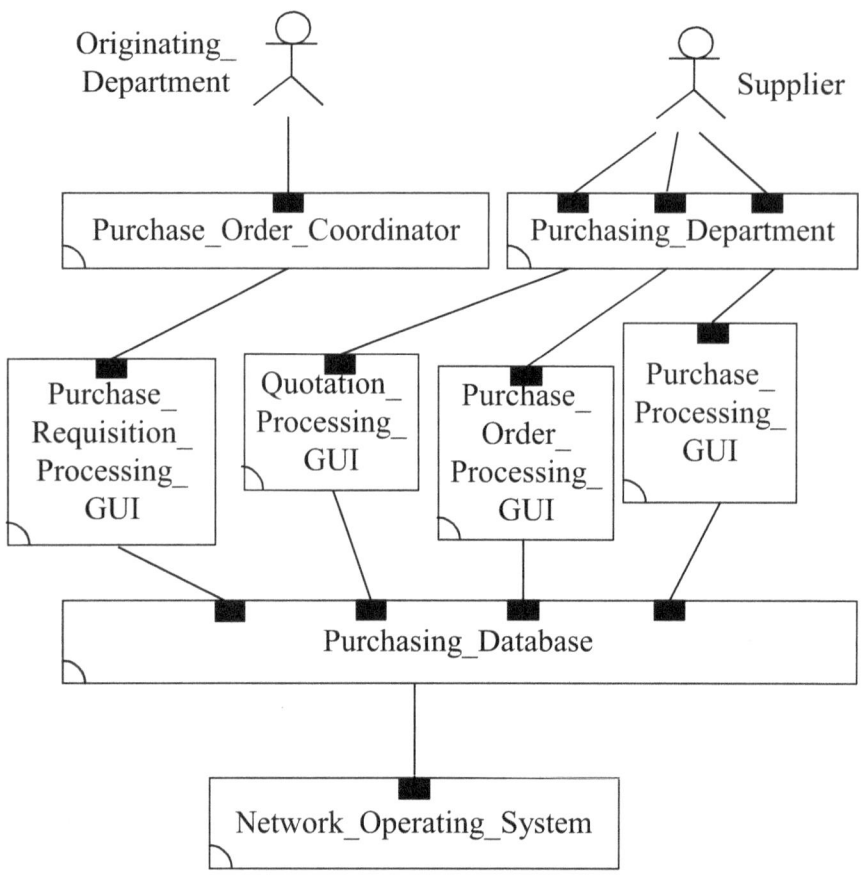

Figure 20-159   Implementation's CCD of the *Purchasing Management* Systems Definition Version 2

20-2-2-4-2 Implementation's Systems Behavior of the Purchasing Management Systems Definition Version 2

The entire implementation's systems behavior of the *purchasing management* systems definition version 2 includes: a) *Implementation's SBCD* and b) *Implementation's IFD* of the *purchasing management* systems definition version 2.

Figure 20-160 shows the implementation's SBCD of the *purchasing management* systems definition version 2 in which interactions among the *Originating_Department, Supplier* actors and the *Purchase_Order_Coordinator, Purchasing_Department, Purchase_Requisition_Processing_GUI, Quotation_Processing_GUI, Purchase_Order_Processing_GUI, Purchase_Processing_GUI, Purchasing_Database, Network_Operating_System* components shall draw forth the *Purchase_Requisition*, *Quotation*, *Purchase_Order* and *Purchase* behaviors.

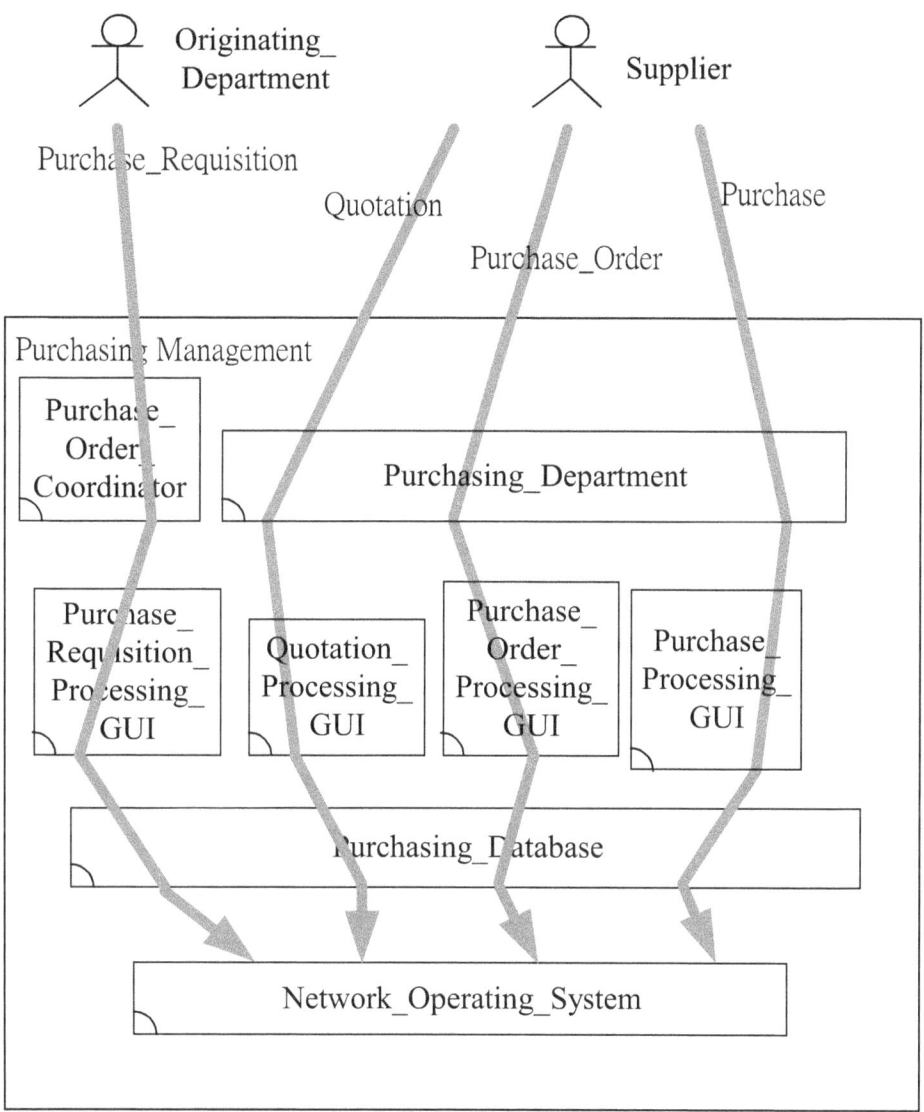

Figure 20-160   Implementation's SBCD of the *Purchasing Management*
Systems Definition Version 2

The overall implementation's behavior of the *purchasing management* systems definition version 2 includes the *Purchase_Requisition*, *Quotation*, *Purchase_Order and Purchase* behaviors. In other words, the *Purchase_Requisition*, *Quotation*, *Purchase_Order and Purchase* behaviors together provide the overall implementation's behavior of the *purchasing management* systems definition version 2.

Be noticed that the *Purchase_Requisition, Quotation, Purchase_Order and Purchase* behaviors are mutually independent of each other. They tend to be executed concurrently [Hoar85, Miln89, Miln99].

The overall implementation's behavior of the *purchasing management* systems definition version 2 includes four individual behaviors: *Purchase_Requisition, Quotation, Purchase_Order and Purchase*. Each individual behavior is represented by an execution path. We use an interaction flow diagram (IFD) to define each one of these execution paths. Figure 20-161 shows the implementation's IFD of the *purchasing management* systems definition version 2 *Purchase_Requisition* behavior. First, actor *Originating_Department* interacts with the *Purchase_Order_Coordinator* component through the *Purchase_Requisition_Verify* operation call interaction, carrying the *Purchase_Requisition_Form* input parameter. Next, component *Purchase_Order_Coordinator* interacts with the *Purchase_Requisition_Processing_GUI* component through the *Purchase_Requisition_Processing_Button_Click* operation call interaction, carrying the *Purchase_Requisition_Form* input parameter. Continuingly, component *Purchase_Requisition_Processing_GUI* interacts with the *Purchasing_Database* component through the *SQL_Purchase_Requisition_Insert* operation call interaction, carrying the *Purchase_Requisition_Query* input parameter. Finally, component *Purchasing_Database* interacts with the *Network_Operating_System* component through the *Infrastructure_Resources_Share* operation call interaction.

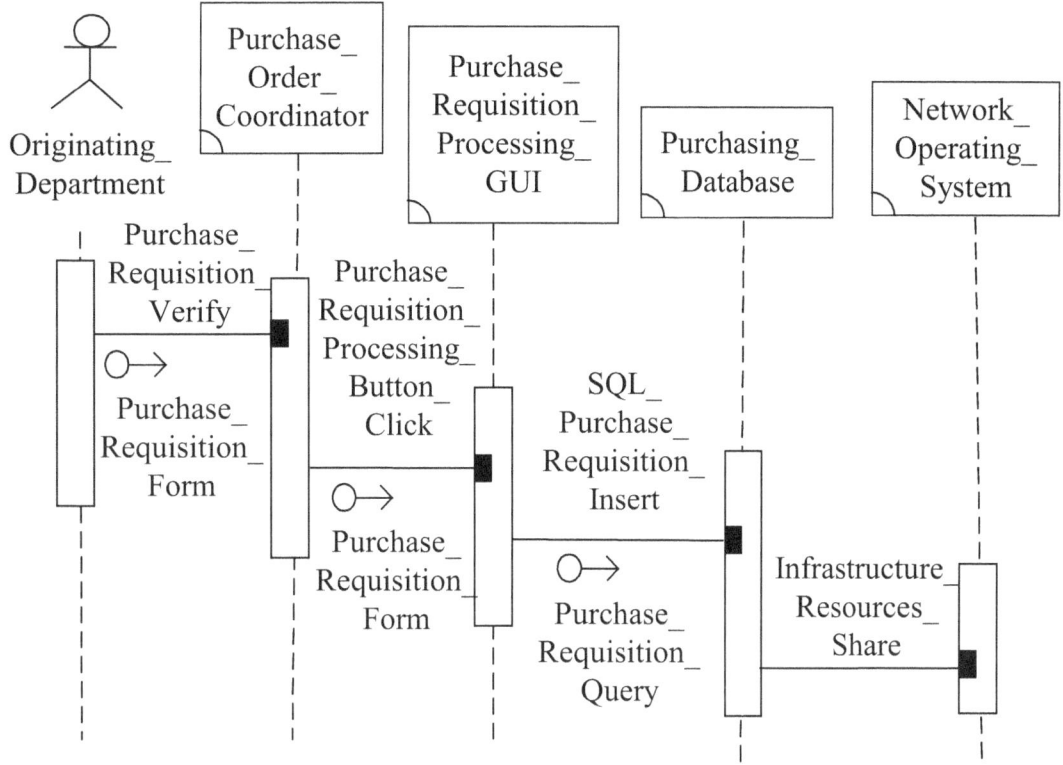

Figure 20-161   Implementation's IFD of the *Purchasing Management Systems Definition Version 2 Purchase_Requisition* Behavior

Figure 20-162 shows the implementation's IFD of the *purchasing management* systems definition version 2 *Quotation* behavior. First, actor *Supplier* interacts with the *Purchasing_Department* component through the *Quotation_Verify* operation call interaction, carrying the *Quotation_Form* input parameter. Next, component *Purchasing_Department* interacts with the *Quotation_Processing_GUI* component through the *Quotation_Processing_Button_Click* operation call interaction, carrying the *Quotation_Form* input parameter. Continuingly, component *Quotation_Processing_GUI* interacts with the *Purchasing_Database* component through the *SQL_Quotation_Insert* operation call interaction, carrying the *Quotation_Query* input parameter. Finally, component *Purchasing_Database* interacts with the *Network_Operating_System* component through the *Infrastructure_Resources_Share* operation call interaction.

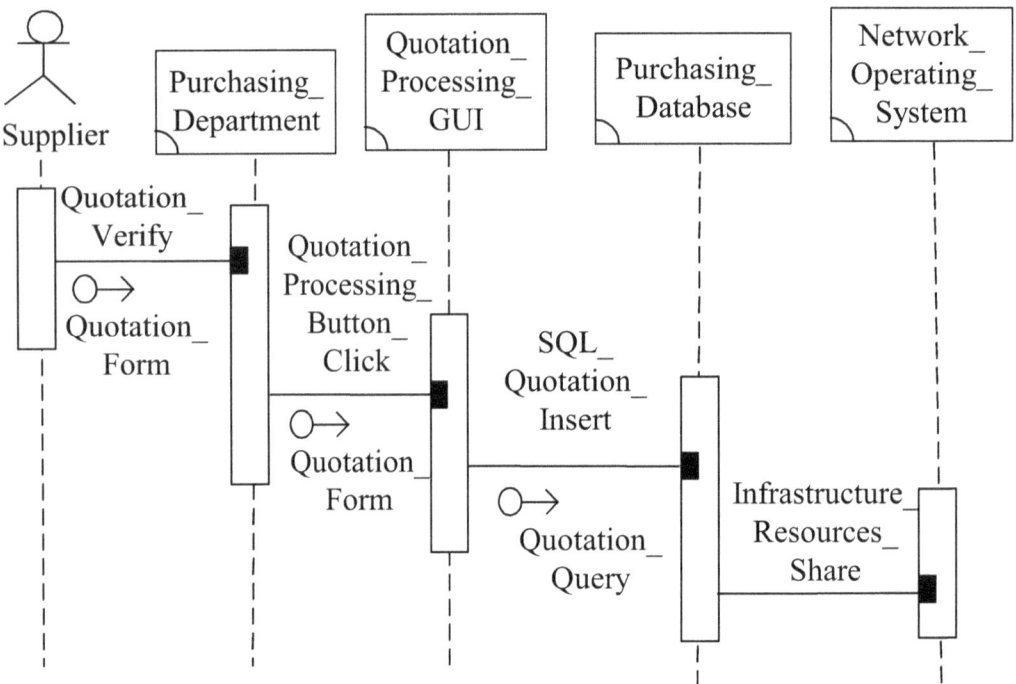

Figure 20-162   Implementation's IFD of the *Purchasing Management
Systems Definition Version 2 Quotation* Behavior

Figure 20-163 shows the implementation's IFD of the *purchasing management* systems definition version 2 *Purchase_Order* behavior. First, actor *Supplier* interacts with the *Purchasing_Department* component through the *Purchase_Order_Verify* operation call interaction. Next, component *Purchasing_Department* interacts with the *Purchase_Order_Processing_GUI* component through the *Purchase_Order_Processing_Button_Click* operation call interaction, carrying the *Purchase_Order_Form* input parameter. Continuingly, component *Purchase_Order_Processing_GUI* interacts with the *Purchasing_Database* component through the *SQL_Purchase_Order_Insert* operation call interaction, carrying the *Purchase_Order_Query* input parameter. Continuingly, component *Purchasing_Database* interacts with the *Network_Operating_System* component through the *Infrastructure_Resources_Share* operation call interaction. Continuingly, component *Purchasing_Department* interacts with the *Purchase_Order_Processing_GUI* component through the *Purchase_Order_Processing_Button_Click* operation return interaction, carrying the *Purchase_Order_Report* output parameter. Finally, actor *Supplier* interacts with the *Purchasing_Department* component through the *Purchase_Order_Verify* operation return interaction, carrying the *Purchase_Order_Report* output parameter.

377

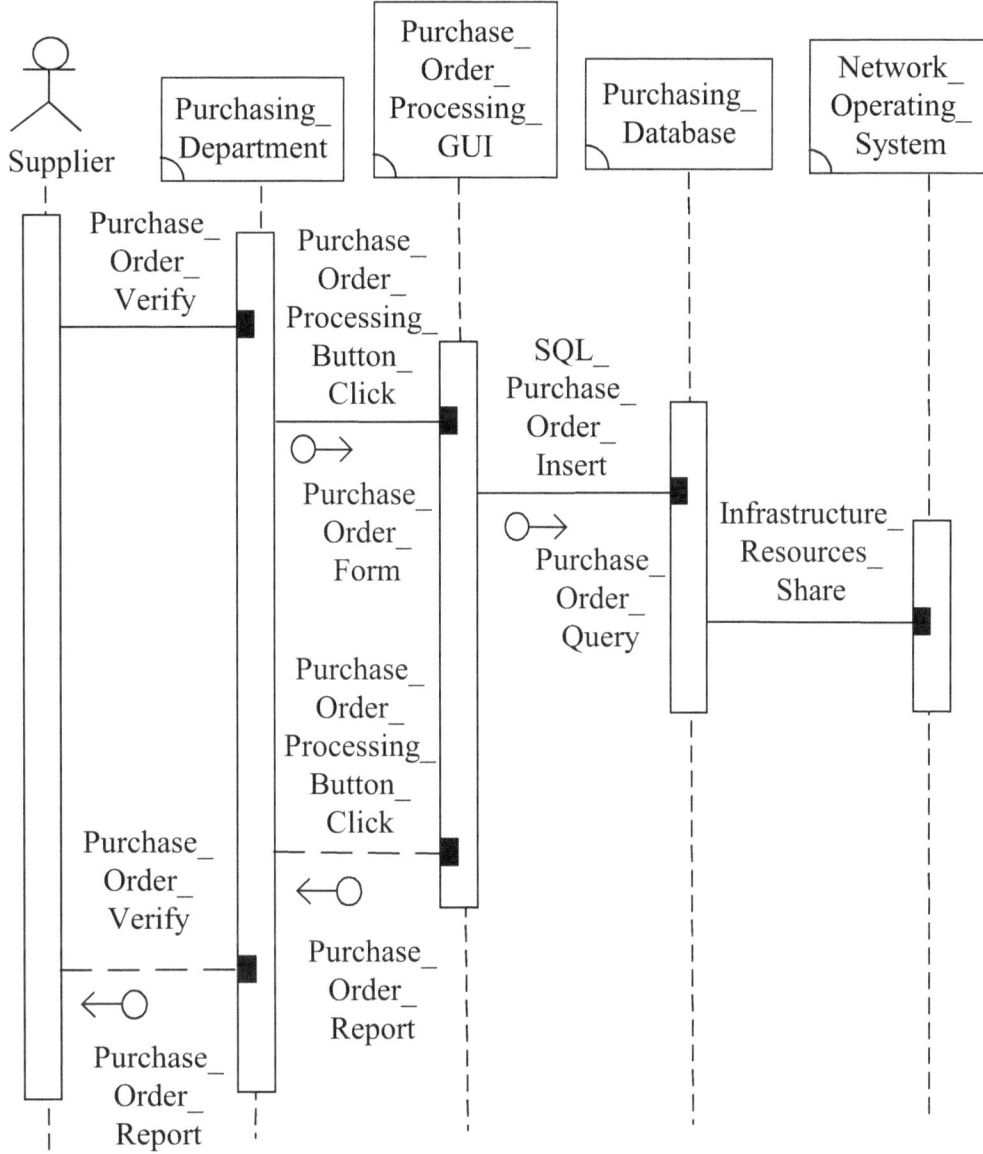

Figure 20-163   Implementation's IFD of the *Purchasing Management Systems* Definition Version 2 *Purchase_Order* Behavior

Figure 20-164 shows the implementation's IFD of the *purchasing management* systems definition version 2 *Purchase* behavior. First, actor *Supplier* interacts with the *Purchasing_Department* component through the *Purchase_Verify* operation call interaction, carrying the *Products_In* and *Invoice* input parameters. Next, component *Purchasing_Department* interacts with the *Purchase_Processing_GUI* component through the *Purchase_Processing_Button_Click* operation call interaction, carrying the *Purchase_Form* input parameter. Continuingly, component *Purchase_Processing_GUI* interacts with the *Purchasing_Database* component through the *SQL_Purchase_Insert* operation call interaction, carrying the

Purchase_Query input parameter. Continuingly, component *Purchasing_Database* interacts with the *Network_Operating_System* component through the *Infrastructure_Resources_Share* operation call interaction. Finally, actor *Supplier* interacts with the *Purchasing_Department* component through the *Purchase_Verify* operation return interaction, carrying the *Products_Out* output parameter.

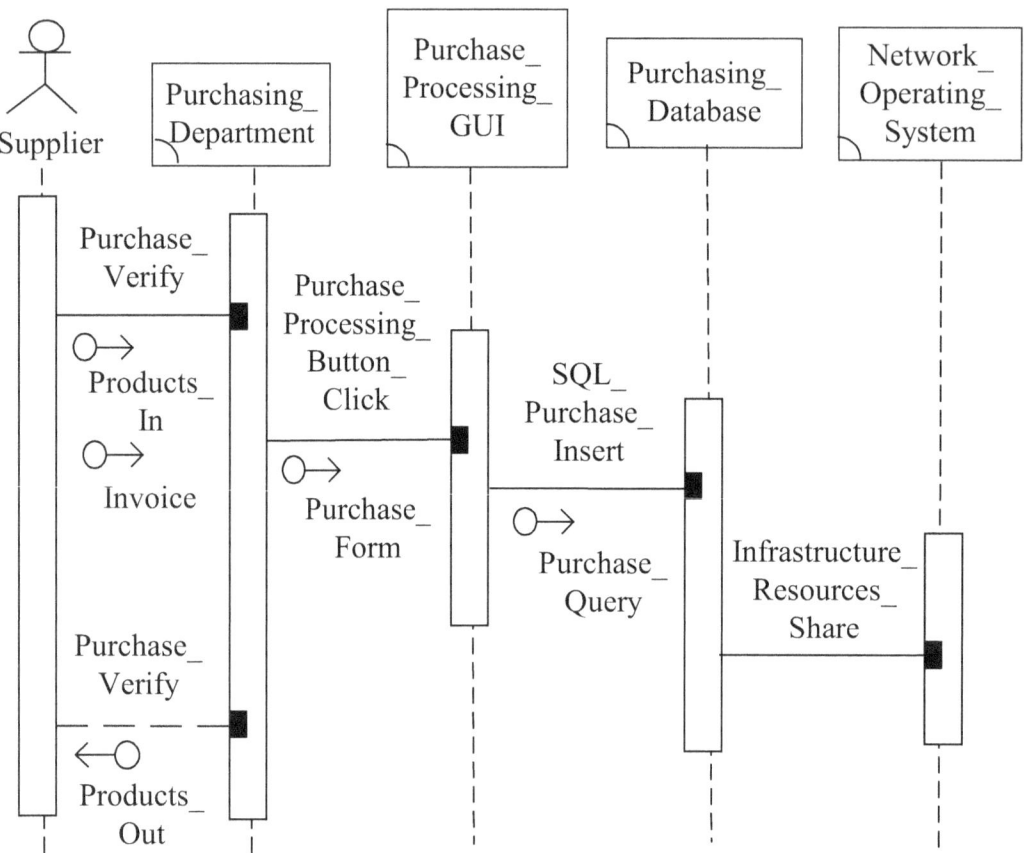

Figure 20-164   Implementation's IFD of the *Purchasing Management* Systems Definition Version 2 *Purchase* Behavior

20-3 Purchasing Management Systems Definition Strategy/Version 3

The *purchasing management* systems motivation model shown in Figure 20-165, being a higher-order system, has the *purchasing management* systems definition strategy 3 as its input and the *purchasing management* systems definition version 3 as its output.

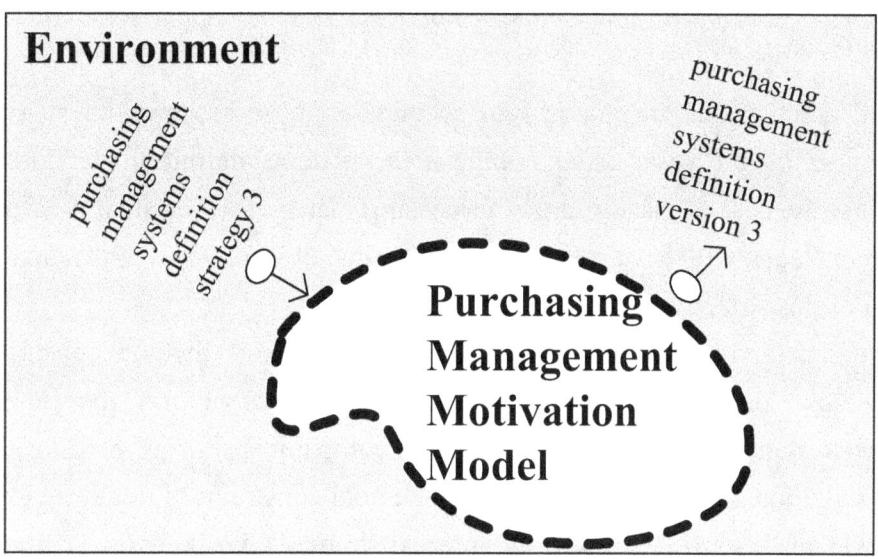

Figure 20-165   *Purchasing Management* Motivation Model is a Higher-Order System

The *purchasing management* systems definition strategy 3 mapped to the *purchasing management* systems definition version 3 can be represented as an ordered pair, as shown in Figure 20-166.

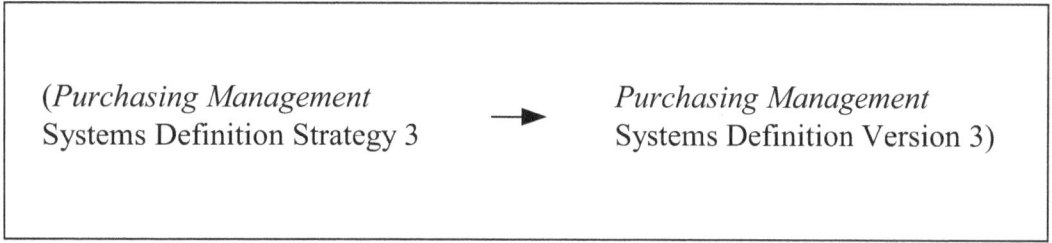

Figure 20-166   An Ordered Pair

20-3-1 Purchasing Management Systems Definition Strategy 3

Here, we use (a) goal drivers, (b) goal assumptions, (c) goal constraints and (d) SWOT analysis, to illustrate the strategic means of the "*purchasing management* systems definition strategy 3".

Goal drivers are up from the policy considerations, the goal driver is kind of why we want to have this *purchasing management* systems definition version 3. The goal drivers of *purchasing management* systems definition version 3 are: currently, a *purchasing management* system with only the *Purchase_Requisition, Quotation, Purchase_Order and Purchase* behaviors is losing its appeal to managers; a *purchasing management* system with the *Purchase_Requisition, Quotation, Purchase_Order, Purchase* and *Collect_Supplier_Data* behaviors will attract more managers.

Goal assumptions are taking into account of those assumptions that have a positive impact on the *purchasing management* systems definition version 3. We assume that if the cost of installing the purchasing management system is affordable, then every manager will have a great desire to install one. This is the major goal assumption of this strategy.

Goal constraints are up from the policy considerations, the goal constraints are related to those restrictions which have a negative impact on the *purchasing management* systems definition version 3. If the company is short of fund to invest in this new installation, then this would become the goal constraint of this strategy.

SWOT analysis is to analyze the internal strengths, weaknesses, opportunities and threats, and so for executing this *purchasing management* systems definition strategy 3. Being a big company, it should be trivial for the company to install this purchasing management system with the *Purchase_Requisition, Quotation, Purchase_Order, Purchase* and *Collect_Supplier_Data* behaviors. This is the internal strength of this company. However, kind of bulky makes the company may not react fast enough to carry out this new installation. This is the internal weakness of this company.

20-3-2 Purchasing Management Systems Definition Version 3

Using the SBC multi-level (hierarchical) view, an architect goes through: a) concept view, b) analysis view, c) design view and d) implementation view for the *purchasing management* systems definition version 3 as shown in Figure 20-167.

381

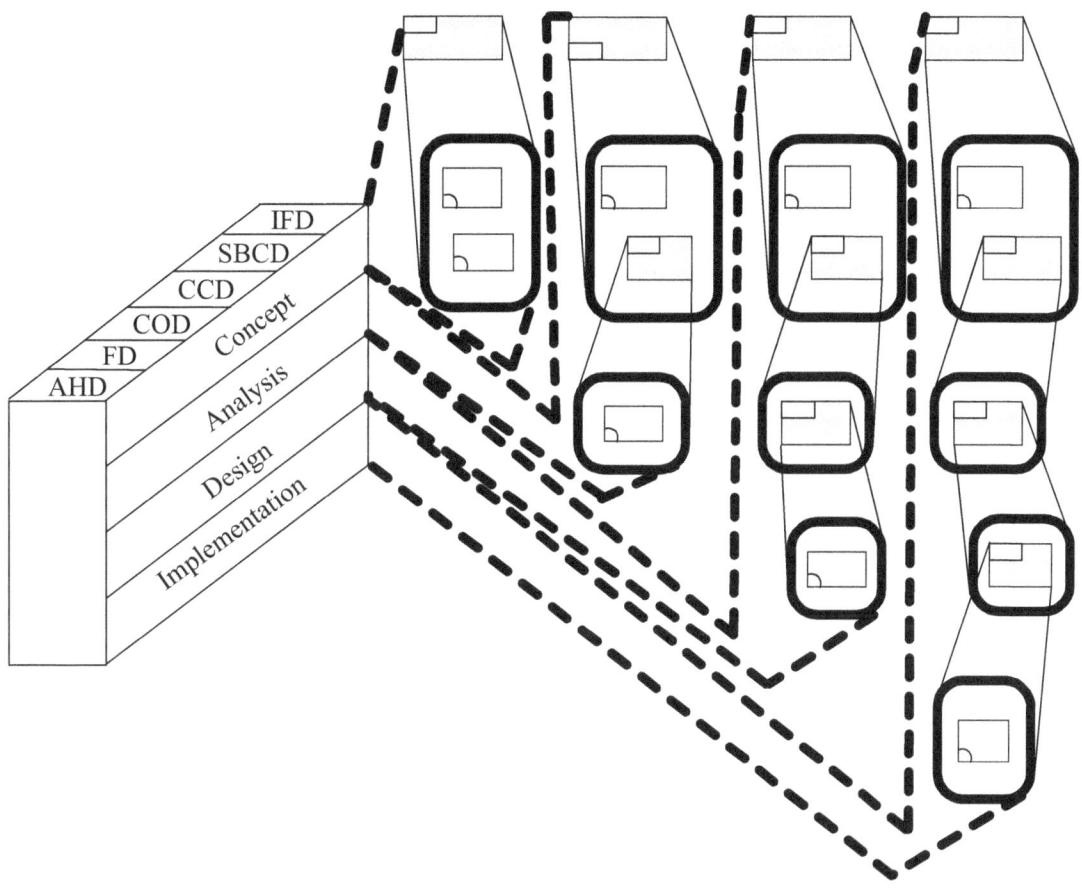

Figure 20-167   *Purchasing Management* Systems Definition Version 3

20-3-2-1 Concept View of the Purchasing Management Systems Definition Version 3

The concept view of the *purchasing management* systems definition version 3 is shown in Figure 20-168.

382

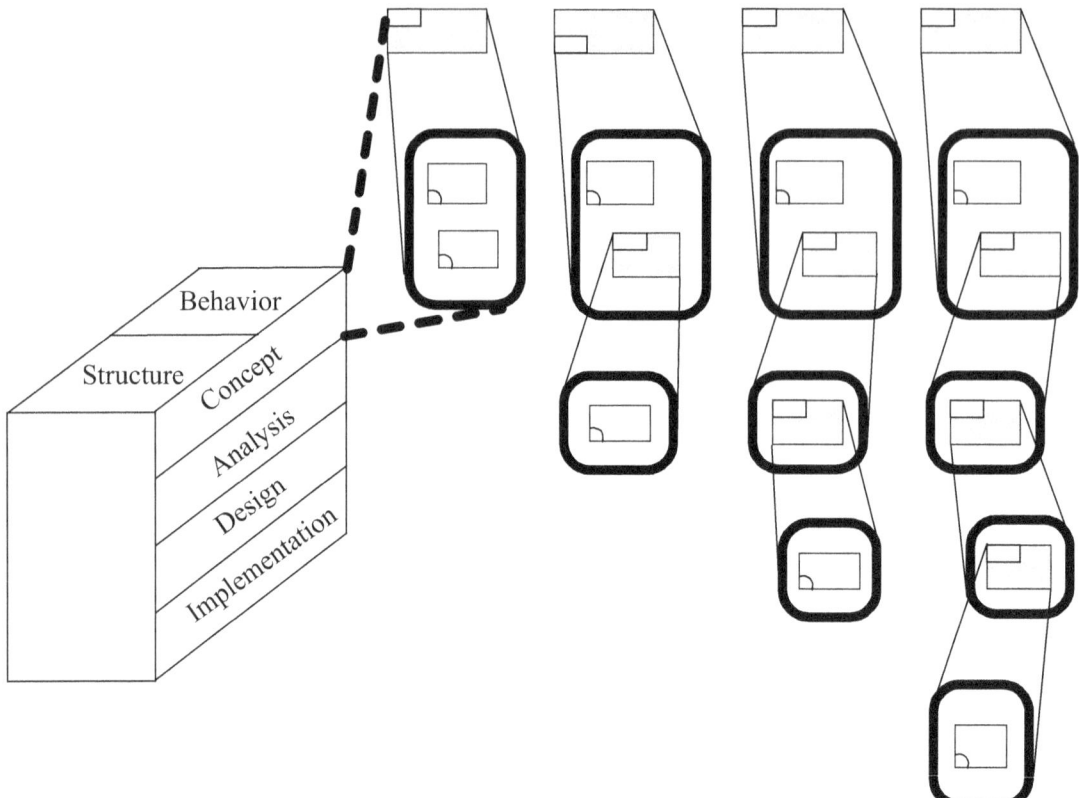

Figure 20-168   Concept View of the *Purchasing Management* Systems Definition Version 3

The concept view of the *purchasing management* systems definition version 3 consists of: a) concept's systems structure of the *purchasing management* systems definition version 3 and b) concept's systems behavior of the *purchasing management* systems definition version 3.

20-3-2-1-1 Concept's Systems Structure of the Purchasing Management Systems Definition Version 3

The entire concept's systems structure of the *purchasing management* systems definition version 3 includes: a) *Concept's AHD*, b) *Concept's FD*, c) *Concept's COD* and d) *Concept's CCD* of the *purchasing management* systems definition version 3.

We first draw the concept's AHD of the *purchasing management* systems definition version 3. As shown in Figure 20-169, *Purchasing Management* is composed of *Purchase_Order_Coordinator*, *Purchasing_Department* and *PM_Subsystem_3*. In the figure, *Purchasing Management* is an aggregated system while *Purchase_Order_Coordinator*, *Purchasing_Department* and *PM_Subsystem_3* are non-aggregated systems.

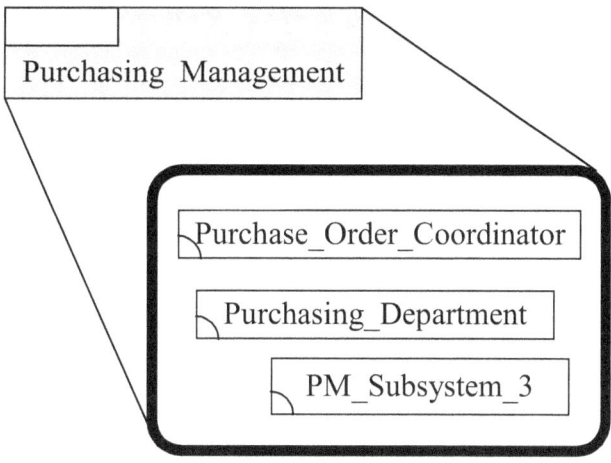

Figure 20-169   Concept's AHD of the *Purchasing Management*
Systems Definition Version 3

Figure 20-170 shows the concept's FD of the *purchasing management* systems definition version 3. In the figure, *Business_Layer* contains the *Purchase_Order_Coordinator*, *Purchasing_Department* and *PM_Subsystem_3* components.

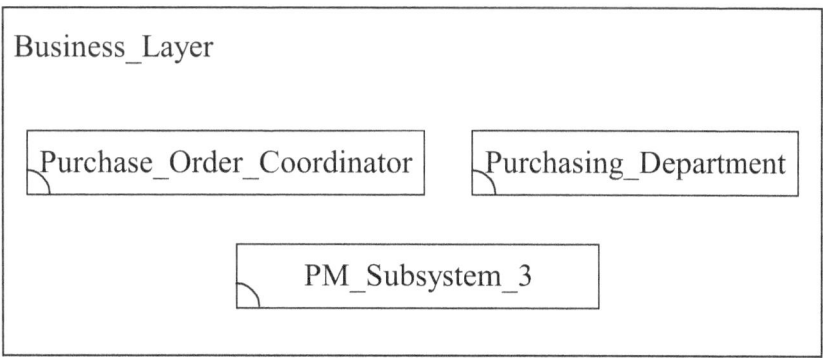

Figure 20-170   Concept's FD of the *Purchasing Management*
Systems Definition Version 3

Figure 20-171 shows the concept's COD of the *purchasing management* systems definition version 3. In the figure, component *Purchase_Order_Coordinator* has one operation: *Purchase_Requisition_Verify*; component *Purchasing_Department* has four operations: *Quotation_Verify*, *Purchase_Order_Verify*, *Purchase_Verify* and *Interview*; component *PM_Subsystem_3* has five operations: *Purchase_Requisition_Processing_Button_Click*,

384

Quotation_Processing_Button_Click,   *Purchase_Order_Processing_Button_Click,*
Purchase_Processing_Button_Click and *Supplier_Data_Processing_Button_Click.*

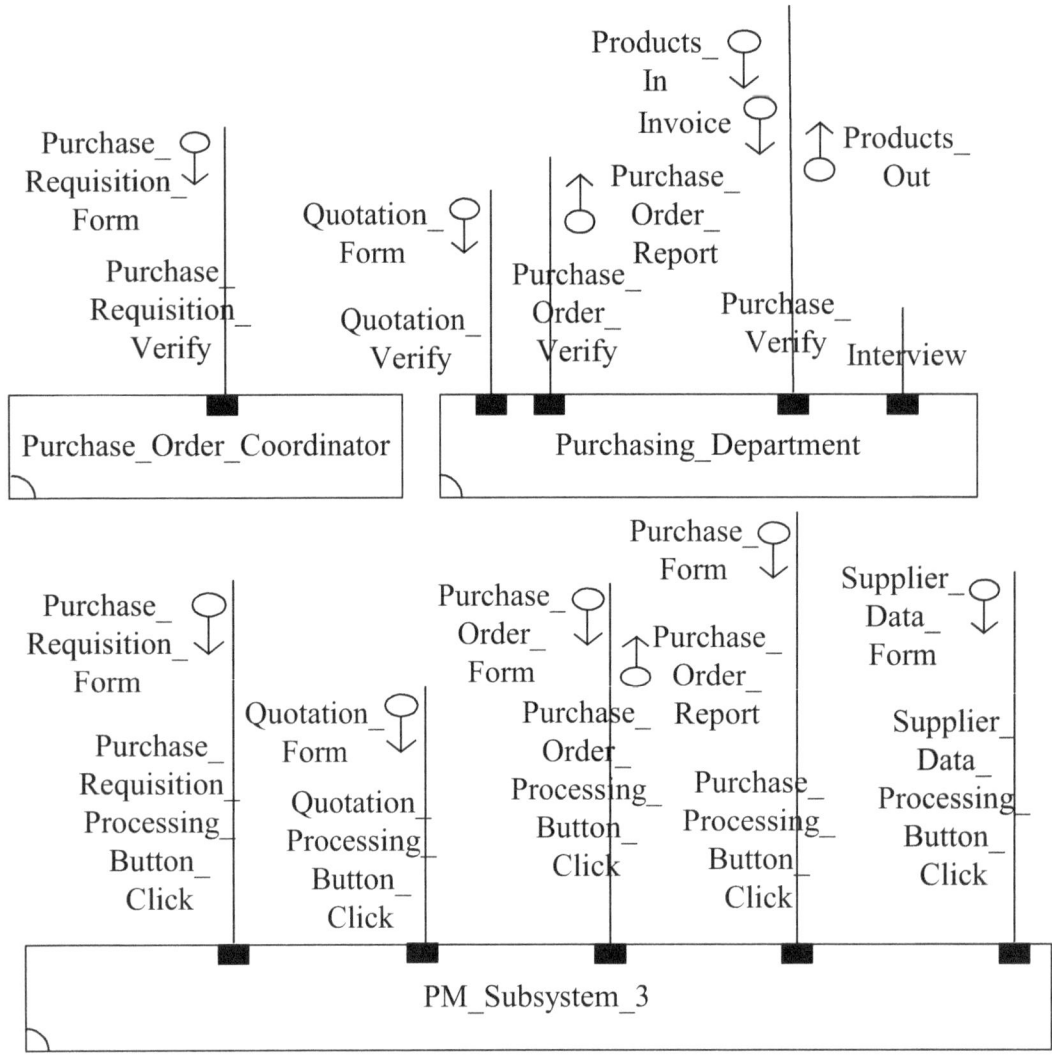

Figure 20-171   Concept's COD of the *Purchasing Management*
Systems Definition Version 3

The   operation   formula   of   *Purchase_Requisition_Verify*   is
Purchase_Requisition_Verify(In Purchase_Requisition_Form). The operation formula
of *Quotation_Verify* is *Quotation_Verify(In Quotation_Form).* The operation formula
of *Purchase_Order_Verify* is *Purchase_Order_Verify(Out Purchase_Order_Report).*
The operation formula of *Purchase_Verify* is *Purchase_Verify(In Products_In,
Invoice; Out Products_Out).* The operation formula of *Interview* is *Interview.* The
operation   formula   of   *Purchase_Requisition_Processing_Button_Click*   is
Purchase_Requisition_Processing_Button_Click(In   Purchase_Requisition_Form).

The operation formula of *Quotation_Processing_Button_Click* is *Quotation_Processing_Button_Click(In Quotation_Form)*. The operation formula of *Purchase_Order_Processing_Button_Click* is *Purchase_Order_Processing_Button_Click(In Purchase_Order_Form; Out Purchase_Order_Report)*. The operation formula of *Purchase_Processing_Button_Click* is *Purchase_Processing_Button_Click(In Purchase_Form)*. The operation formula of *Supplier_Data_Processing_Button_Click* is *Supplier_Data_Processing_Button_Click(In Supplier_Data_Form)*.

Figure 20-172 shows the composite data type specification of the *Purchase_Requisition_Form* input parameter occurring in the *Purchase_Requisition_Verify(In Purchase_Requisition_Form)* and *Purchase_Requisition_Processing_Button_Click(In Purchase_Requisition_Form)* operation formulas.

Parameter	*Purchase_Requisition_Form*
Data Type	TABLE of Date : Text OD : Text ProductNo : Text Quantity : Integer End TABLE ;
Instances	**Purchase Requisition Form** Date: 2011/10/17 Originating_Department : Sales Dept. ProductNo Quantity __A00001(Pen)_____300_____ __A00002(Mouse)_____400_____ __A00003(Camera)_____500_____

Figure 20-172   Composite Data Type Specification

386

Figure 20-173 shows the composite data type specification of the *Quotation_Form* input parameter occurring in the *Quotation_Verify(In Quotation_Form)* and *Quotation_Processing_Button_Click(In Quotation_Form)* operation formulas.

Parameter	*Quotation_Form*
Data Type	TABLE of Date : Text SupplierName: Text ProductNo : Text Quantity : Integer UnitPrice : Real Total : Real End TABLE ;
Instances	**Quotation Form** Date: 2011/10/25 SupplierName : Johnson Corp. ProductNo Quantity UnitPrice A00001(Pen) 300 100.00 _A00002(Mouse)____400_____200.00__ _A00003(Camera)___500_____300.00__ Total : 260,000.00

Figure 20-173   Composite Data Type Specification

Figure 20-174 shows the composite data type specification of the *Purchase_Order_Report* output parameter occurring in the *Purchase_Order_Verify(Out Purchase_Order_Report)* and *Purchase_Order_Processing_Button_Click(In Purchase_Order_Form; Out Purchase_Order_Report)* operation formulas.

Parameter	*Purchase_Order_Report*
Data Type	TABLE of Date : Text SupplierName: Text ProductNo : Text Quantity : Integer End TABLE ;
Instances	Date : 20111118 SupplierName : Johnson Corp. <table><tr><th>ProductNo</th><th>Quantity</th></tr><tr><td>A00001(Pen)</td><td>300</td></tr><tr><td>A00002(Mouse)</td><td>400</td></tr><tr><td>A00003(Camera)</td><td>500</td></tr></table>

Figure 20-174   Composite Data Type Specification

Figure 20-175 shows the primitive data type specification of the *Products_In* input parameter occurring in the *Purchase_Verify(In Products_In, Invoice; Out Products_Out)* operation formula.

Parameter	Data Type	Instances
Products_In	Physical Object	Pen, Mouse, Camera

Figure 20-175   Primitive Data Type Specification

Figure 20-176 shows the composite data type specification of the *Invoice* input parameter occurring in the *Purchase_Verify(In Products_In, Invoice; Out Products_Out)* operation formula.

388

Parameter	*Invoice*

Data Type	TABLE of
	  Date : Text
	  SupplierName: Text
	  ProductNo : Text
	  Quantity : Integer
	  UnitPrice : Real
	  Total : Real
	End TABLE ;

Instances	Date : 20111130 SupplierName : Johnson Corp. | ProductNo	Quantity	UnitPrice
---	---	---
A00001(Pen)	300	100.00
A00002(Mouse)	400	200.00
A00003(Camera)	500	300.00	 Total : 260,000.00

Figure 20-176   Composite Data Type Specification

Figure 20-177 shows the primitive data type specification of the *Products_Out* output parameter occurring in the *Purchase_Verify(In Products_In, Invoice; Out Products_Out)* operation formula.

Parameter	Data Type	Instances
Products_Out	Physical Object	Pen, Mouse, Camera

Figure 20-177   Primitive Data Type Specification

Figure 20-178 shows the composite data type specification of the *Purchase_Order_Form* input parameter occurring in the *Purchase_Order_Processing_Button_Click(In Purchase_Order_Form; Out Purchase_Order_Report)* operation formula.

Parameter	*Purchase_Order_Form*

Data Type	TABLE of
	  Date : Text
	  SupplierName: Text
	  ProductNo : Text
	  Quantity : Integer
	End TABLE ;

Instances	**Purchase Order Form** Date: 2011/11/18 SupplierName : Johnson Corp. ProductNo Quantity __A00001(Pen) ____300 ___ __A00002(Mouse)___400__ __A00003(Camera)___500__

Figure 20-178   Composite Data Type Specification

Figure 20-179 shows the composite data type specification of the *Purchase_Form* input parameter occurring in the *Purchase_Processing_Button_Click(In Purchase_Form)* operation formula.

Parameter	*Purchase_Form*

Data Type	TABLE of
	  Date : Text
	  SupplierName: Text
	  ProductNo : Text
	  Quantity : Integer
	  UnitPrice : Real
	  ReturnQuantity : Integer
	  Total : Real
	End TABLE ;

Instances	**Purchase Form** Date: 2011/12/12 SupplierName : Johnson Corp. ProductNo Quantity UnitPrice ReturnQuantity A00001(Pen) 300 100.00 0 A00002(Mouse) 390 200.00 10 A00003(Camera)500 300.00 0 Total : 258,000.00

Figure 20-179   Composite Data Type Specification

Figure 20-180 shows the composite data type specification of the *Supplier_Data_Form* input parameter occurring in the *Supplier_Data_Processing_Button_Click(In Supplier_Data_Form)* operation formula.

Parameter	*Supplier_Data_Form*

Data Type	TABLE of
	  SupplierName:Text
	  Address :Text
	  PhoneNumber:Text
	  FaxNumber:Text
	  E-mail : Text
	  Rank : Text
	End TABLE ;

Instances	**Supplier Data Form** SupplierName : Johnson Corp. Address : 1232 Fair Circle, Austin, TX PhoneNumber : 512-463-8472 FaxNumber : 512-463-8499 E-mail : Johnson1122@gmail.com Rank : B

Figure 20-180   Composite Data Type Specification

Figure 20-181 shows the concept's CCD of the *purchasing management* systems definition version 3. In the figure, actor *Originating_Department* has one connection with the *Purchase_Order_Coordinator* component; actor *Supplier* has four connections with the *Purchasing_Department* component; component *Purchase_Order_Coordinator* has one connection with the *PM_Subsystem_3* component; component *Purchasing_Department* has four connections with the *PM_Subsystem_3* component.

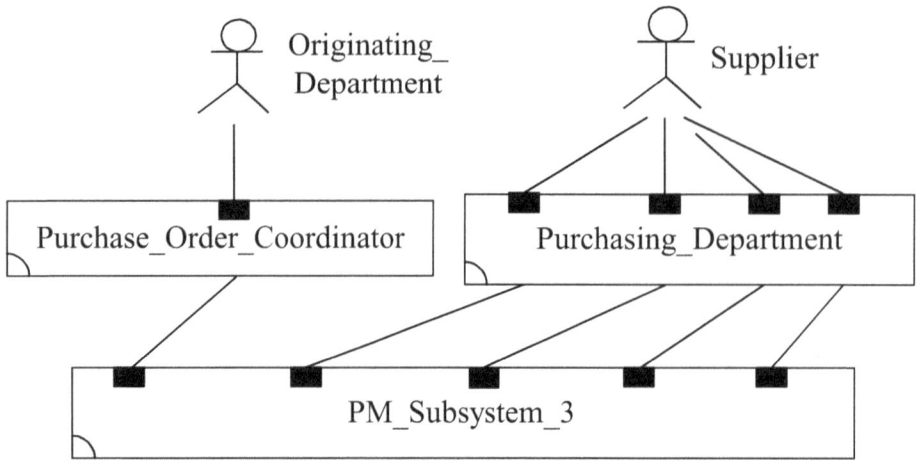

Figure 20-181   Concept's CCD of the *Purchasing Management*
Systems Definition Version 3

20-3-2-1-2 Concept's Systems Behavior of the Purchasing Management Systems Definition Version 3

The entire concept's systems behavior of the *purchasing management* systems definition version 3 includes: a) *Concept's SBCD* and b) *Concept's IFD* of the *purchasing management* systems definition version 3.

Figure 20-182 shows the concept's SBCD of the *purchasing management* systems definition version 3 in which interactions among the *Originating_Department*, *Supplier* actors and the *Purchase_Order_Coordinator*, *Purchasing_Department*, *PM_Subsystem_3* components shall draw forth the *Purchase_Requisition*, *Quotation*, *Purchase_Order*, *Purchase* and *Collect_Supplier_Data* behaviors.

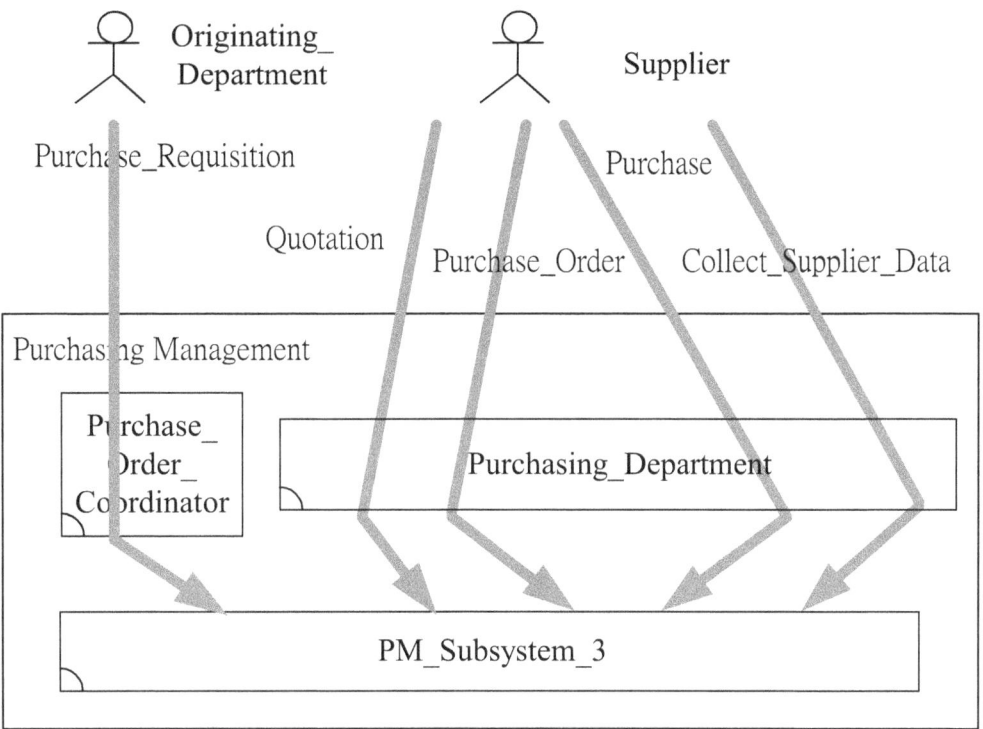

Figure 20-182   Concept's SBCD of the *Purchasing Management*
Systems Definition Version 3

The overall concept's behavior of the *purchasing management* systems definition version 3 includes the *Purchase_Requisition, Quotation, Purchase_Order, Purchase* and *Collect_Supplier_Data* behaviors. In other words, the *Purchase_Requisition, Quotation, Purchase_Order, Purchase* and *Collect_Supplier_Data* behaviors together provide the overall concept's behavior of the *purchasing management* systems definition version 3.

Be noticed that the *Purchase_Requisition, Quotation, Purchase_Order, Purchase* and *Collect_Supplier_Data* behaviors are mutually independent of each other. They tend to be executed concurrently [Hoar85, Miln89, Miln99].

The overall concept's behavior of the *purchasing management* systems definition version 3 includes five individual behaviors: *Purchase_Requisition, Quotation, Purchase_Order, Purchase* and *Collect_Supplier_Data*. Each individual behavior is represented by an execution path. We use an interaction flow diagram (IFD) to define each one of these execution paths. Figure 20-183 shows the concept's IFD of the *purchasing management* systems definition version 3 *Purchase_Requisition* behavior. First, actor *Originating_Department* interacts with the *Purchase_Order_Coordinator* component through the

Purchase_Requisition_Verify operation call interaction, carrying the *Purchase_Requisition_Form* input parameter. Finally, component *Purchase_Order_Coordinator* interacts with the *PM_Subsystem_3* component through the *Purchase_Requisition_Processing_Button_Click* operation call interaction, carrying the *Purchase_Requisition_Form* input parameter.

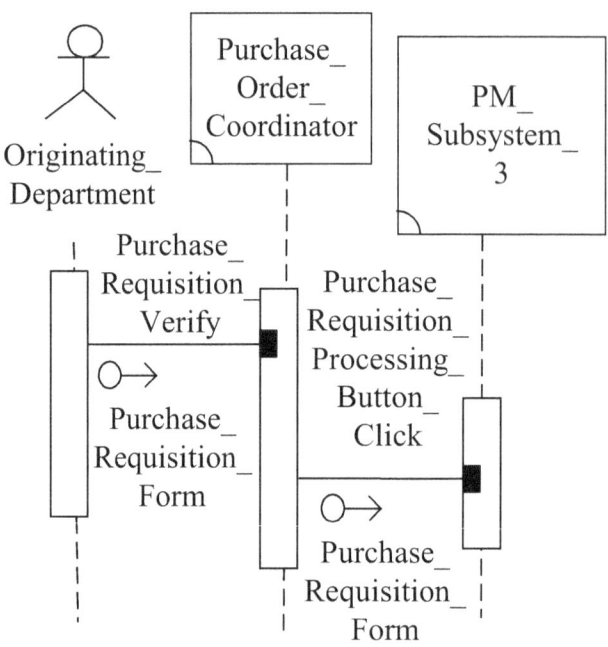

Figure 20-183   Concept's IFD of the *Purchasing Management*
Systems Definition Version 3 *Purchase_Requisition* Behavior

Figure 20-184 shows the concept's IFD of the *purchasing management* systems definition version 3 *Quotation* behavior. First, actor *Supplier* interacts with the *Purchasing_Department* component through the *Quotation_Verify* operation call interaction, carrying the *Quotation_Form* input parameter. Finally, component *Purchasing_Department* interacts with the *PM_Subsystem_3* component through the *Quotation_Processing_Button_Click* operation call interaction, carrying the *Quotation_Form* input parameter.

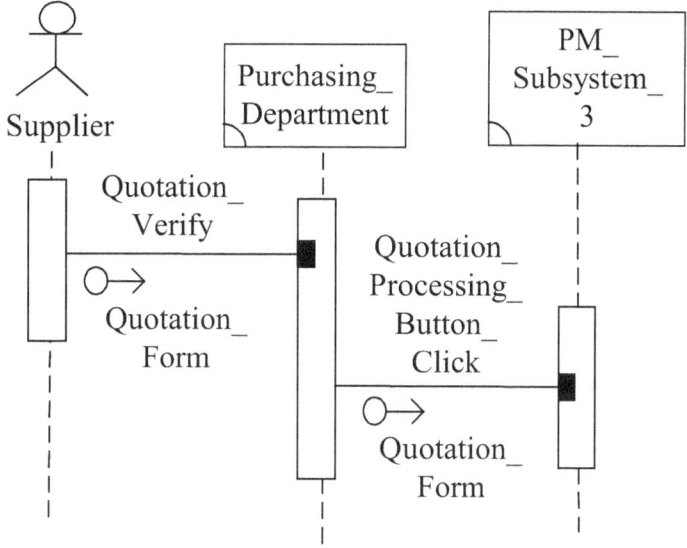

Figure 20-184   Concept's IFD of the *Purchasing Management Systems* Definition Version 3 *Quotation* Behavior

Figure 20-185 shows the concept's IFD of the *purchasing management systems* definition version 3 *Purchase_Order* behavior. First, actor *Supplier* interacts with the *Purchasing_Department* component through the *Purchase_Order_Verify* operation call interaction. Next, component *Purchasing_Department* interacts with the *PM_Subsystem_3* component through the *Purchase_Order_Processing_Button_Click* operation call interaction, carrying the *Purchase_Order_Form* input parameter and *Purchase_Order_Report* output parameter. Finally, actor *Supplier* interacts with the *Purchasing_Department* component through the *Purchase_Order_Verify* operation return interaction, carrying the *Purchase_Order_Report* output parameter.

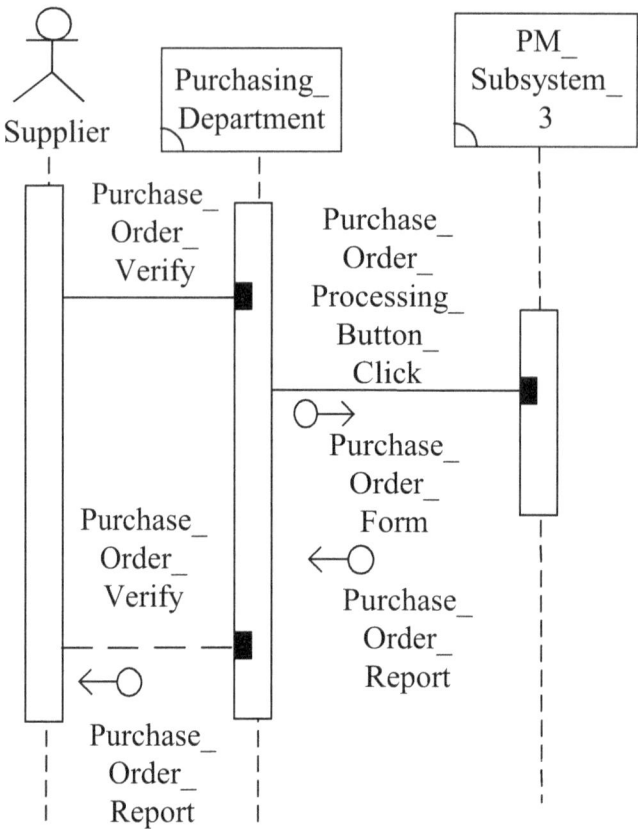

Figure 20-185   Concept's IFD of the *Purchasing Management*
Systems Definition Version 3 *Purchase_Order* Behavior

Figure 20-186 shows the concept's IFD of the *purchasing management* systems definition version 3 *Purchase* behavior. First, actor *Supplier* interacts with the *Purchasing_Department* component through the *Purchase_Verify* operation call interaction, carrying the *Products_In* and *Invoice* input parameters. Next, component *Purchasing_Department* interacts with the *PM_Subsystem_3* component through the *Purchase_Processing_Button_Click* operation call interaction, carrying the *Purchase_Form* input parameter. Finally, actor *Supplier* interacts with the *Purchasing_Department* component through the *Purchase_Verify* operation return interaction, carrying the *Products_Out* output parameter.

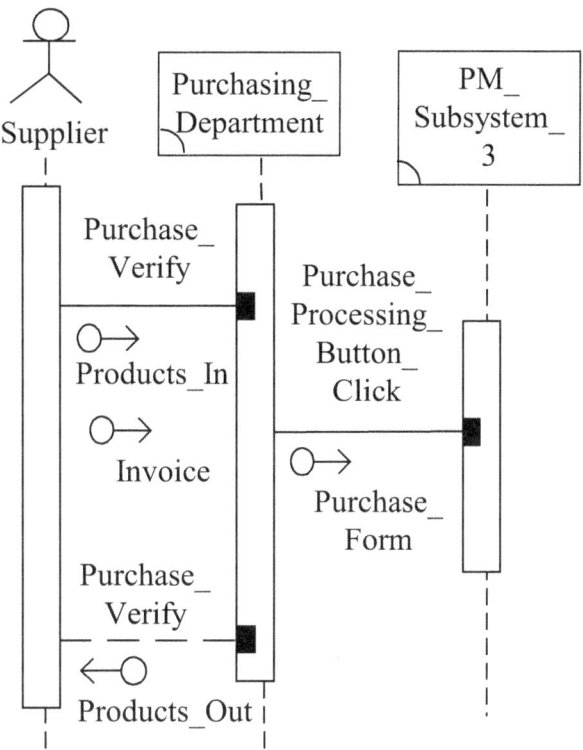

Figure 20-186   Concept's IFD of the *Purchasing Management Systems Definition* Version 3 *Purchase* Behavior

Figure 20-187 shows the concept's IFD of the *purchasing management* systems definition version 3 *Collect_Supplier_Data* behavior. First, actor *Supplier* interacts with the *Purchasing_Department* component through the *Interview* operation call interaction. Finally, component *Purchasing_Department* interacts with the *PM_Subsystem_3* component through the *Supplier_Data_Processing_Button_Click* operation call interaction, carrying the *Supplier_Data_Form* input parameter.

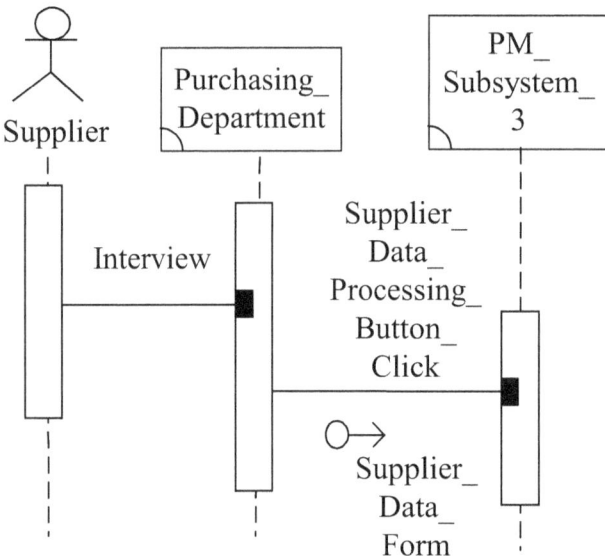

Figure 20-187   Concept's IFD of the *Purchasing Management Systems Definition Version 3 Collect_Supplier_Data* Behavior

20-3-2-2 Analysis View of the Purchasing Management Systems Definition Version 3

The analysis view of the *purchasing management* systems definition version 3 is shown in Figure 20-188.

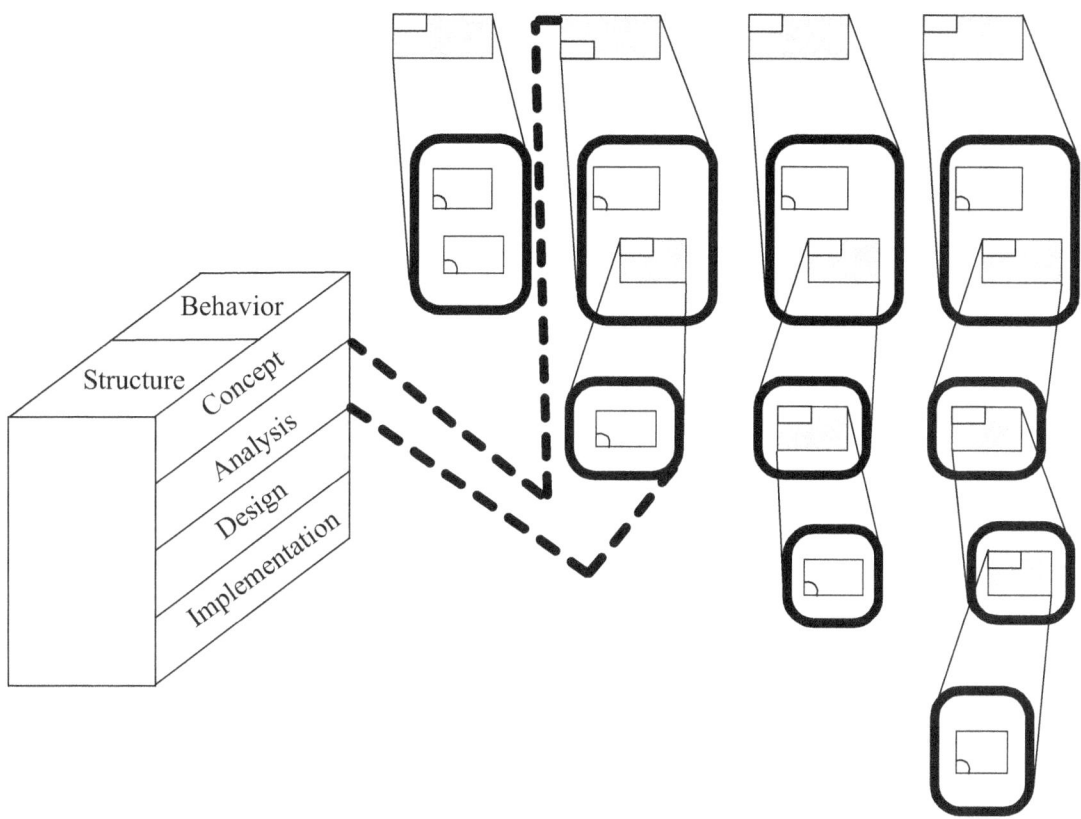

Figure 20-188   Analysis View of the *Purchasing Management* Systems Definition Version 3

The analysis view of the *purchasing management* systems definition version 3 consists of: a) analysis' systems structure of the *purchasing management* systems definition version 3 and b) analysis' systems behavior of the *purchasing management* systems definition version 3.

20-3-2-2-1 Analysis' Systems Structure of the Purchasing Management Systems Definition Version 3

The entire analysis' systems structure of the *purchasing management* systems definition version 3 includes: a) *Analysis' AHD*, b) *Analysis' FD*, c) *Analysis' COD* and d) *Analysis' CCD* of the *purchasing management* systems definition version 3.

We first draw the analysis' AHD of the *purchasing management* systems definition version 3. As shown in Figure 20-189, *Purchasing Management* is composed of *Purchase_Order_Coordinator*, *Purchasing_Department* and *PM_Subsystem_3*; *PM_Subsystem_3* is composed of *Purchase_Requisition_Processing_GUI*, *Quotation_Processing_GUI*, *Purchase_Order_Processing_GUI*, *Purchase_Processing_GUI*,

Supplier_Data_Processing_GUI and *PM_Subsystem_2*. In the figure, *Purchasing Management* and *PM_Subsystem_3* are aggregated systems while *Purchase_Order_Coordinator*, *Purchasing_Department*, *Purchase_Requisition_Processing_GUI*, *Quotation_Processing_GUI*, *Purchase_Order_Processing_GUI*, *Purchase_Processing_GUI*, *Supplier_Data_Processing_GUI* and *PM_Subsystem_2* are non-aggregated systems.

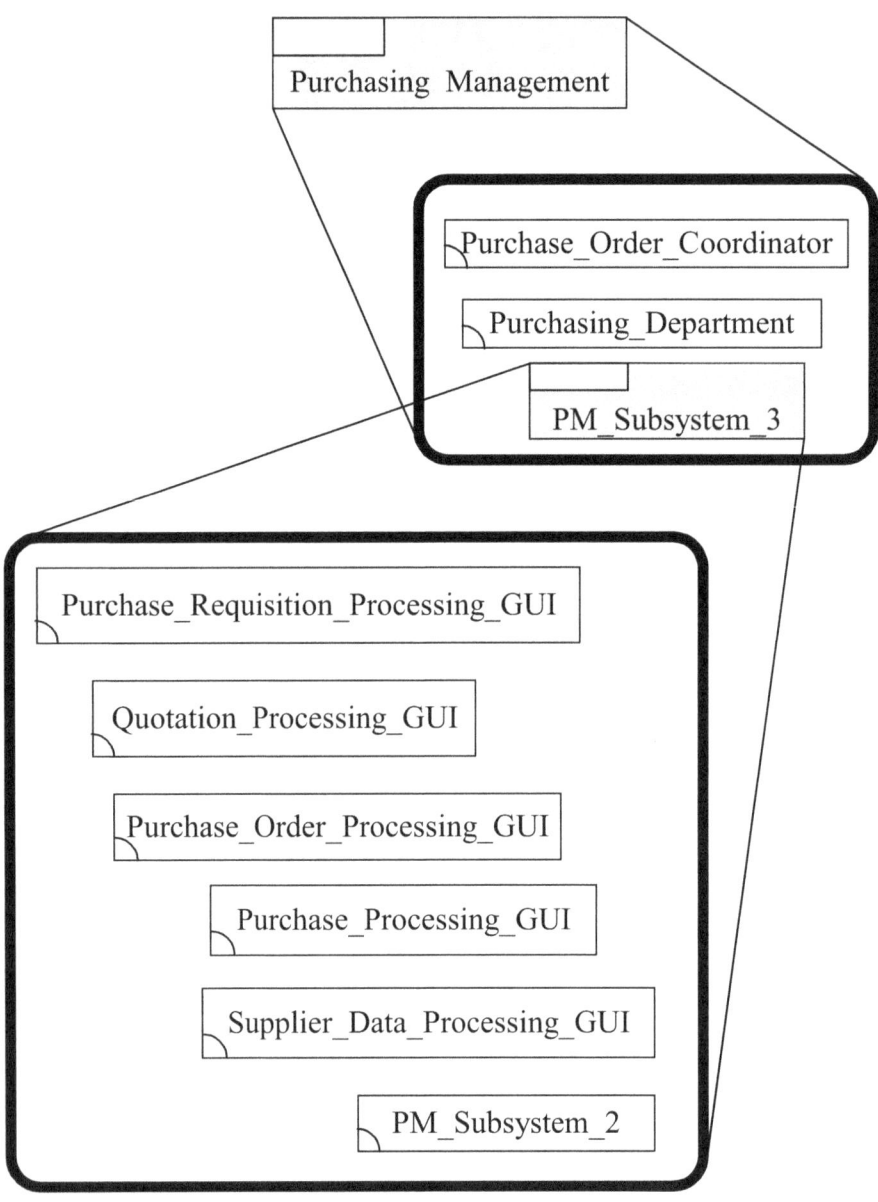

Figure 20-189   Analysis' AHD of the *Purchasing Management*
Systems Definition Version 3

Figure 20-190 shows the analysis' FD of the *purchasing management* systems definition version 3. In the figure, *Business_Layer* contains the *Purchase_Order_Coordinator* and *Purchasing_Department* components; *Application_Layer* contains the *Purchase_Requisition_Processing_GUI*, *Quotation_Processing_GUI*, *Purchase_Order_Processing_GUI*, *Purchase_Processing_GUI*, *Supplier_Data_Processing_GUI* and *PM_Subsystem_2* components.

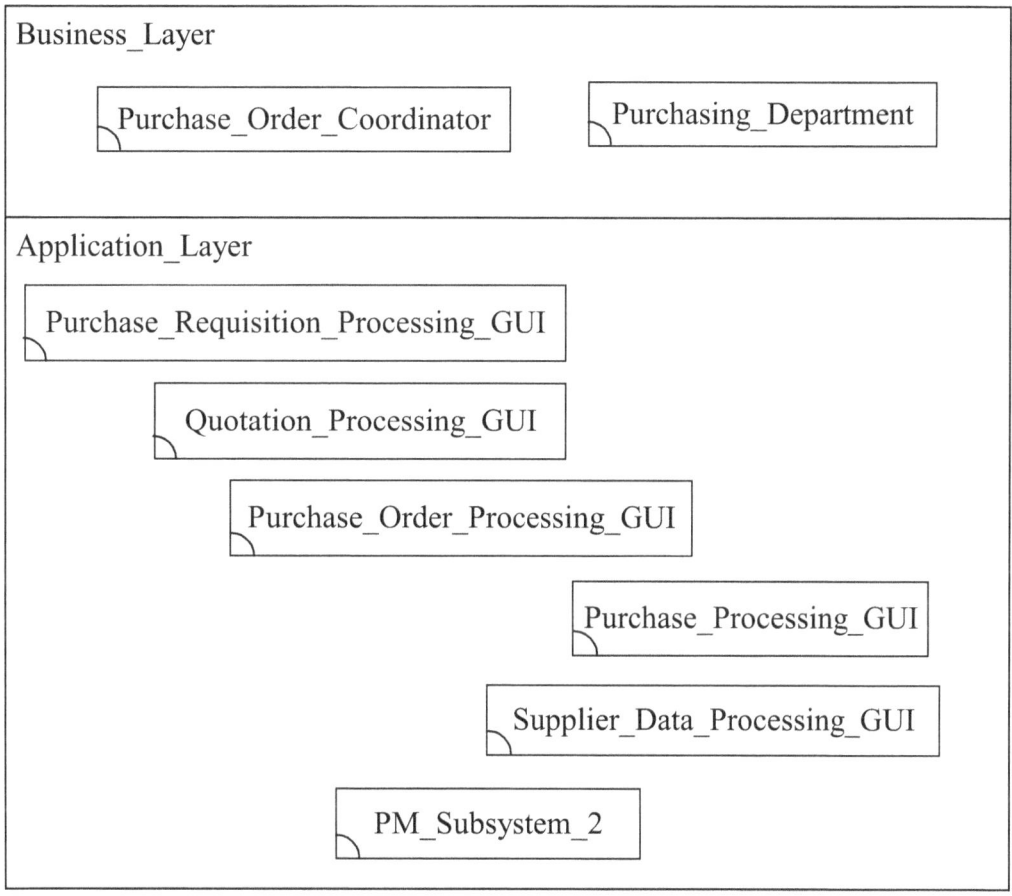

Figure 20-190   Analysis' FD of the *Purchasing Management*
Systems Definition Version 3

Figure 20-191 shows the analysis' COD of the *purchasing management* systems definition version 3. In the figure, component *Purchase_Order_Coordinator* has one operation: *Purchase_Requisition_Verify*; component *Purchasing_Department* has four operations: *Quotation_Verify*, *Purchase_Order_Verify*, *Purchase_Verify* and *Interview*; component *Purchase_Requisition_Processing_GUI* has one operation: *Purchase_Requisition_Processing_Button_Click*;   component *Quotation_Processing_GUI* has one operation: *Quotation_Processing_Button_Click*;

component *Purchase_Order_Processing_GUI* has one operation: *Purchase_Order_Processing_Button_Click*; component *Purchase_Processing_GUI* has one operation: *Purchase_Processing_Button_Click*; component *Supplier_Data_Processing_GUI* has one operation: *Supplier_Data_Processing_Button_Click*; component *PM_Subsystem_2* has five operations: *SQL_Purchase_Requisition_Insert*, *SQL_Quotation_Insert*, *SQL_Purchase_Order_Insert*, *SQL_Purchase_Insert* and *SQL_Supplier_Data_Insert*.

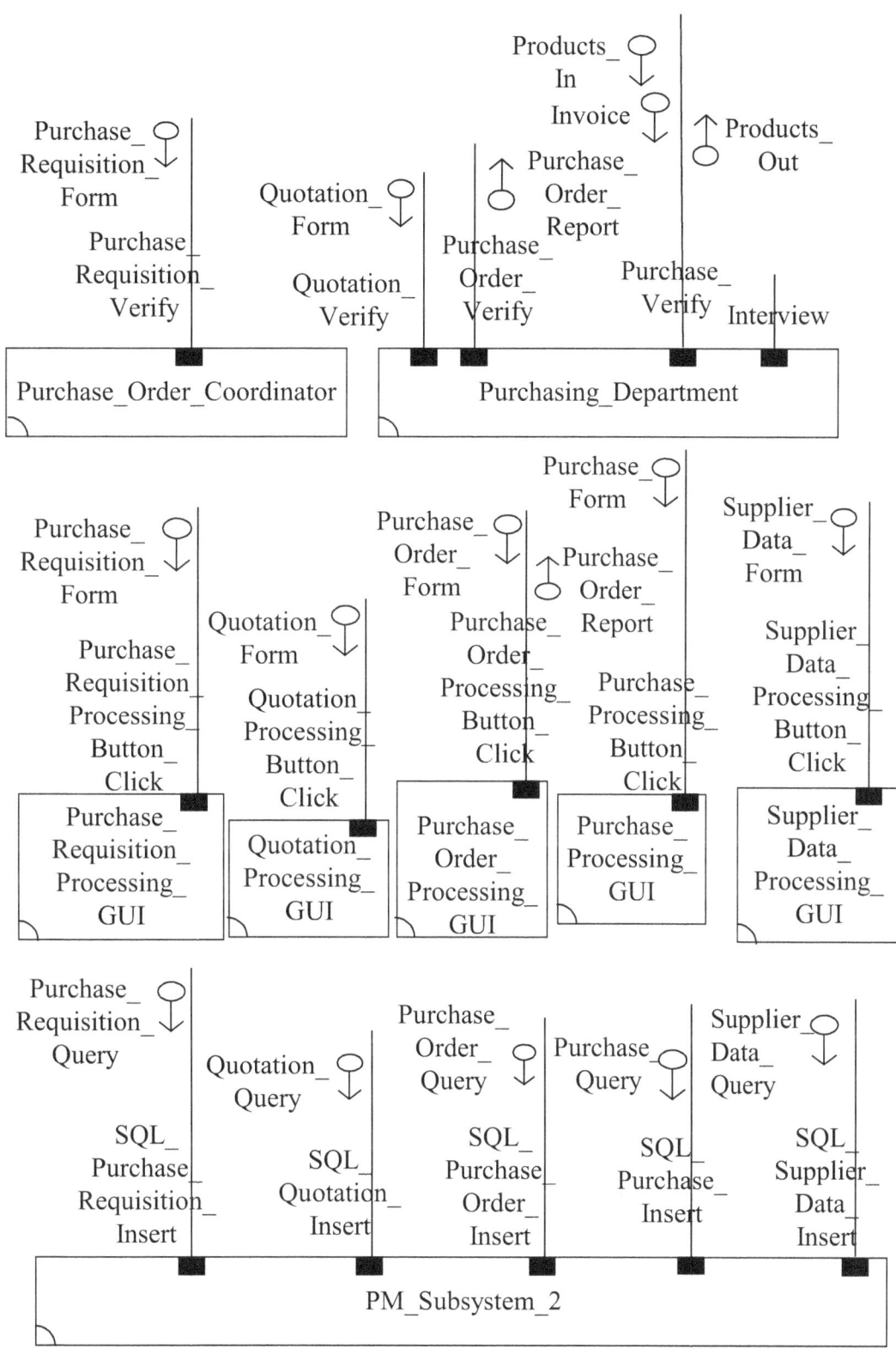

Figure 20-191   Analysis' COD of the *Purchasing Management*
Systems Definition Version 3

The operation formula of *Purchase_Requisition_Verify* is *Purchase_Requisition_Verify(In Purchase_Requisition_Form)*. The operation formula of *Quotation_Verify* is *Quotation_Verify(In Quotation_Form)*. The operation formula of *Purchase_Order_Verify* is *Purchase_Order_Verify(Out Purchase_Order_Report)*. The operation formula of *Purchase_Verify* is *Purchase_Verify(In Products_In, Invoice; Out Products_Out)*. The operation formula of *Interview* is *Interview*. The operation formula of *Purchase_Requisition_Processing_Button_Click* is *Purchase_Requisition_Processing_Button_Click(In Purchase_Requisition_Form)*. The operation formula of *Quotation_Processing_Button_Click* is *Quotation_Processing_Button_Click(In Quotation_Form)*. The operation formula of *Purchase_Order_Processing_Button_Click* is *Purchase_Order_Processing_Button_Click(In Purchase_Order_Form; Out Purchase_Order_Report)*. The operation formula of *Purchase_Processing_Button_Click* is *Purchase_Processing_Button_Click(In Purchase_Form)*. The operation formula of *Supplier_Data_Processing_Button_Click* is *Supplier_Data_Processing_Button_Click(In Supplier_Data_Form)*. The operation formula of *SQL_Purchase_Requisition_Insert* is *SQL_Purchase_Requisition_Insert(In Purchase_Requisition_Query)*. The operation formula of *SQL_Quotation_Insert* is *SQL_Quotation_Insert(In Quotation_Query)*. The operation formula of *SQL_Purchase_Order_Insert* is *SQL_Purchase_Order_Insert (In Purchase_Order_Query)*. The operation formula of *SQL_Purchase_Insert* is *SQL_Purchase_Insert (In Purchase_Query)*. The operation formula of *SQL_Supplier_Data_Insert* is *SQL_Supplier_Data_Insert(In Supplier_Data_Query)*.

Figure 20-192 shows the composite data type specification of the *Purchase_Requisition_Form* input parameter occurring in the *Purchase_Requisition_Verify(In Purchase_Requisition_Form)* and *Purchase_Requisition_Processing_Button_Click(In Purchase_Requisition_Form)* operation formulas.

Parameter	*Purchase_Requisition_Form*

Data Type	TABLE of
	  Date : Text
	  OD : Text
	  ProductNo : Text
	  Quantity : Integer
	End TABLE ;

Instances	**Purchase Requisition Form** Date: 2011/10/17 Originating_Department : Sales Dept. ProductNo Quantity __A00001(Pen)_____300_____ __A00002(Mouse)_____400_____ __A00003(Camera)_____500_____

Figure 20-192   Composite Data Type Specification

Figure 20-193 shows the composite data type specification of the *Quotation_Form* input parameter occurring in the *Quotation_Verify(In Quotation_Form)* and *Quotation_Processing_Button_Click(In Quotation_Form)* operation formulas.

Parameter	*Quotation_Form*

Data Type	TABLE of
	  Date : Text
	  SupplierName: Text
	  ProductNo : Text
	  Quantity : Integer
	  UnitPrice : Real
	  Total : Real
	End TABLE ;

Instances	**Quotation Form** Date: 2011/10/25 SupplierName : Johnson Corp. ProductNo Quantity UnitPrice A00001(Pen) 300 100.00 A00002(Mouse) 400 200.00 A00003(Camera) 500 300.00 Total : 260,000.00

Figure 20-193   Composite Data Type Specification

Figure 20-194 shows the composite data type specification of the *Purchase_Order_Report* output parameter occurring in the *Purchase_Order_Verify(Out Purchase_Order_Report)* and *Purchase_Order_Processing_Button_Click(In Purchase_Order_Form; Out Purchase_Order_Report)* operation formulas.

Parameter	*Purchase_Order_Report*
Data Type	TABLE of Date : Text SupplierName: Text ProductNo : Text Quantity : Integer End TABLE ;
Instances	Date : 20111118 SupplierName : Johnson Corp. | ProductNo	Quantity
A00001(Pen)	300
A00002(Mouse)	400
A00003(Camera)	500	

Figure 20-194   Composite Data Type Specification

Figure 20-195 shows the primitive data type specification of the *Products_In* input parameter occurring in the *Purchase_Verify(In Products_In, Invoice; Out Products_Out)* operation formula.

Parameter	Data Type	Instances
Products_In	Physical Object	Pen, Mouse, Camera

Figure 20-195   Primitive Data Type Specification

Figure 20-196 shows the composite data type specification of the *Invoice* input parameter occurring in the *Purchase_Verify(In Products_In, Invoice; Out Products_Out)* operation formula.

Parameter	*Invoice*
Data Type	TABLE of Date : Text SupplierName: Text ProductNo : Text Quantity : Integer UnitPrice : Real Total : Real End TABLE ;
Instances	Date : 20111130 SupplierName : Johnson Corp. | ProductNo	Quantity	UnitPrice
---	---	---
A00001(Pen)	300	100.00
A00002(Mouse)	400	200.00
A00003(Camera)	500	300.00	 Total : 260,000.00

Figure 20-196   Composite Data Type Specification

Figure 20-197 shows the primitive data type specification of the *Products_Out* output parameter occurring in the *Purchase_Verify(In Products_In, Invoice; Out Products_Out)* operation formula.

Parameter	Data Type	Instances
Products_Out	Physical Object	Pen, Mouse, Camera

Figure 20-197   Primitive Data Type Specification

Figure 20-198 shows the composite data type specification of the *Purchase_Order_Form* input parameter occurring in the *Purchase_Order_Processing_Button_Click(In Purchase_Order_Form; Out Purchase_Order_Report)* operation formula.

Parameter	*Purchase_Order_Form*

Data Type	TABLE of
	  Date : Text
	  SupplierName: Text
	  ProductNo : Text
	  Quantity : Integer
	End TABLE ;

Instances	**Purchase Order Form** Date: 2011/11/18 SupplierName : Johnson Corp. ProductNo Quantity __A00001(Pen) ____300 ___ __A00002(Mouse)____400___ __A00003(Camera)___500___

Figure 20-198   Composite Data Type Specification

Figure 20-199 shows the composite data type specification of the *Purchase_Form* input parameter occurring in the *Purchase_Processing_Button_Click(In Purchase_Form)* operation formula.

Parameter	*Purchase_Form*

Data Type	TABLE of
	  Date : Text
	  SupplierName: Text
	  ProductNo : Text
	  Quantity : Integer
	  UnitPrice : Real
	  ReturnQuantity : Integer
	  Total : Real
	End TABLE ;

Instances	**Purchase Form** Date: 2011/12/12 SupplierName : Johnson Corp. ProductNo Quantity UnitPrice ReturnQuantity A00001(Pen) 300 100.00 0 A00002(Mouse) 390 200.00 10 A00003(Camera)500 300.00 0 Total : 258,000.00

Figure 20-199   Composite Data Type Specification

Figure 20-200 shows the composite data type specification of the *Supplier_Data_Form* input parameter occurring in the *Supplier_Data_Processing_Button_Click(In Supplier_Data_Form)* operation formula.

Parameter	*Supplier_Data_Form*

Data Type	TABLE of
	  SupplierName:Text
	  Address :Text
	  PhoneNumber:Text
	  FaxNumber:Text
	  E-mail : Text
	  Rank : Text
	End TABLE ;

Instances	**Supplier Data Form** SupplierName : Johnson Corp. Address : 1232 Fair Circle, Austin, TX PhoneNumber : 512-463-8472 FaxNumber : 512-463-8499 E-mail : Johnson1122@gmail.com Rank : B

Figure 20-200   Composite Data Type Specification

Figure 20-201 shows the composite data type specification of the *Purchase_Requisition_Query* input parameter occurring in the *SQL_Purchase_Requisition_Insert(In   Purchase_Requisition_Query)* operation formula.

Parameter	*Purchase_Requisition_Query*
Data Type	TABLE of 　Date : Text 　OD : Text 　ProductNo : Text 　Quantity : Integer End TABLE ;
Instances	<table><tr><th>Date</th><th colspan="2">Originating_Department :</th></tr><tr><td>20111017</td><td colspan="2">Sales Dept.</td></tr><tr><th>ProductNo</th><th>Quantity</th></tr><tr><td>A00001(Pen)</td><td>300</td></tr><tr><td>A00002(Mouse)</td><td>400</td></tr><tr><td>A00003(Camera)</td><td>500</td></tr></table>

Figure 20-201　　Composite Data Type Specification

Figure 20-202 shows the composite data type specification of the *Quotation_Query* input parameter occurring in the *SQL_Quotation_Insert(In Quotation_Query)* operation formula.

Parameter	*Quotation_Query*

Data Type	TABLE of
	  Date : Text
	  SupplierName: Text
	  ProductNo : Text
	  Quantity : Integer
	  UnitPrice : Real
	  Total : Real
	End TABLE ;

Instances	

Date	SupplierName	Total
20111025	Johnson Corp.	260,000.00

ProductNo	Quantity	UnitPrice
A00001(Pen)	300	100.00
A00002(Mouse)	400	200.00
A00003(Camera)	500	300.00

Figure 20-202   Composite Data Type Specification

Figure 20-203 shows the composite data type specification of the *Purchase_Order_Query* input parameter occurring in the *SQL_Purchase_Order_Insert(In Purchase_Order_Query)* operation formula.

414

Parameter	*Purchase_Order_Query*
Data Type	TABLE of Date : Text SupplierName: Text ProductNo : Text Quantity : Integer End TABLE ;
Instances	

Date	SupplierName
20111118	Johnson Corp.

ProductNo	Quantity
A00001(Pen)	300
A00002(Mouse)	400
A00003(Camera)	500

Figure 20-203   Composite Data Type Specification

Figure 20-204 shows the composite data type specification of the *Purchase_Query* input parameter occurring in the *SQL_Purchase_Insert (In Purchase_Query)* operation formula.

Parameter	*Purchase_Query*

Data Type	TABLE of
	  Date : Text
	  SupplierName: Text
	  ProductNo : Text
	  Quantity : Integer
	  UnitPrice : Real
	  ReturnQuantity : Integer
	  Total : Real
	End TABLE ;

Instances	<table><tr><th>Date</th><th>SupplierName</th><th>Total</th></tr><tr><td>20111212</td><td>Johnson Corp.</td><td>258,000.00</td></tr></table> <table><tr><th>ProductNo</th><th>Quantity</th><th>UnitPrice</th><th>ReturnQuantity</th></tr><tr><td>A00001(Pen)</td><td>300</td><td>100.00</td><td>0</td></tr><tr><td>A00002(Mouse)</td><td>390</td><td>200.00</td><td>10</td></tr><tr><td>A00003(Camera)</td><td>500</td><td>300.00</td><td>0</td></tr></table>

Figure 20-204   Composite Data Type Specification

Figure 20-205 shows the composite data type specification of the *Supplier_Data_Query* input parameter occurring in the *SQL_Supplier_Data_Insert(In Supplier_Data_Query)* operation formula.

Parameter	*Supplier_Data_Query*
Data Type	TABLE of SupplierName:Text Address :Text PhoneNumber:Text FaxNumber:Text E-mail : Text Rank : Text End TABLE ;
Instances	<table><tr><td>SupplierName</td><td>Johnson Corp.</td></tr><tr><td>Address</td><td>1232 Fair Circle, Austin, TX</td></tr><tr><td>PhoneNumber</td><td>512-463-8472</td></tr><tr><td>FaxNumber</td><td>512-463-8499</td></tr><tr><td>E-mail</td><td>Johnson1122@gmail.com</td></tr><tr><td>Rank</td><td>B</td></tr></table>

Figure 20-205   Composite Data Type Specification

Figure 20-206 shows the analysis' CCD of the *purchasing management* systems definition version 3. In the figure, actor *Originating_Department* has a connection with the *Purchase_Order_Coordinator* component; actor *Supplier* has three connections with the *Purchasing_Department* component; component *Purchase_Order_Coordinator* has a connection with the *Purchase_Requisition_Processing_GUI* component; component *Purchasing_Department* has a connection with each one of the *Quotation_Processing_GUI*, *Purchase_Order_Processing_GUI*, *Purchase_Processing_GUI* and *Supplier_Data_Processing_GUI* components; each one of the *Purchase_Requisition_Processing_GUI*, *Quotation_Processing_GUI*, *Purchase_Order_Processing_GUI*, *Purchase_Processing_GUI* and *Supplier_Data_Processing_GUI* components has a connection with the *PM_Subsystem_2* component.

417

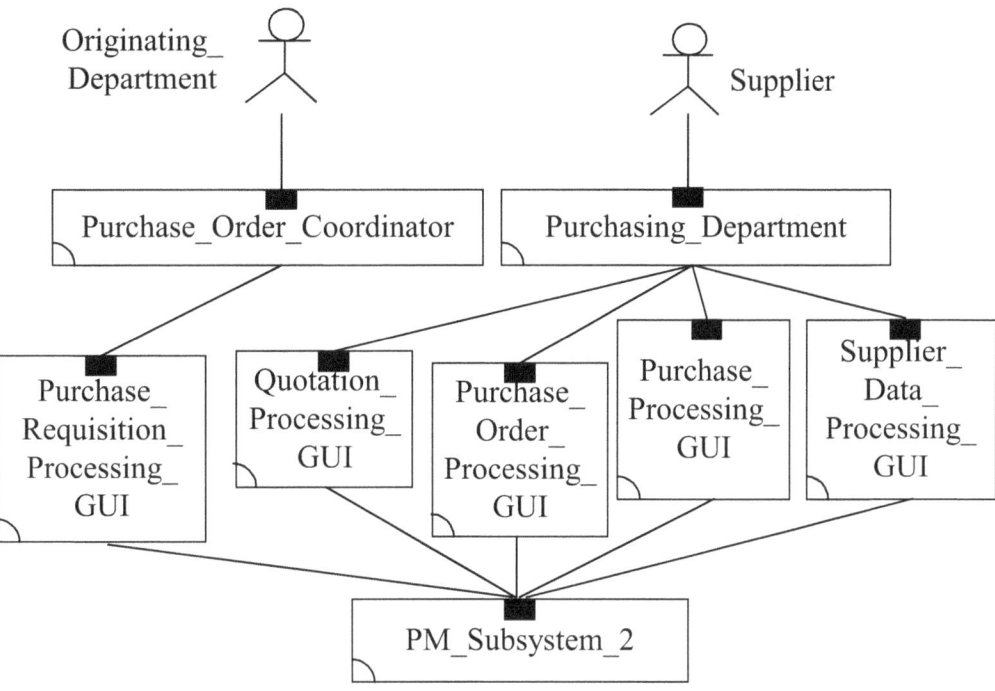

Figure 20-206   Analysis' CCD of the *Purchasing Management Systems Definition Version 3*

20-3-2-2-2 Analysis' Systems Behavior of the Purchasing Management Systems Definition Version 3

The entire analysis' systems behavior of the *purchasing management* systems definition version 3 includes: a) *Analysis' SBCD* and b) *Analysis' IFD* of the *purchasing management* systems definition version 3.

Figure 20-207 shows the analysis' SBCD of the *purchasing management* systems definition version 3 in which interactions among the *Originating_Department*, *Supplier* actors and the *Purchase_Order_Coordinator*, *Purchasing_Department*, *Purchase_Requisition_Processing_GUI*, *Quotation_Processing_GUI*, *Purchase_Order_Processing_GUI*, *Purchase_Processing_GUI*, *Supplier_Data_Processing_GUI*, *PM_Subsystem_2* components shall draw forth the *Purchase_Requisition, Quotation, Purchase_Order* and *Purchase* behaviors.

418

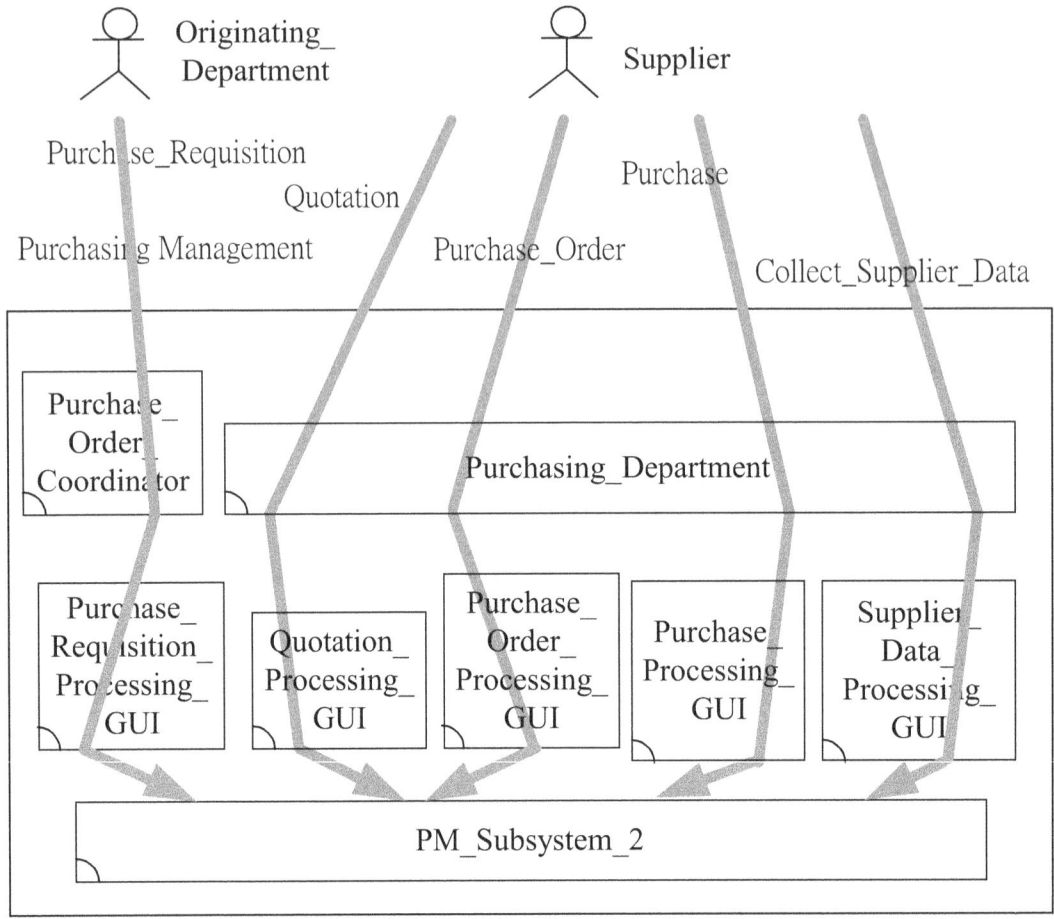

Figure 20-207   Analysis' SBCD of the *Purchasing Management*
Systems Definition Version 3

The overall analysis' behavior of the *purchasing management* systems definition version 3 includes the *Purchase_Requisition*, *Quotation*, *Purchase_Order*, *Purchase* and *Collect_Supplier_Data* behaviors. In other words, the *Purchase_Requisition*, *Quotation*, *Purchase_Order*, *Purchase* and *Collect_Supplier_Data* behaviors together provide the overall analysis' behavior of the *purchasing management* systems definition version 3.

Be noticed that the *Purchase_Requisition*, *Quotation*, *Purchase_Order*, *Purchase* and *Collect_Supplier_Data* behaviors are mutually independent of each other. They tend to be executed concurrently [Hoar85, Miln89, Miln99].

The overall analysis' behavior of the *purchasing management* systems definition version 3 includes five individual behaviors: *Purchase_Requisition*, *Quotation*, *Purchase_Order*, *Purchase* and *Collect_Supplier_Data*. Each individual behavior is represented by an execution path. We use an interaction flow diagram

419

(IFD) to define each one of these execution paths. Figure 20-208 shows the analysis' IFD of the *purchasing management* systems definition version 3 *Purchase_Requisition* behavior. First, actor *Originating_Department* interacts with the *Purchase_Order_Coordinator* component through the *Purchase_Requisition_Verify* operation call interaction, carrying the *Purchase_Requisition_Form* input parameter. Next, component *Purchase_Order_Coordinator* interacts with the *Purchase_Requisition_Processing_GUI* component through the *Purchase_Requisition_Processing_Button_Click* operation call interaction, carrying the *Purchase_Requisition_Form* input parameter. Finally, component *Purchase_Requisition_Processing_GUI* interacts with the *PM_Subsystem_2* component through the *SQL_Purchase_Requisition_Insert* operation call interaction, carrying the *Purchase_Requisition_Query* input parameter.

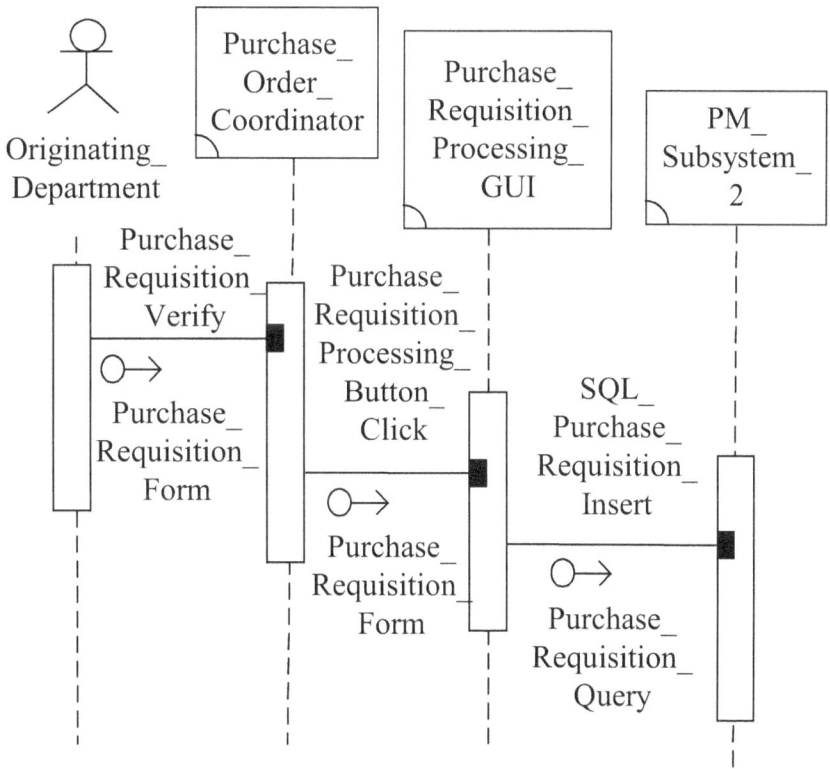

Figure 20-208   Analysis' IFD of the *Purchasing Management* Systems Definition Version 3 *Purchase_Requisition* Behavior

Figure 20-209 shows the analysis' IFD of the *purchasing management* systems definition version 3 *Quotation* behavior. First, actor *Supplier* interacts with the *Purchasing_Department* component through the *Quotation_Verify* operation call interaction, carrying the *Quotation_Form* input parameter. Next, component

Purchasing_Department interacts with the *Quotation_Processing_GUI* component through the *Quotation_Processing_Button_Click* operation call interaction, carrying the *Quotation_Form* input parameter. Finally, component *Quotation_Processing_GUI* interacts with the *PM_Subsystem_2* component through the *SQL_Quotation_Insert* operation call interaction, carrying the *Quotation_Query* input parameter.

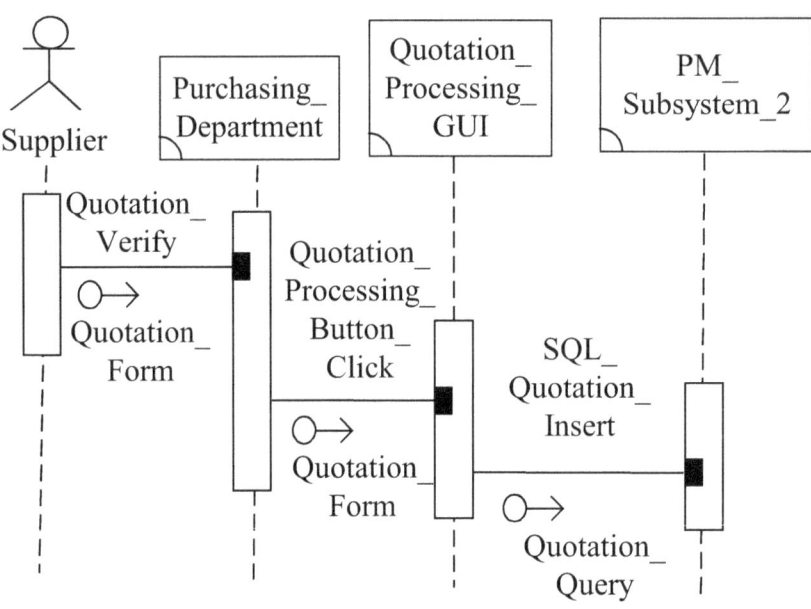

Figure 20-209   Analysis' IFD of the *Purchasing Management*
Systems Definition Version 3 *Quotation* Behavior

Figure 20-210 shows the analysis' IFD of the *purchasing management* systems definition version 3 *Purchase_Order* behavior. First, actor *Supplier* interacts with the *Purchasing_Department* component through the *Purchase_Order_Verify* operation call interaction. Next, component *Purchasing_Department* interacts with the *Purchase_Order_Processing_GUI* component through the *Purchase_Order_Processing_Button_Click* operation call interaction, carrying the *Purchase_Order_Form* input parameter. Continuingly, component *Purchase_Order_Processing_GUI* interacts with the *PM_Subsystem_2* component through the *SQL_Purchase_Order_Insert* operation call interaction, carrying the *Purchase_Order_Query* input parameter. Continuingly, component *Purchasing_Department* interacts with the *Purchase_Order_Processing_GUI* component through the *Purchase_Order_Processing_Button_Click* operation return interaction, carrying the *Purchase_Order_Report* output parameter. Finally, actor *Supplier* interacts with the *Purchasing_Department* component through the *Purchase_Order_Verify* operation return interaction, carrying the

Purchase_Order_Report output parameter.

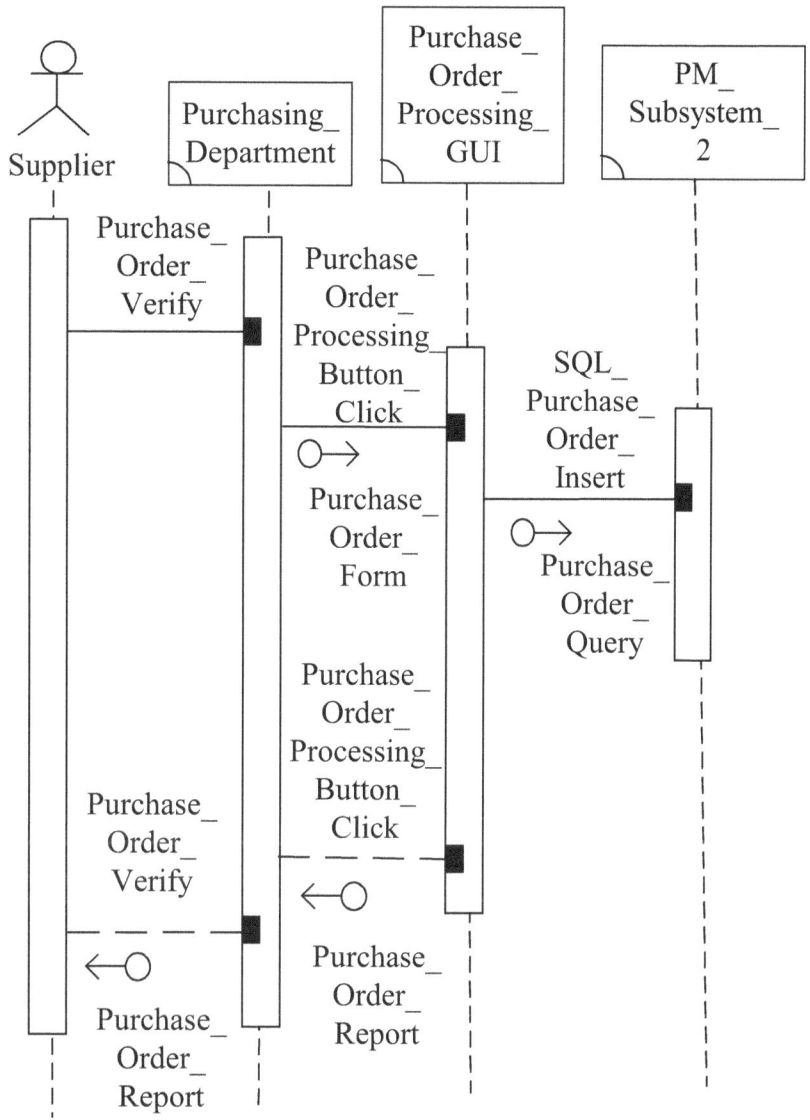

Figure 20-210   Analysis' IFD of the *Purchasing Management*
Systems Definition Version 3 *Purchase_Order* Behavior

Figure 20-211 shows the analysis' IFD of the *purchasing management* systems definition version 3 *Purchase* behavior. First, actor *Supplier* interacts with the *Purchasing_Department* component through the *Purchase_Verify* operation call interaction, carrying the *Products_In* and *Invoice* input parameters. Next, component *Purchasing_Department* interacts with the *Purchase_Processing_GUI* component through the *Purchase_Processing_Button_Click* operation call interaction, carrying the *Purchase_Form* input parameter. Continuingly, component *Purchase_Processing_GUI* interacts with the *PM_Subsystem_2* component through the *SQL_Purchase_Insert* operation call interaction, carrying the *Purchase_Query* input parameter. Finally, actor *Supplier* interacts with the *Purchasing_Department* component through the *Purchase_Verify* operation return interaction, carrying the *Products_Out* output parameter.

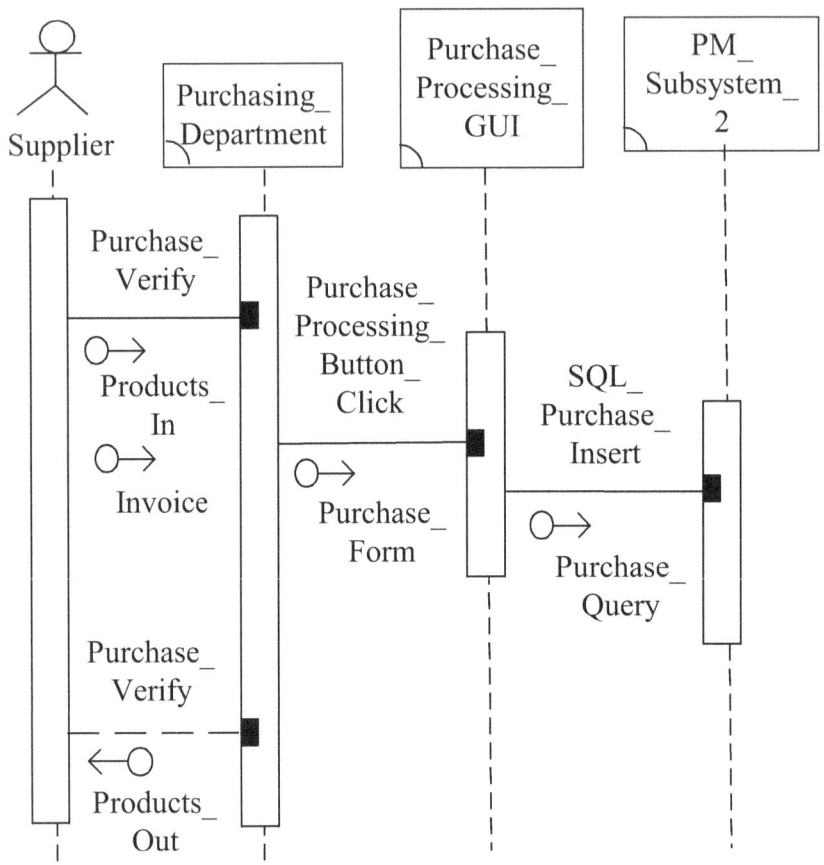

Figure 20-211   Analysis' IFD of the *Purchasing Management* Systems Definition Version 3 *Purchase* Behavior

Figure 20-212 shows the analysis' IFD of the *purchasing management* systems definition version 3 *Collect_Supplier_Data* behavior. First, actor *Supplier* interacts with the *Purchasing_Department* component through the *Interview* operation call interaction. Next, component *Purchasing_Department* interacts with the *Supplier_Data_Processing_GUI* component through the *Supplier_Data_Processing_Button_Click* operation call interaction, carrying the *Supplier_Data_Form* input parameter. Finally, component *Supplier_Data_Processing_GUI* interacts with the *PM_Subsystem_2* component through the *SQL_Supplier_Data_Insert* operation call interaction, carrying the *Supplier_Data_Query* input parameter.

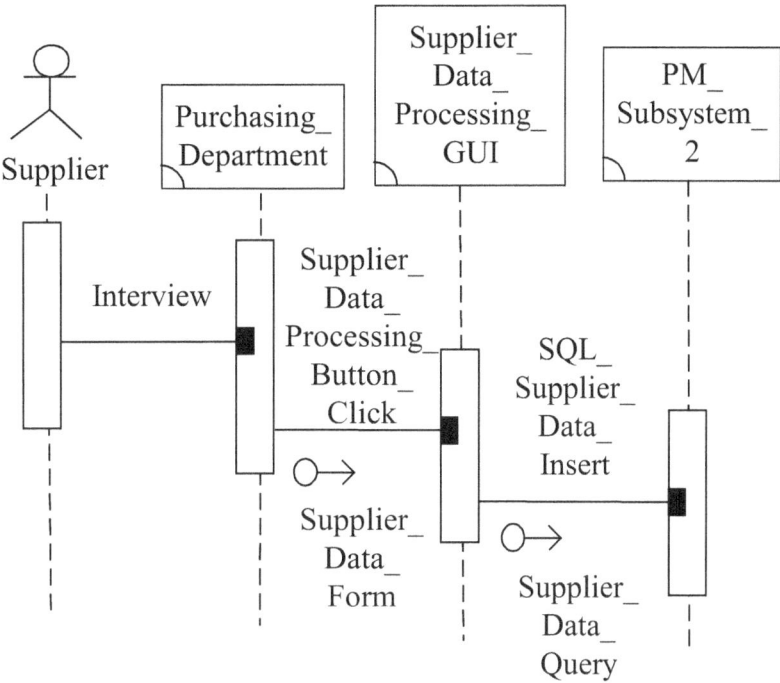

Figure 20-212   Analysis' IFD of the *Purchasing Management Systems Definition Version 3 Collect_Supplier_Data* Behavior

20-3-2-3 Design View of the Purchasing Management Systems Definition Version 3

The design view of the *purchasing management* systems definition version 3 is shown in Figure 20-213.

424

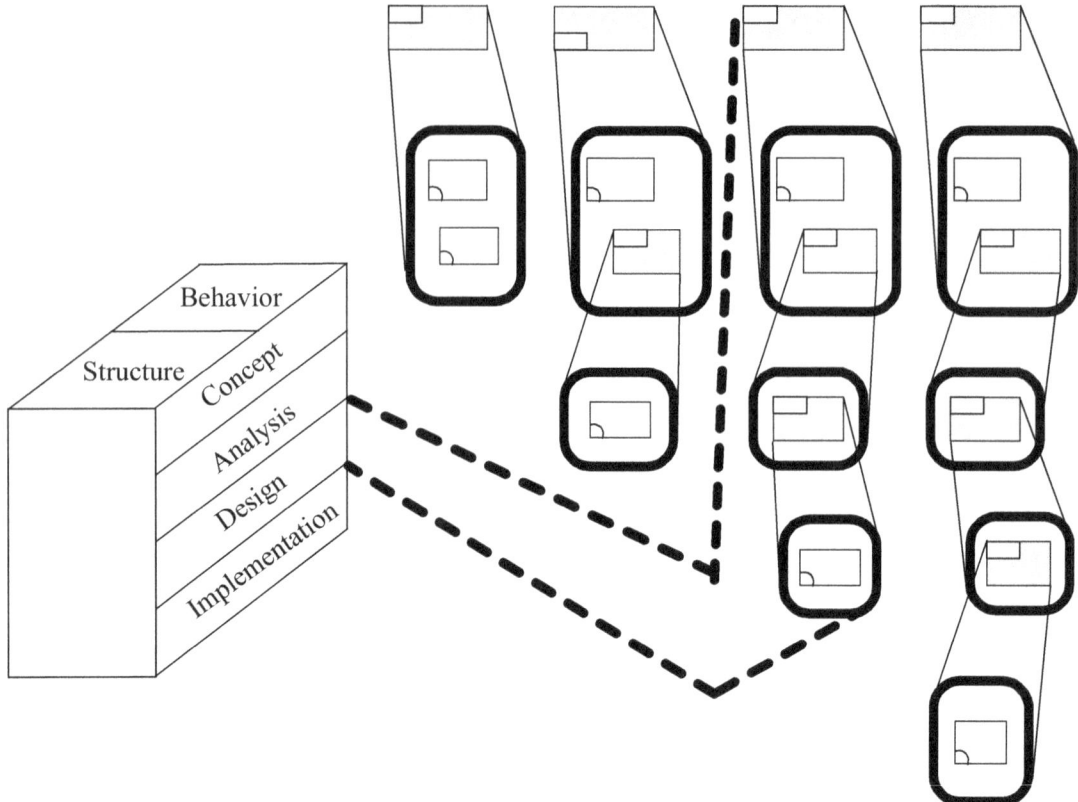

Figure 20-213   Design View of the *Purchasing Management* Systems Definition Version 3

The design view of the *purchasing management* systems definition version 3 consists of: a) design's systems structure of the *purchasing management* systems definition version 3 and b) design's systems behavior of the *purchasing management* systems definition version 3.

20-3-2-3-1 Design's Systems Structure of the Purchasing Management Systems Definition Version 3

The entire design's systems structure of the *purchasing management* systems definition version 3 includes: a) *Design's AHD*, b) *Design's FD*, c) *Design's COD* and d) *Design's CCD* of the *purchasing management* systems definition version 3.

We first draw the design's AHD of the *purchasing management* systems definition version 3. As shown in Figure 20-214, *Purchasing Management* is composed of *Purchase_Order_Coordinator*, *Purchasing_Department* and *PM_Subsystem_3*; *PM_Subsystem_3* is composed of *Purchase_Requisition_Processing_GUI*, *Quotation_Processing_GUI*, *Purchase_Order_Processing_GUI*, *Purchase_Processing_GUI* and *PM_Subsystem_2*; *PM_Subsystem_2* is composed of *Purchasing_Database* and *PM_Subsystem_1*. In the figure, *Purchasing Management*, *PM_Subsystem_3* and *PM_Subsystem_2* are aggregated systems while *Purchase_Order_Coordinator*, *Purchasing_Department*,

Purchase_Requisition_Processing_GUI, *Quotation_Processing_GUI*, *Purchase_Order_Processing_GUI*, *Purchase_Processing_GUI*, *Supplier_Data_Processing_GUI*, *Purchasing_Database* and *PM_Subsystem_1* are non-aggregated systems.

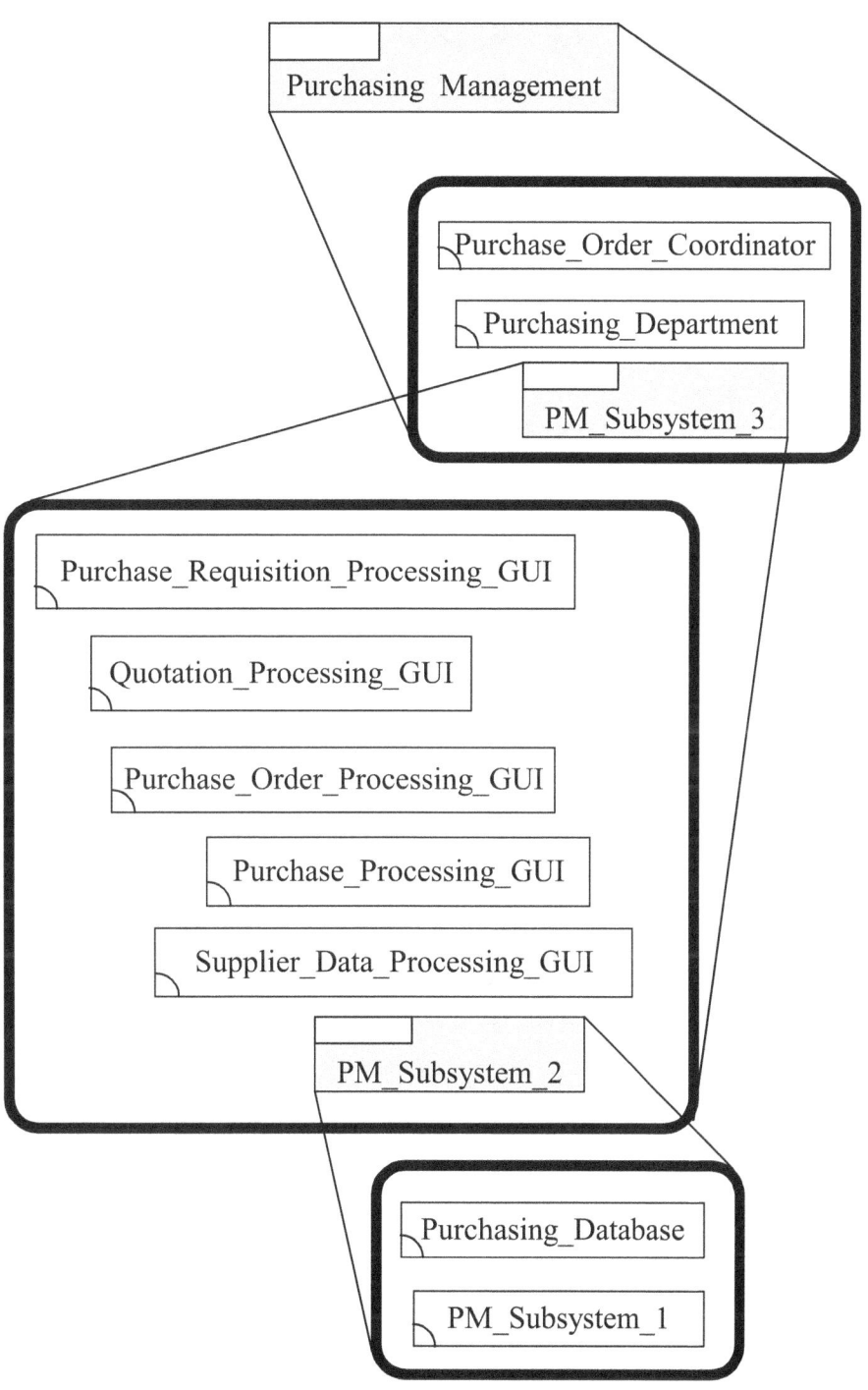

Figure 20-214   Design's AHD of the *Purchasing Management*
Systems Definition Version 3

Figure 20-215 shows the design's FD of the *purchasing management* systems definition version 3. In the figure, *Business_Layer* contains the *Purchase_Order_Coordinator* and *Purchasing_Department* components; *Application_Layer* contains the *Purchase_Requisition_Processing_GUI*, *Quotation_Processing_GUI*, *Purchase_Order_Processing_GUI*, *Purchase_Processing_GUI* and *Supplier_Data_Processing_GUI* components; *Data_Layer* contains the *Purchasing_Database* and *PM_Subsystem_1* components.

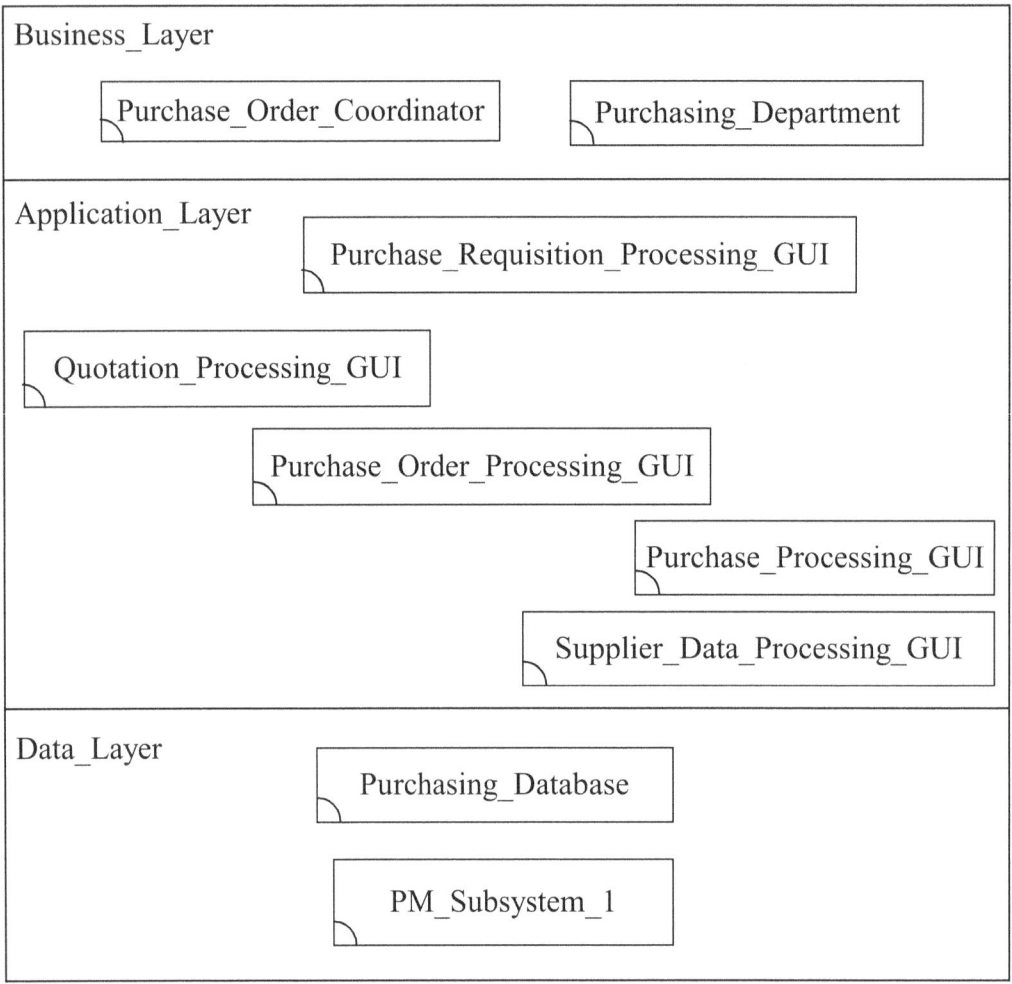

Figure 20-215   Design's FD of the *Purchasing Management*
Systems Definition Version 3

Figure 20-216 shows the design's COD of the *purchasing management* systems definition version 3. In the figure, component *Purchase_Order_Coordinator* has one operation: *Purchase_Requisition_Verify*; component *Purchasing_Department* has four operations: *Quotation_Verify*, *Purchase_Order_Verify*, *Purchase_Verify* and

Interview; component *Purchase_Requisition_Processing_GUI* has one operation: *Purchase_Requisition_Processing_Button_Click*; component *Quotation_Processing_GUI* has one operation: *Quotation_Processing_Button_Click*; component *Purchase_Order_Processing_GUI* has one operation: *Purchase_Order_Processing_Button_Click*; component *Purchase_Processing_GUI* has one operation: *Purchase_Processing_Button_Click*; component *Supplier_Data_Processing_GUI* has one operation: *Supplier_Data_Processing_Button_Click*; component *Purchasing_Database* has five operations: *SQL_Purchase_Requisition_Insert*, *SQL_Quotation_Insert*, *SQL_Purchase_Order_Insert*, *SQL_Purchase_Insert* and *SQL_Supplier_Data_Insert*; component *PM_Subsystem_1* has one operation: *Infrastructure_Resources_Share*.

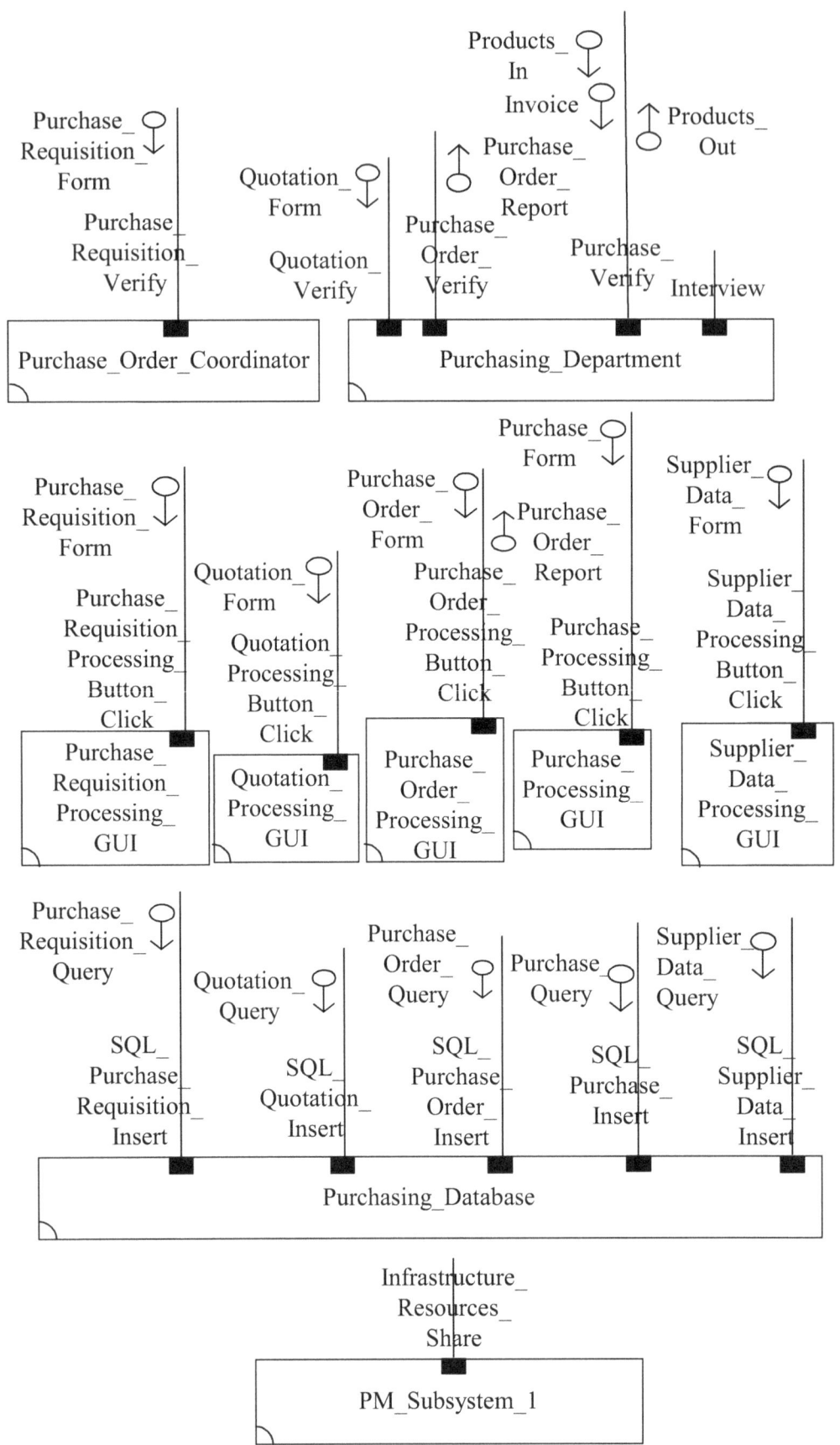

Figure 20-216   Design's COD of the *Purchasing Management Systems Definition Version 3*

The operation formula of *Purchase_Requisition_Verify* is *Purchase_Requisition_Verify(In Purchase_Requisition_Form)*. The operation formula of *Quotation_Verify* is *Quotation_Verify(In Quotation_Form)*. The operation formula of *Purchase_Order_Verify* is *Purchase_Order_Verify(Out Purchase_Order_Report)*. The operation formula of *Purchase_Verify* is *Purchase_Verify(In Products_In, Invoice; Out Products_Out)*. The operation formula of *Interview* is *Interview*. The operation formula of *Purchase_Requisition_Processing_Button_Click* is *Purchase_Requisition_Processing_Button_Click(In Purchase_Requisition_Form)*. The operation formula of *Quotation_Processing_Button_Click* is *Quotation_Processing_Button_Click(In Quotation_Form)*. The operation formula of *Purchase_Order_Processing_Button_Click* is *Purchase_Order_Processing_Button_Click(In Purchase_Order_Form; Out Purchase_Order_Report)*. The operation formula of *Purchase_Processing_Button_Click* is *Purchase_Processing_Button_Click(In Purchase_Form)*. The operation formula of *Supplier_Data_Processing_Button_Click* is *Supplier_Data_Processing_Button_Click(In Supplier_Data_Form)*. The operation formula of *SQL_Purchase_Requisition_Insert* is *SQL_Purchase_Requisition_Insert(In Purchase_Requisition_Query)*. The operation formula of *SQL_Quotation_Insert* is *SQL_Quotation_Insert(In Quotation_Query)*. The operation formula of *SQL_Purchase_Order_Insert* is *SQL_Purchase_Order_Insert (In Purchase_Order_Query)*. The operation formula of *SQL_Purchase_Insert* is *SQL_Purchase_Insert (In Purchase_Query)*. The operation formula of *SQL_Supplier_Data_Insert* is *SQL_Supplier_Data_Insert(In Supplier_Data_Query)*. The operation formula of *Infrastructure_Resources_Share* is *Infrastructure_Resources_Share*.

Figure 20-217 shows the composite data type specification of the *Purchase_Requisition_Form* input parameter occurring in the *Purchase_Requisition_Verify(In Purchase_Requisition_Form)* and *Purchase_Requisition_Processing_Button_Click(In Purchase_Requisition_Form)* operation formulas.

Parameter	*Purchase_Requisition_Form*

Data Type	TABLE of
	  Date : Text
	  OD : Text
	  ProductNo : Text
	  Quantity : Integer
	End TABLE ;

Instances	**Purchase Requisition Form** Date: 2011/10/17 Originating_Department : Sales Dept. ProductNo Quantity __A00001(Pen)_____300_____ __A00002(Mouse)_____400_____ __A00003(Camera)_____500_____

Figure 20-217   Composite Data Type Specification

Figure 20-218 shows the composite data type specification of the *Quotation_Form* input parameter occurring in the *Quotation_Verify(In Quotation_Form)* and *Quotation_Processing_Button_Click(In Quotation_Form)* operation formulas.

Parameter	*Quotation_Form*

Data Type	TABLE of
	  Date : Text
	  SupplierName: Text
	  ProductNo : Text
	  Quantity : Integer
	  UnitPrice : Real
	  Total : Real
	End TABLE ;

Instances	**Quotation Form** Date: 2011/10/25 SupplierName : Johnson Corp. ProductNo Quantity UnitPrice A 0 0 0 0 1 (P e n) 3 0 0 1 0 0 . 0 0 _A00002(Mouse)____400____200.00__ _A00003(Camera)___500____300.00__ Total : 260,000.00

Figure 20-218   Composite Data Type Specification

Figure 20-219 shows the composite data type specification of the *Purchase_Order_Report* output parameter occurring in the *Purchase_Order_Verify(Out Purchase_Order_Report)* and *Purchase_Order_Processing_Button_Click(In Purchase_Order_Form; Out Purchase_Order_Report)* operation formulas.

432

Parameter	Purchase_Order_Report

Data Type	TABLE of
	  Date : Text
	  SupplierName: Text
	  ProductNo : Text
	  Quantity : Integer
	End TABLE ;

Instances	Date : 20111118 SupplierName : Johnson Corp. | ProductNo	Quantity
A00001(Pen)	300
A00002(Mouse)	400
A00003(Camera)	500	

Figure 20-219   Composite Data Type Specification

Figure 20-220 shows the primitive data type specification of the *Products_In* input parameter occurring in the *Purchase_Verify(In Products_In, Invoice; Out Products_Out)* operation formula.

Parameter	Data Type	Instances
Products_In	Physical Object	Pen, Mouse, Camera

Figure 20-220   Primitive Data Type Specification

Figure 20-221 shows the composite data type specification of the *Invoice* input parameter occurring in the *Purchase_Verify(In Products_In, Invoice; Out Products_Out)* operation formula.

Parameter	*Invoice*

Data Type	TABLE of
	  Date : Text
	  SupplierName: Text
	  ProductNo : Text
	  Quantity : Integer
	  UnitPrice : Real
	  Total : Real
	End TABLE ;

Instances	Date : 20111130 SupplierName : Johnson Corp. ProductNo / Quantity / UnitPrice: A00001(Pen) — 300 — 100.00 A00002(Mouse) — 400 — 200.00 A00003(Camera) — 500 — 300.00 Total : 260,000.00

Figure 20-221   Composite Data Type Specification

Figure 20-222 shows the primitive data type specification of the *Products_Out* output parameter occurring in the *Purchase_Verify(In Products_In, Invoice; Out Products_Out)* operation formula.

Parameter	Data Type	Instances
Products_Out	Physical Object	Pen, Mouse, Camera

Figure 20-222   Primitive Data Type Specification

Figure 20-223 shows the composite data type specification of the *Purchase_Order_Form* input parameter occurring in the *Purchase_Order_Processing_Button_Click(In Purchase_Order_Form; Out Purchase_Order_Report)* operation formula.

434

Parameter	*Purchase_Order_Form*

Data Type	TABLE of
	  Date : Text
	  SupplierName: Text
	  ProductNo : Text
	  Quantity : Integer
	End TABLE ;

Instances	**Purchase Order Form** Date: 2011/11/18 SupplierName : Johnson Corp. ProductNo Quantity __A00001(Pen) ___300 ___ __A00002(Mouse)___400___ __A00003(Camera)___500___

Figure 20-223   Composite Data Type Specification

Figure 20-224 shows the composite data type specification of the *Purchase_Form* input parameter occurring in the *Purchase_Processing_Button_Click(In Purchase_Form)* operation formula.

Parameter	*Purchase_Form*

Data Type	TABLE of
	  Date : Text
	  SupplierName: Text
	  ProductNo : Text
	  Quantity : Integer
	  UnitPrice : Real
	  ReturnQuantity : Integer
	  Total : Real
	End TABLE ;

Instances	**Purchase Form** Date: 2011/12/12 SupplierName : Johnson Corp. ProductNo Quantity UnitPrice ReturnQuantity A00001(Pen) 300 100.00 0 A00002(Mouse) 390 200.00 10 A00003(Camera)500 300.00 0 Total : 258,000.00

Figure 20-224   Composite Data Type Specification

Figure 20-225 shows the composite data type specification of the *Supplier_Data_Form* input parameter occurring in the *Supplier_Data_Processing_Button_Click(In Supplier_Data_Form)* operation formula.

Parameter	*Supplier_Data_Form*
Data Type	TABLE of SupplierName:Text Address :Text PhoneNumber:Text FaxNumber:Text E-mail : Text Rank : Text End TABLE ;
Instances	**Supplier Data Form** SupplierName : Johnson Corp. Address : 1232 Fair Circle, Austin, TX PhoneNumber : 512-463-8472 FaxNumber : 512-463-8499 E-mail : Johnson1122@gmail.com Rank : B

Figure 20-225   Composite Data Type Specification

Figure 20-226 shows the composite data type specification of the *Purchase_Requisition_Query* input parameter occurring in the *SQL_Purchase_Requisition_Insert(In Purchase_Requisition_Query)* operation formula.

Parameter	*Purchase_Requisition_Query*

Data Type	TABLE of
	  Date : Text
	  OD : Text
	  ProductNo : Text
	  Quantity : Integer
	End TABLE ;

Instances	<table><tr><td>Date</td><td colspan="2">Originating_Department :</td></tr><tr><td>20111017</td><td colspan="2">Sales Dept.</td></tr><tr><td>ProductNo</td><td>Quantity</td></tr><tr><td>A00001(Pen)</td><td>300</td></tr><tr><td>A00002(Mouse)</td><td>400</td></tr><tr><td>A00003(Camera)</td><td>500</td></tr></table>

Figure 20-226　　Composite Data Type Specification

Figure 20-227 shows the composite data type specification of the *Quotation_Query* input parameter occurring in the *SQL_Quotation_Insert(In Quotation_Query)* operation formula.

438

Parameter	*Quotation_Query*

Data Type	TABLE of
	  Date : Text
	  SupplierName: Text
	  ProductNo : Text
	  Quantity : Integer
	  UnitPrice : Real
	  Total : Real
	End TABLE ;

Instances	<table><tr><th>Date</th><th>SupplierName</th><th>Total</th></tr><tr><td>20111025</td><td>Johnson Corp.</td><td>260,000.00</td></tr></table> <table><tr><th>ProductNo</th><th>Quantity</th><th>UnitPrice</th></tr><tr><td>A00001(Pen)</td><td>300</td><td>100.00</td></tr><tr><td>A00002(Mouse)</td><td>400</td><td>200.00</td></tr><tr><td>A00003(Camera)</td><td>500</td><td>300.00</td></tr></table>

Figure 20-227   Composite Data Type Specification

Figure 20-228 shows the composite data type specification of the *Purchase_Order_Query* input parameter occurring in the *SQL_Purchase_Order_Insert(In Purchase_Order_Query)* operation formula.

Parameter	*Purchase_Order_Query*

Data Type	TABLE of
	  Date : Text
	  SupplierName: Text
	  ProductNo : Text
	  Quantity : Integer
	End TABLE ;

Instances	<table><tr><td>Date</td><td>SupplierName</td></tr><tr><td>20111118</td><td>Johnson Corp.</td></tr></table> <table><tr><td>ProductNo</td><td>Quantity</td></tr><tr><td>A00001(Pen)</td><td>300</td></tr><tr><td>A00002(Mouse)</td><td>400</td></tr><tr><td>A00003(Camera)</td><td>500</td></tr></table>

Figure 20-228   Composite Data Type Specification

Figure 20-229 shows the composite data type specification of the *Purchase_Query* input parameter occurring in the *SQL_Purchase_Insert (In Purchase_Query)* operation formula.

Parameter	*Purchase_Query*

Data Type	TABLE of
	  Date : Text
	  SupplierName: Text
	  ProductNo : Text
	  Quantity : Integer
	  UnitPrice : Real
	  ReturnQuantity : Integer
	  Total : Real
	End TABLE ;

Instances	

Date	SupplierName	Total
20111212	Johnson Corp.	258,000.00

ProductNo	Quantity	UnitPrice	ReturnQuantity
A00001(Pen)	300	100.00	0
A00002(Mouse)	390	200.00	10
A00003(Camera)	500	300.00	0

Figure 20-229   Composite Data Type Specification

Figure 20-230 shows the composite data type specification of the *Supplier_Data_Query* input parameter occurring in the *SQL_Supplier_Data_Insert(In Supplier_Data_Query)* operation formula.

Parameter	*Supplier_Data_Query*
Data Type	TABLE of SupplierName:Text Address :Text PhoneNumber:Text FaxNumber:Text E-mail : Text Rank : Text End TABLE ;
Instances	

SupplierName	Johnson Corp.
Address	1232 Fair Circle, Austin, TX
PhoneNumber	512-463-8472
FaxNumber	512-463-8499
E-mail	Johnson1122@gmail.com
Rank	B

Figure 20-230   Composite Data Type Specification

Figure 20-231 shows the design's CCD of the *purchasing management* systems definition version 3. In the figure, actor *Originating_Department* has a connection with the *Purchase_Order_Coordinator* component; actor *Supplier* has four connections with the *Purchasing_Department* component; component *Purchase_Order_Coordinator* has a connection with the *Purchase_Requisition_Processing_GUI* component; component *Purchasing_Department* has a connection with each one of the *Quotation_Processing_GUI*, *Purchase_Order_Processing_GUI*, *Purchase_Processing_GUI* and *Supplier_Data_Processing_GUI* components; each one of the *Purchase_Requisition_Processing_GUI*, *Quotation_Processing_GUI*, *Purchase_Order_Processing_GUI*, *Purchase_Processing_GUI* and *Supplier_Data_Processing_GUI* components has a connection with the *Purchasing_Database* component; component *Purchasing_Database* has a connection with the *PM_Subsystem_1* component.

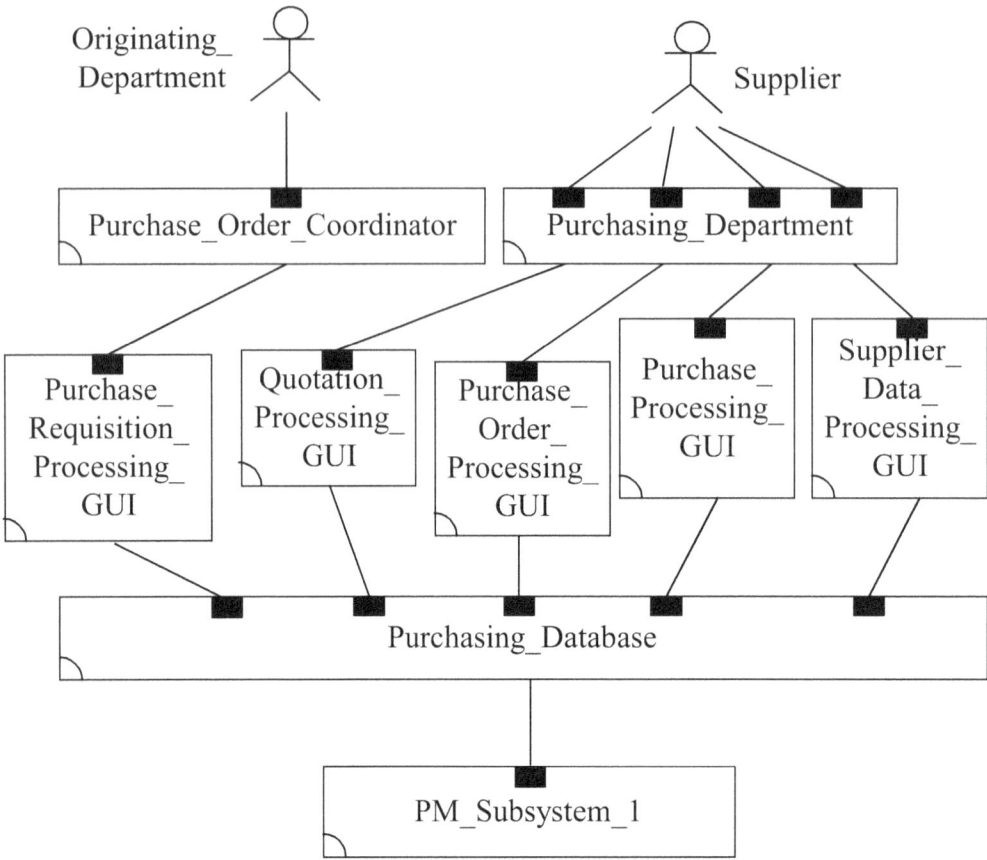

Figure 20-231   Design's CCD of the *Purchasing Management Systems Definition Version 3*

20-3-2-3-2 Design's Systems Behavior of the Purchasing Management Systems Definition Version 3

The entire design's systems behavior of the *purchasing management* systems definition version 3 includes: a) *Design's SBCD* and b) *Design's IFD* of the *purchasing management* systems definition version 3.

Figure 20-232 shows the design's SBCD of the *purchasing management* systems definition version 3 in which interactions among the *Originating_Department*, *Supplier* actors and the *Purchase_Order_Coordinator*, *Purchasing_Department*, *Purchase_Requisition_Processing_GUI*, *Quotation_Processing_GUI*, *Purchase_Order_Processing_GUI*, *Purchase_Processing_GUI*, *Supplier_Data_Processing_GUI*, *Purchasing_Database*, *PM_Subsystem_1* components shall draw forth the *Purchase_Requisition*, *Quotation*, *Purchase_Order*, *Purchase* and *Collect_Supplier_Data* behaviors.

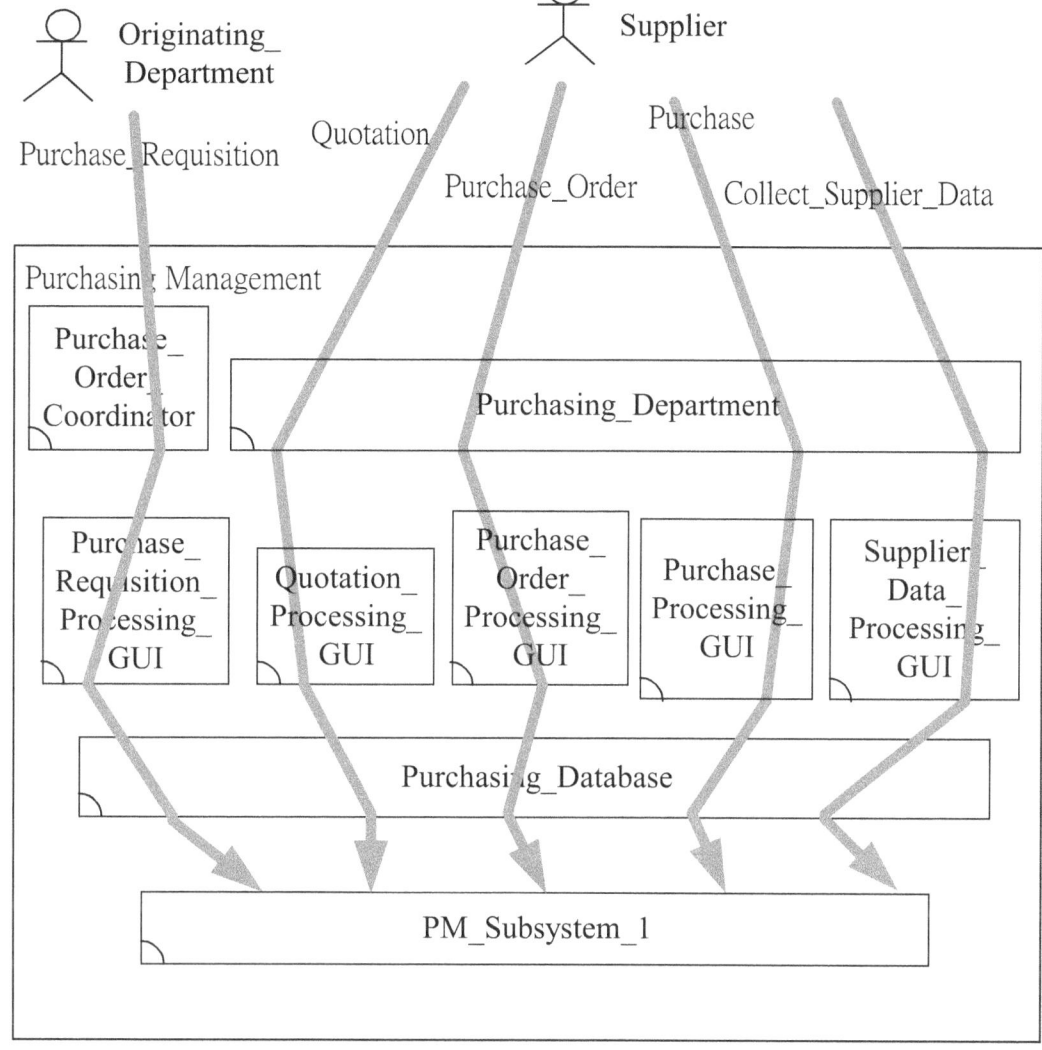

Figure 20-232   Design's SBCD of the *Purchasing Management*
Systems Definition Version 3

The overall design's behavior of the *purchasing management* systems definition version 3 includes the *Purchase_Requisition*, *Quotation*, *Purchase_Order*, *Purchase* and *Collect_Supplier_Data* behaviors. In other words, the *Purchase_Requisition*, *Quotation*, *Purchase_Order*, *Purchase* and *Collect_Supplier_Data* behaviors together provide the overall design's behavior of the *purchasing management* systems definition version 3.

Be noticed that the *Purchase_Requisition*, *Quotation*, *Purchase_Order*, *Purchase* and *Collect_Supplier_Data* behaviors are mutually independent of each other. They tend to be executed concurrently [Hoar85, Miln89, Miln99].

The overall design's behavior of the *purchasing management* systems definition version 3 includes five individual behaviors: *Purchase_Requisition*,

Quotation, *Purchase_Order*, *Purchase* and *Collect_Supplier_Data*. Each individual behavior is represented by an execution path. We use an interaction flow diagram (IFD) to define each one of these execution paths. Figure 20-233 shows the design's IFD of the *purchasing management* systems definition version 3 *Purchase_Requisition* behavior. First, actor *Originating_Department* interacts with the *Purchase_Order_Coordinator* component through the *Purchase_Requisition_Verify* operation call interaction, carrying the *Purchase_Requisition_Form* input parameter. Next, component *Purchase_Order_Coordinator* interacts with the *Purchase_Requisition_Processing_GUI* component through the *Purchase_Requisition_Processing_Button_Click* operation call interaction, carrying the *Purchase_Requisition_Form* input parameter. Continuingly, component *Purchase_Requisition_Processing_GUI* interacts with the *Purchasing_Database* component through the *SQL_Purchase_Requisition_Insert* operation call interaction, carrying the *Purchase_Requisition_Query* input parameter. Finally, component *Purchasing_Database* interacts with the *PM_Subsystem_1* component through the *Infrastructure_Resources_Share* operation call interaction.

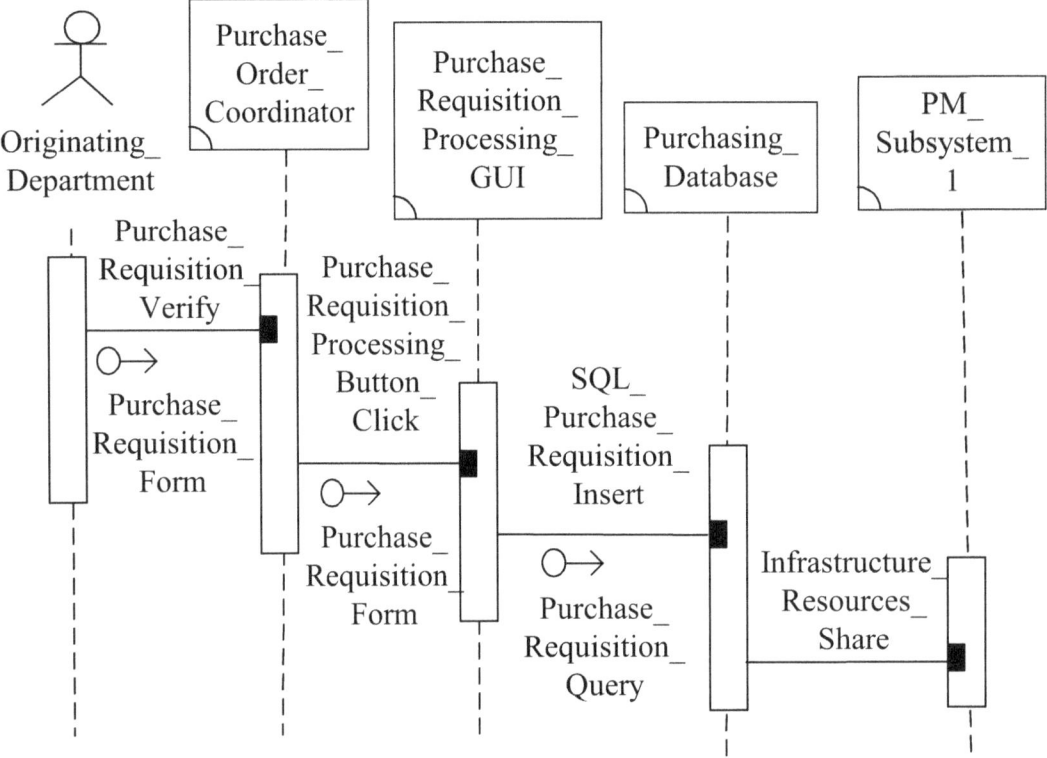

Figure 20-233   Design's IFD of the *Purchasing Management*
Systems Definition Version 3 *Purchase_Requisition* Behavior

Figure 20-234 shows the design's IFD of the *purchasing management* systems definition version 3 *Quotation* behavior. First, actor *Supplier* interacts with the *Purchasing_Department* component through the *Quotation_Verify* operation call interaction, carrying the *Quotation_Form* input parameter. Next, component *Purchasing_Department* interacts with the *Quotation_Processing_GUI* component through the *Quotation_Processing_Button_Click* operation call interaction, carrying the *Quotation_Form* input parameter. Continuingly, component *Quotation_Processing_GUI* interacts with the *Purchasing_Database* component through the *SQL_Quotation_Insert* operation call interaction, carrying the *Quotation_Query* input parameter. Finally, component *Purchasing_Database* interacts with the *PM_Subsystem_1* component through the *Infrastructure_Resources_Share* operation call interaction.

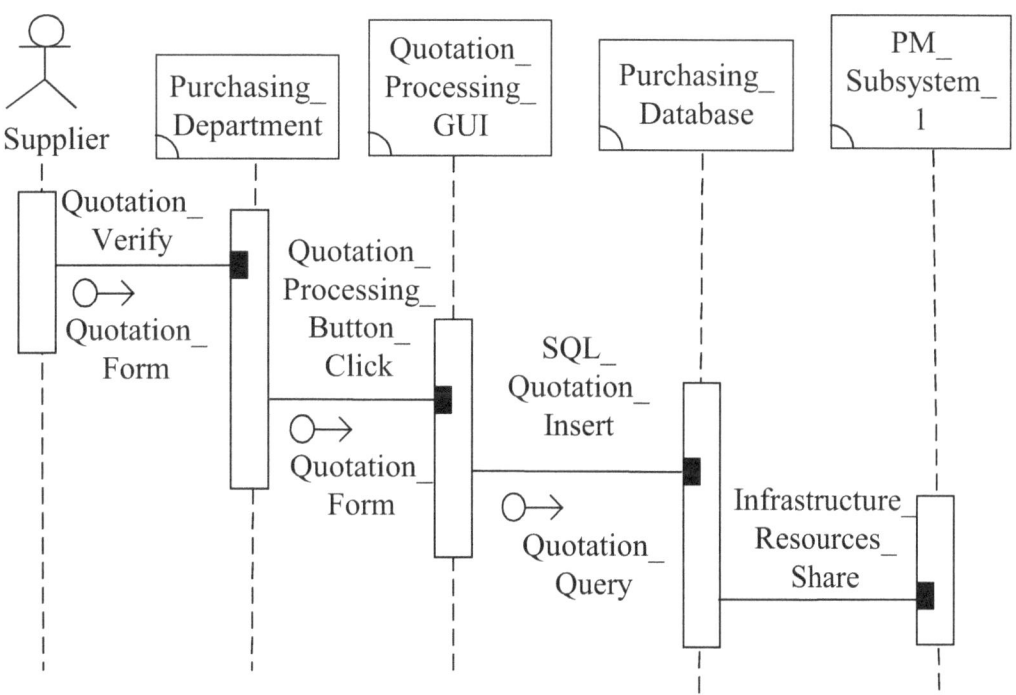

Figure 20-234   Design's IFD of the *Purchasing Management*
Systems Definition Version 3 *Quotation* Behavior

Figure 20-235 shows the design's IFD of the *purchasing management* systems definition version 3 *Purchase_Order* behavior. First, actor *Supplier* interacts with the *Purchasing_Department* component through the *Purchase_Order_Verify* operation call interaction. Next, component *Purchasing_Department* interacts with the *Purchase_Order_Processing_GUI* component through the *Purchase_Order_Processing_Button_Click* operation call interaction, carrying the

Purchase_Order_Form input parameter. Continuingly, component *Purchase_Order_Processing_GUI* interacts with the *Purchasing_Database* component through the *SQL_Purchase_Order_Insert* operation call interaction, carrying the *Purchase_Order_Query* input parameter. Continuingly, component *Purchasing_Database* interacts with the *PM_Subsystem_1* component through the *Infrastructure_Resources_Share* operation call interaction. Continuingly, component *Purchasing_Department* interacts with the *Purchase_Order_Processing_GUI* component through the *Purchase_Order_Processing_Button_Click* operation return interaction, carrying the *Purchase_Order_Report* output parameter. Finally, actor *Supplier* interacts with the *Purchasing_Department* component through the *Purchase_Order_Verify* operation return interaction, carrying the *Purchase_Order_Report* output parameter.

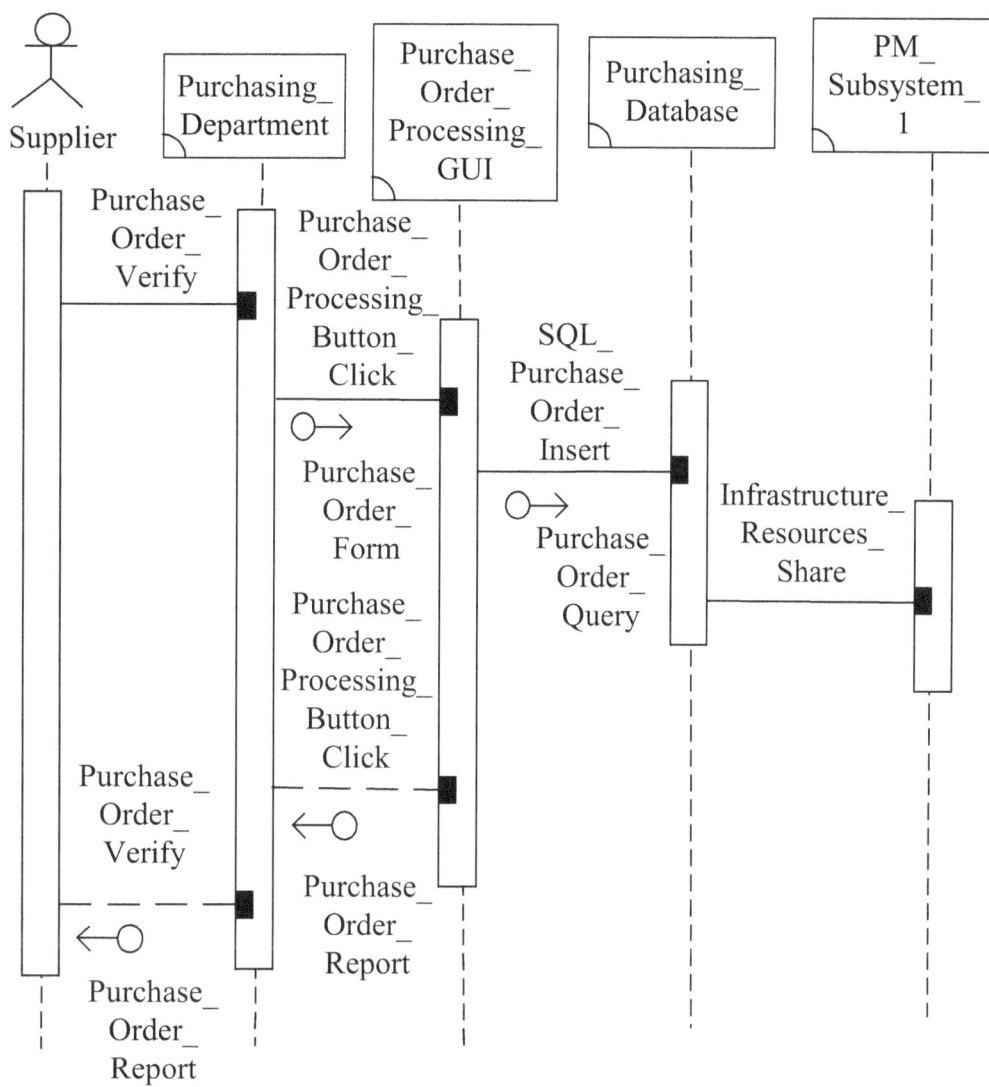

Figure 20-235   Design's IFD of the *Purchasing Management*
Systems Definition Version 3 *Purchase_Order* Behavior

Figure 20-236 shows the design's IFD of the *purchasing management* systems definition version 3 *Purchase* behavior. First, actor *Supplier* interacts with the *Purchasing_Department* component through the *Purchase_Verify* operation call interaction, carrying the *Products_In* and *Invoice* input parameters. Next, component *Purchasing_Department* interacts with the *Purchase_Processing_GUI* component through the *Purchase_Processing_Button_Click* operation call interaction, carrying the *Purchase_Form* input parameter. Continuingly, component *Purchase_Processing_GUI* interacts with the *Purchasing_Database* component through the *SQL_Purchase_Insert* operation call interaction, carrying the *Purchase_Query* input parameter. Continuingly, component *Purchasing_Database* interacts with the *PM_Subsystem_1* component through the *Infrastructure_Resources_Share* operation call interaction. Finally, actor *Supplier* interacts with the *Purchasing_Department* component through the *Purchase_Verify* operation return interaction, carrying the *Products_Out* output parameter.

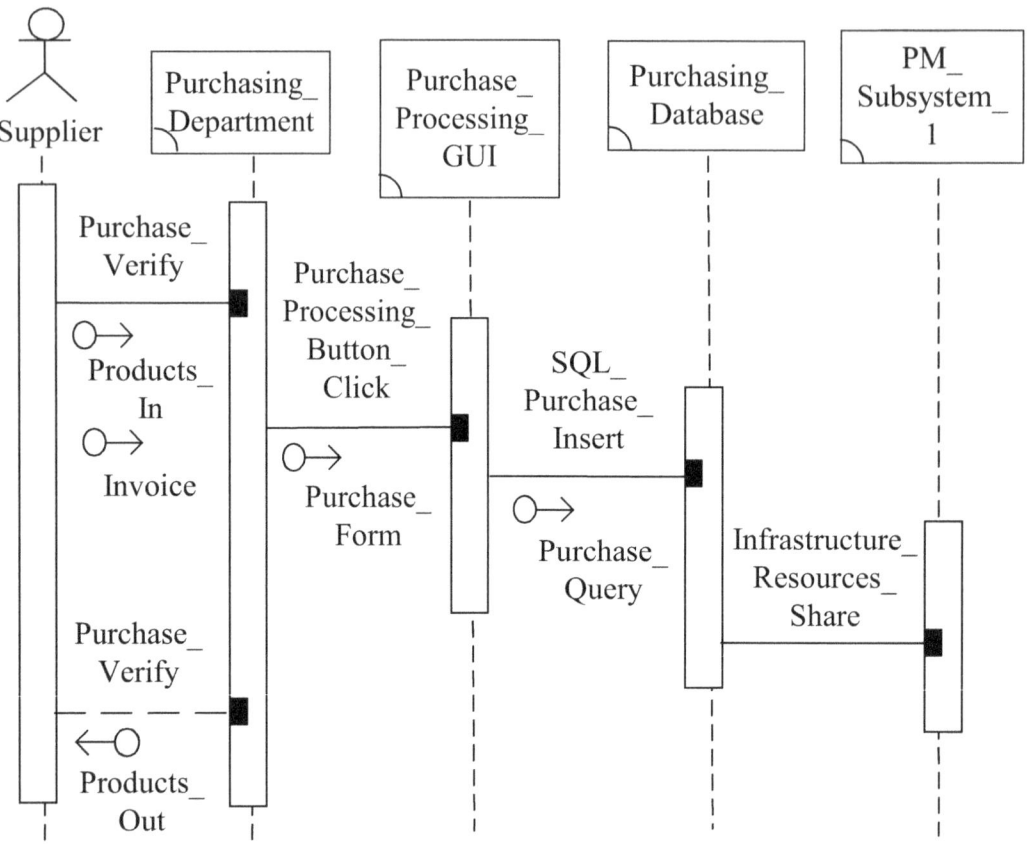

Figure 20-236   Design's IFD of the *Purchasing Management*
Systems Definition Version 3 *Purchase* Behavior

Figure 20-237 shows the design's IFD of the *purchasing management* systems definition version 3 *Collect_Supplier_Data* behavior. First, actor *Supplier* interacts with the *Purchasing_Department* component through the *Interview* operation call interaction. Next, component *Purchasing_Department* interacts with the *Supplier_Data_Processing_GUI* component through the *Supplier_Data_Processing_Button_Click* operation call interaction, carrying the *Supplier_Data_Form* input parameter. Continuingly, component *Supplier_Data_Processing_GUI* interacts with the *Purchasing_Database* component through the *SQL_Supplier_Data_Insert* operation call interaction, carrying the *Supplier_Data_Query* input parameter. Finally, component *Purchasing_Database* interacts with the *PM_Subsystem_1* component through the *Infrastructure_Resources_Share* operation call interaction.

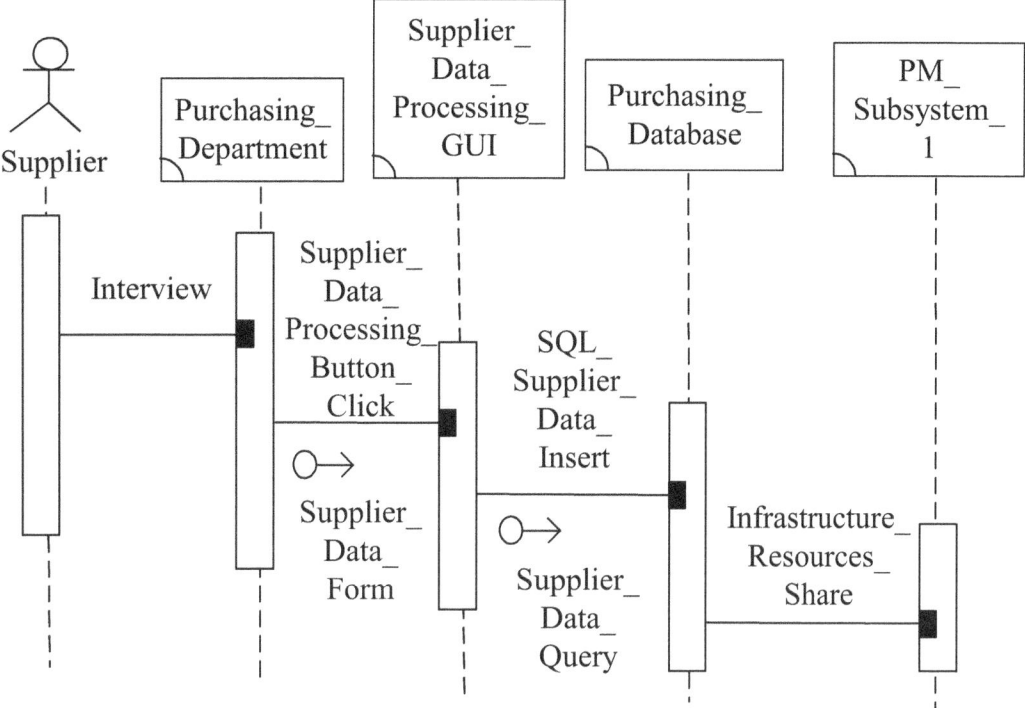

Figure 20-237   Design's IFD of the *Purchasing Management Systems Definition Version 3 Collect_Supplier_Data* Behavior

20-3-2-4 Implementation View of the Purchasing Management Systems Definition Version 3

The implementation view of the *purchasing management* systems definition version 3 is shown in Figure 20-238.

450

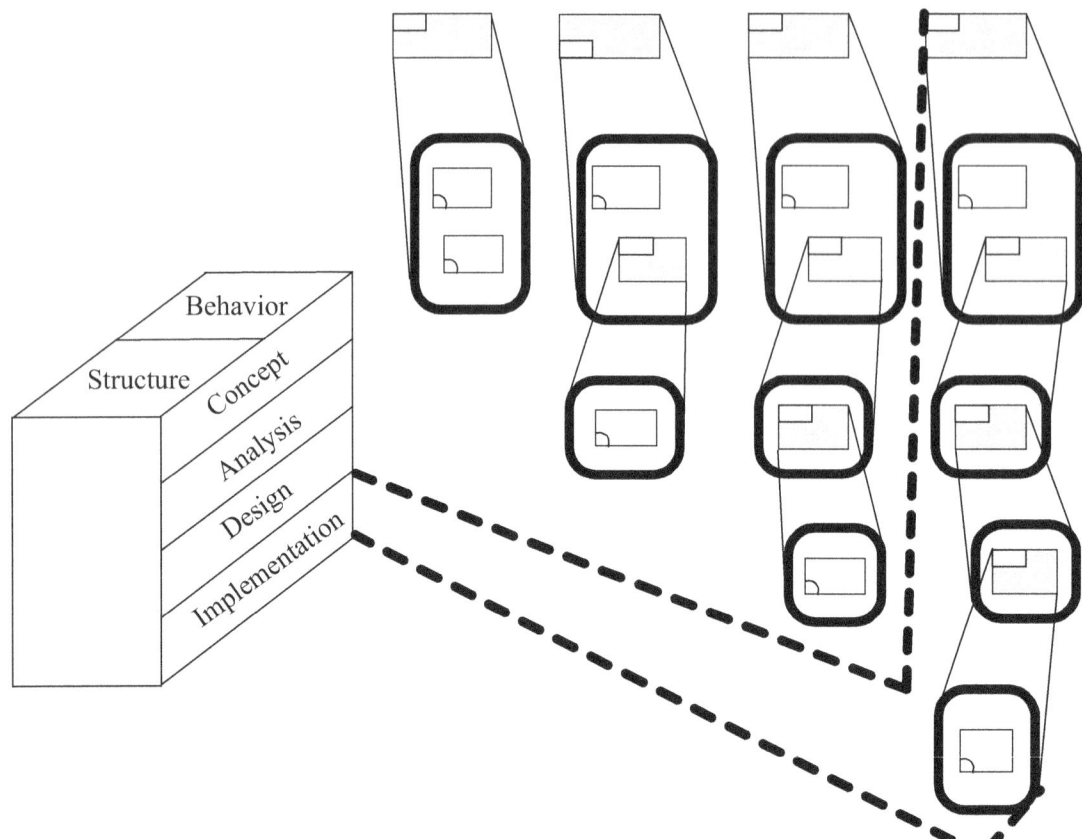

Figure 20-238   Implementation View of the *Purchasing Management*
Systems Definition Version 3

The implementation view of the *purchasing management* systems definition version 3 consists of: a) implementation's systems structure of the *purchasing management* systems definition version 3 and b) implementation's systems behavior of the *purchasing management* systems definition version 3.

20-3-2-4-1 Implementation's Systems Structure of the Purchasing Management Systems Definition Version 3

The entire implementation's systems structure of the *purchasing management* systems definition version 3 includes: a) *Implementation's AHD*, b) *Implementation's FD*, c) *Implementation's COD* and d) *Implementation's CCD* of the *purchasing management* systems definition version 3.

We first draw the implementation's AHD of the *purchasing management* systems definition version 3. As shown in Figure 20-239, *Purchasing Management* is composed of *Purchase_Order_Coordinator*, *Purchasing_Department* and *PM_Subsystem_3*; *PM_Subsystem_3* is composed of *Purchase_Requisition_Processing_GUI*, *Quotation_Processing_GUI*, *Purchase_Order_Processing_GUI*, *Purchase_Processing_GUI* and *PM_Subsystem_2*;

451

PM_Subsystem_2 is composed of *Purchasing_Database* and *PM_Subsystem_1*; *PM_Subsystem_1* is composed of *Network_Operating_System*. In the figure, *Purchasing Management, PM_Subsystem_3, PM_Subsystem_2 and PM_Subsystem_1* are aggregated systems while *Purchase_Order_Coordinator*, *Purchasing_Department*, *Purchase_Requisition_Processing_GUI*, *Quotation_Processing_GUI*, *Purchase_Order_Processing_GUI*, *Purchase_Processing_GUI*, *Purchasing_Database* and *Network_Operating_System* are non-aggregated systems.

452

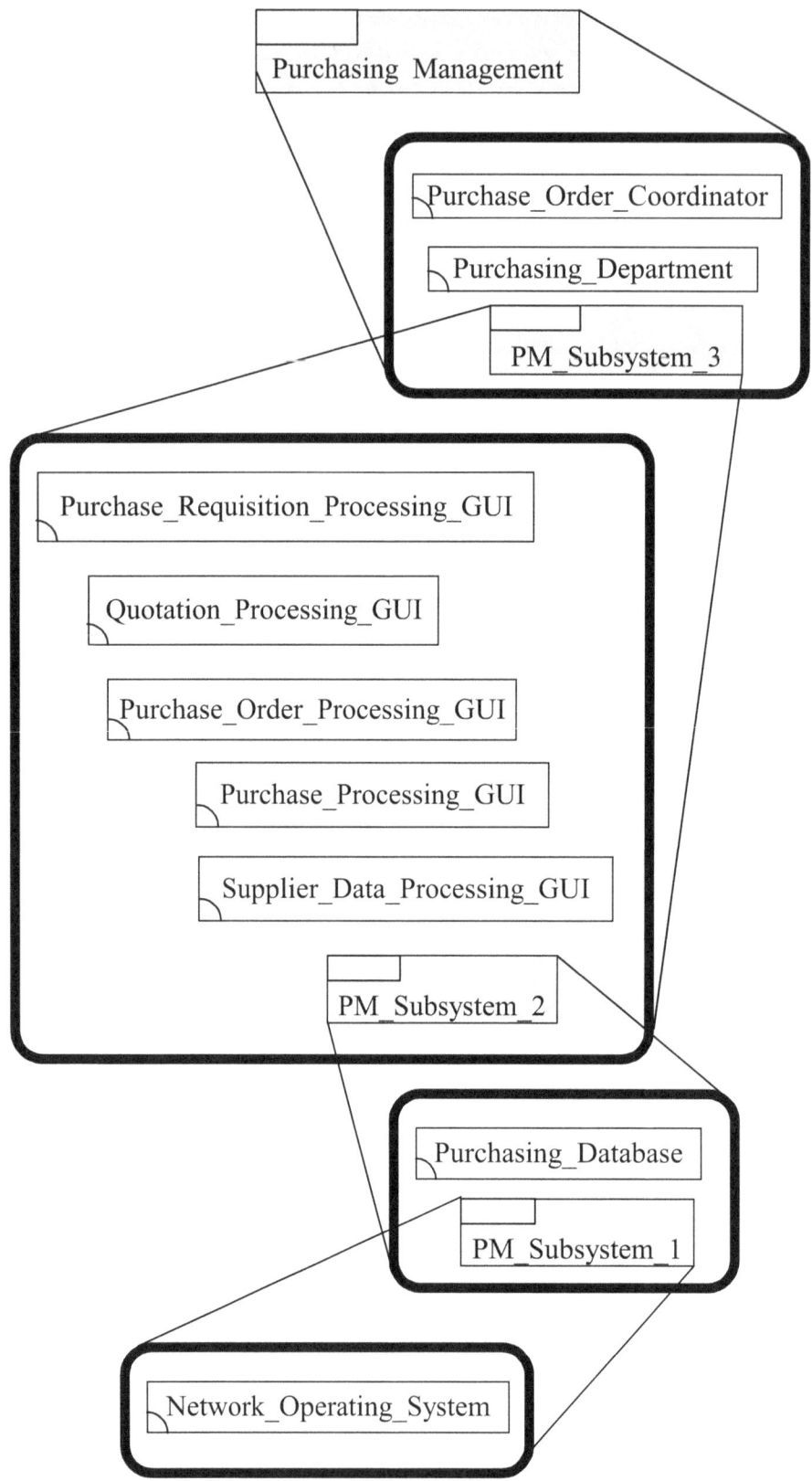

Figure 20-239   Implementation's AHD of the *Purchasing Management*
Systems Definition Version 3

Figure 20-240 shows the implementation's FD of the *purchasing management* systems definition version 3. In the figure, *Business_Layer* contains the *Purchase_Order_Coordinator* and *Purchasing_Department* components; *Application_Layer* contains the *Purchase_Requisition_Processing_GUI*, *Quotation_Processing_GUI*, *Purchase_Order_Processing_GUI*, *Purchase_Processing_GUI* and *Supplier_Data_Processing_GUI* components; *Data_Layer* contains the *Purchasing_Database* component; *Technology_Layer* contains the *Network_Operating_System* component.

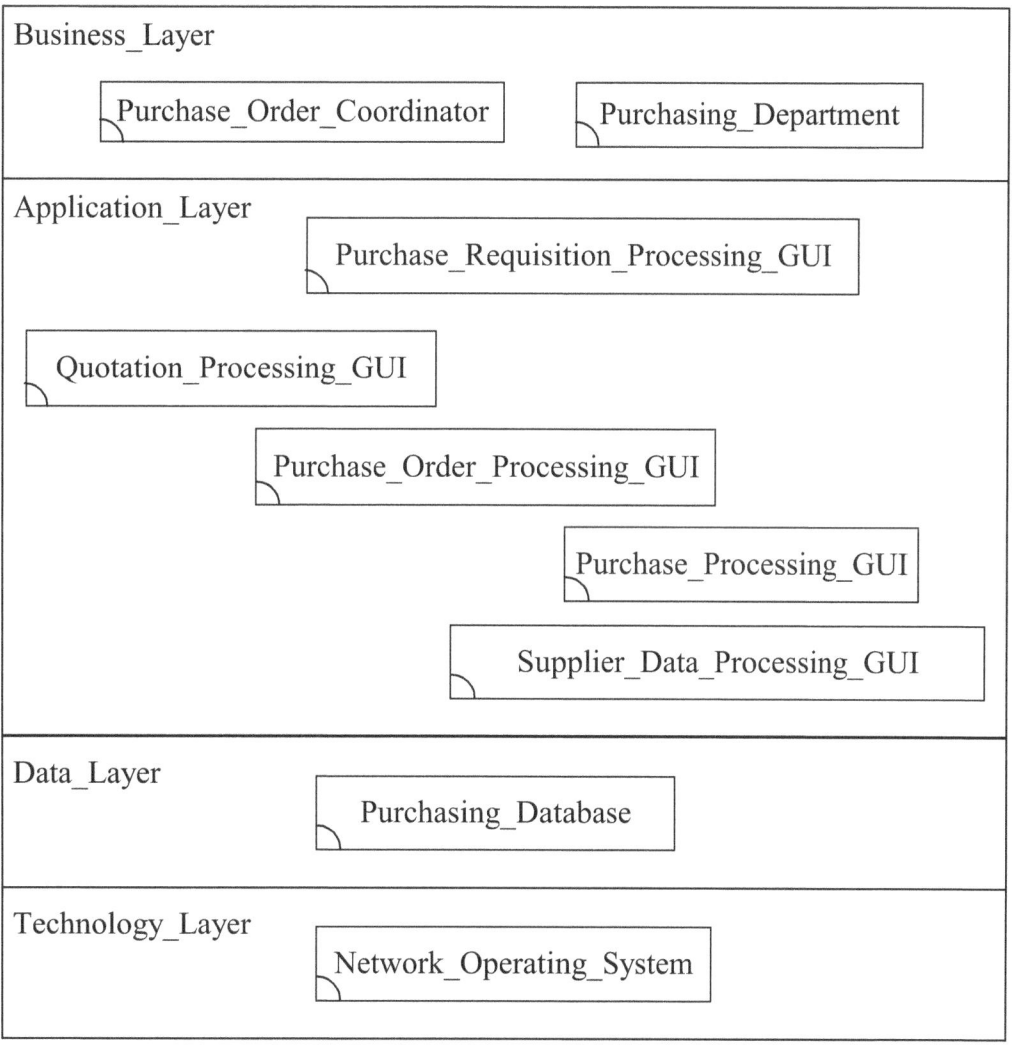

Figure 20-240   Implementation's FD of the *Purchasing Management Systems Definition Version 3*

Figure 20-241 shows the implementation's COD of the *purchasing management* systems definition version 3. In the figure, component *Purchase_Order_Coordinator* has one operation: *Purchase_Requisition_Verify*;

component *Purchasing_Department* has four operations: *Quotation_Verify*, *Purchase_Order_Verify*, *Purchase_Verify* and *Interview*; component *Purchase_Requisition_Processing_GUI* has one operation: *Purchase_Requisition_Processing_Button_Click*; component *Quotation_Processing_GUI* has one operation: *Quotation_Processing_Button_Click*; component *Purchase_Order_Processing_GUI* has one operation: *Purchase_Order_Processing_Button_Click*; component *Purchase_Processing_GUI* has one operation: *Purchase_Processing_Button_Click*; component *Supplier_Data_Processing_GUI* has one operation: *Supplier_Data_Processing_Button_Click*; component *Purchasing_Database* has five operations: *SQL_Purchase_Requisition_Insert*, *SQL_Quotation_Insert*, *SQL_Purchase_Order_Insert*, *SQL_Purchase_Insert* and *SQL_Supplier_Data_Insert*; component *Network_Operating_System* has one operation: *Infrastructure_Resources_Share*.

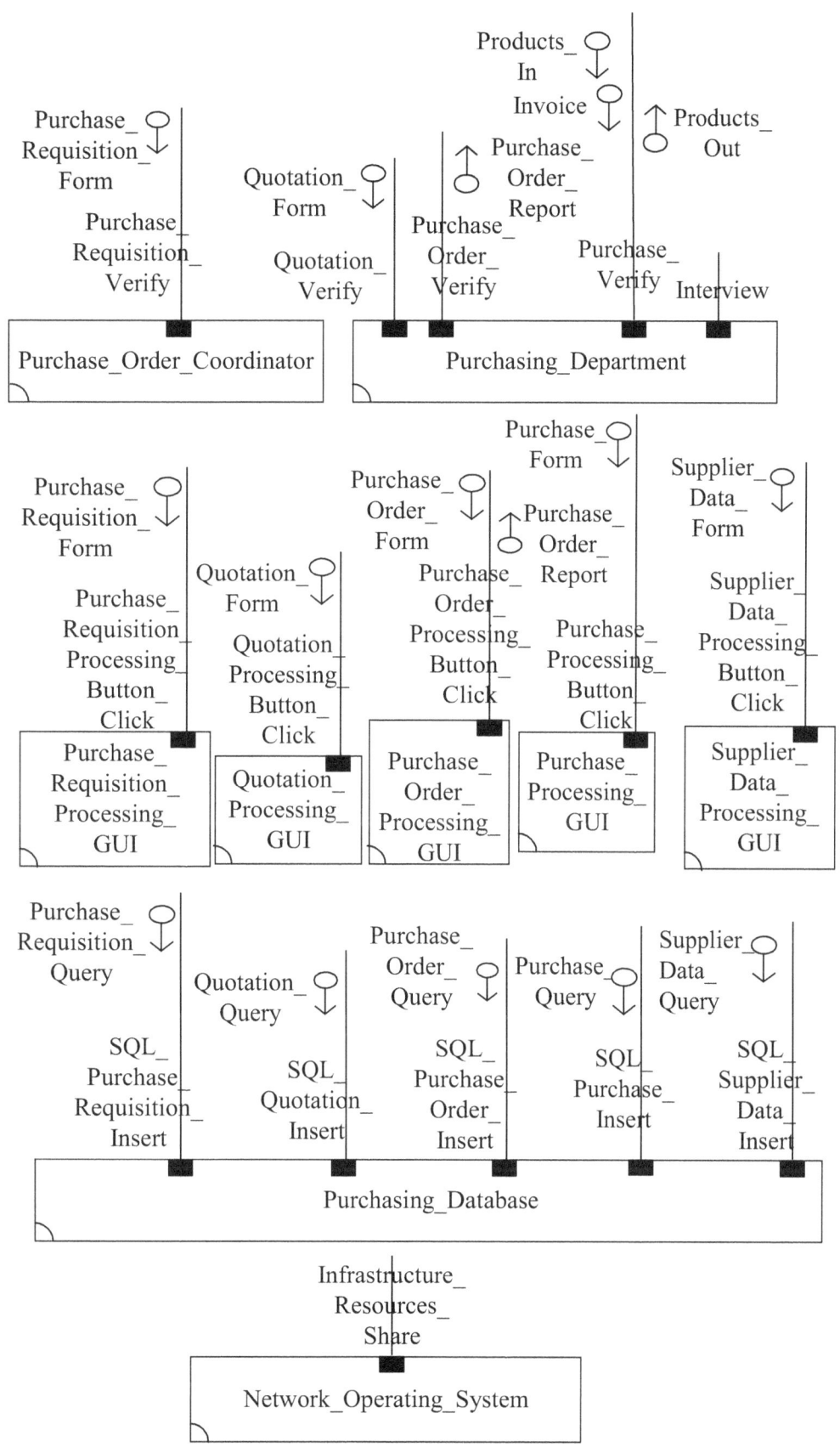

Figure 20-241   Implementation's COD of the *Purchasing Management Systems Definition Version 3*

The operation formula of *Purchase_Requisition_Verify* is *Purchase_Requisition_Verify(In Purchase_Requisition_Form)*. The operation formula of *Quotation_Verify* is *Quotation_Verify(In Quotation_Form)*. The operation formula of *Purchase_Order_Verify* is *Purchase_Order_Verify(Out Purchase_Order_Report)*. The operation formula of *Purchase_Verify* is *Purchase_Verify(In Products_In, Invoice; Out Products_Out)*. The operation formula of *Interview* is *Interview*. The operation formula of *Purchase_Requisition_Processing_Button_Click* is *Purchase_Requisition_Processing_Button_Click(In Purchase_Requisition_Form)*. The operation formula of *Quotation_Processing_Button_Click* is *Quotation_Processing_Button_Click(In Quotation_Form)*. The operation formula of *Purchase_Order_Processing_Button_Click* is *Purchase_Order_Processing_Button_Click(In Purchase_Order_Form; Out Purchase_Order_Report)*. The operation formula of *Purchase_Processing_Button_Click* is *Purchase_Processing_Button_Click(In Purchase_Form)*. The operation formula of *Supplier_Data_Processing_Button_Click* is *Supplier_Data_Processing_Button_Click(In Supplier_Data_Form)*. The operation formula of *SQL_Purchase_Requisition_Insert* is *SQL_Purchase_Requisition_Insert(In Purchase_Requisition_Query)*. The operation formula of *SQL_Quotation_Insert* is *SQL_Quotation_Insert(In Quotation_Query)*. The operation formula of *SQL_Purchase_Order_Insert* is *SQL_Purchase_Order_Insert (In Purchase_Order_Query)*. The operation formula of *SQL_Purchase_Insert* is *SQL_Purchase_Insert (In Purchase_Query)*. The operation formula of *SQL_Supplier_Data_Insert* is *SQL_Supplier_Data_Insert(In Supplier_Data_Query)*. The operation formula of *Infrastructure_Resources_Share* is *Infrastructure_Resources_Share*.

Figure 20-242 shows the composite data type specification of the *Purchase_Requisition_Form* input parameter occurring in the *Purchase_Requisition_Verify(In Purchase_Requisition_Form)* and *Purchase_Requisition_Processing_Button_Click(In Purchase_Requisition_Form)* operation formulas.

457

Parameter	*Purchase_Requisition_Form*

Data Type	TABLE of
	  Date : Text
	  OD : Text
	  ProductNo : Text
	  Quantity : Integer
	End TABLE ;

Instances	**Purchase Requisition Form** Date: 2011/10/17 Originating_Department : Sales Dept. ProductNo Quantity __A00001(Pen)_____300_____ __A00002(Mouse)_____400_____ __A00003(Camera)_____500_____

Figure 20-242   Composite Data Type Specification

Figure 20-243 shows the composite data type specification of the *Quotation_Form* input parameter occurring in the *Quotation_Verify(In Quotation_Form)* and *Quotation_Processing_Button_Click(In Quotation_Form)* operation formulas.

458

Parameter	*Quotation_Form*

Data Type	TABLE of
	  Date : Text
	  SupplierName: Text
	  ProductNo : Text
	  Quantity : Integer
	  UnitPrice : Real
	  Total : Real
	End TABLE ;

Instances	**Quotation Form** Date: 2011/10/25 SupplierName : Johnson Corp. ProductNo Quantity UnitPrice A00001(Pen) 300 100.00 A00002(Mouse) 400 200.00 A00003(Camera) 500 300.00 Total : 260,000.00

Figure 20-243   Composite Data Type Specification

Figure 20-244 shows the composite data type specification of the *Purchase_Order_Report* output parameter occurring in the *Purchase_Order_Verify(Out Purchase_Order_Report)* and *Purchase_Order_Processing_Button_Click(In Purchase_Order_Form; Out Purchase_Order_Report)* operation formulas.

459

Parameter	*Purchase_Order_Report*

Data Type	TABLE of
	  Date : Text
	  SupplierName: Text
	  ProductNo : Text
	  Quantity : Integer
	End TABLE ;

Instances	Date : 20111118 SupplierName : Johnson Corp. | ProductNo	Quantity
A00001(Pen)	300
A00002(Mouse)	400
A00003(Camera)	500	

Figure 20-244   Composite Data Type Specification

Figure 20-245 shows the primitive data type specification of the *Products_In* input parameter occurring in the *Purchase_Verify(In Products_In, Invoice; Out Products_Out)* operation formula.

Parameter	Data Type	Instances
Products_In	Physical Object	Pen, Mouse, Camera

Figure 20-245   Primitive Data Type Specification

Figure 20-246 shows the composite data type specification of the *Invoice* input parameter occurring in the *Purchase_Verify(In Products_In, Invoice; Out Products_Out)* operation formula.

Parameter	*Invoice*

Data Type	TABLE of
	  Date : Text
	  SupplierName: Text
	  ProductNo : Text
	  Quantity : Integer
	  UnitPrice : Real
	  Total : Real
	End TABLE ;

Instances	Date : 20111130 SupplierName : Johnson Corp. { ProductNo / Quantity / UnitPrice table }

Instances detail:

Date : 20111130
SupplierName : Johnson Corp.

ProductNo	Quantity	UnitPrice
A00001(Pen)	300	100.00
A00002(Mouse)	400	200.00
A00003(Camera)	500	300.00

Total : 260,000.00

Figure 20-246   Composite Data Type Specification

Figure 20-247 shows the primitive data type specification of the *Products_Out* output parameter occurring in the *Purchase_Verify(In Products_In, Invoice; Out Products_Out)* operation formula.

Parameter	Data Type	Instances
Products_Out	Physical Object	Pen, Mouse, Camera

Figure 20-247   Primitive Data Type Specification

Figure 20-248 shows the composite data type specification of the *Purchase_Order_Form* input parameter occurring in the *Purchase_Order_Processing_Button_Click(In Purchase_Order_Form; Out Purchase_Order_Report)* operation formula.

461

Parameter	*Purchase_Order_Form*

Data Type	TABLE of
	  Date : Text
	  SupplierName: Text
	  ProductNo : Text
	  Quantity : Integer
	End TABLE ;

Instances	**Purchase Order Form** Date: 2011/11/18 SupplierName : Johnson Corp. ProductNo Quantity __A00001(Pen) ____300 ___ __A00002(Mouse)___400__ __A00003(Camera)___500___

Figure 20-248   Composite Data Type Specification

Figure 20-249 shows the composite data type specification of the *Purchase_Form* input parameter occurring in the *Purchase_Processing_Button_Click(In Purchase_Form)* operation formula.

462

Parameter	*Purchase_Form*

Data Type	TABLE of
	  Date : Text
	  SupplierName: Text
	  ProductNo : Text
	  Quantity : Integer
	  UnitPrice : Real
	  ReturnQuantity : Integer
	  Total : Real
	End TABLE ;

Instances	**Purchase Form** Date: 2011/12/12 SupplierName : Johnson Corp. ProductNo Quantity UnitPrice ReturnQuantity A00001(Pen) 300 100.00 0 A00002(Mouse) 390 200.00 10 A00003(Camera)500 300.00 0 Total : 258,000.00

Figure 20-249   Composite Data Type Specification

Figure 20-250 shows the composite data type specification of the *Supplier_Data_Form* input parameter occurring in the *Supplier_Data_Processing_Button_Click(In Supplier_Data_Form)* operation formula.

Parameter	*Supplier_Data_Form*

Data Type	TABLE of
	  SupplierName:Text
	  Address :Text
	  PhoneNumber:Text
	  FaxNumber:Text
	  E-mail : Text
	  Rank : Text
	End TABLE ;

Instances	**Supplier Data Form** SupplierName : Johnson Corp. Address : 1232 Fair Circle, Austin, TX PhoneNumber : 512-463-8472 FaxNumber : 512-463-8499 E-mail : Johnson1122@gmail.com Rank : B

Figure 20-250   Composite Data Type Specification

Figure 20-251 shows the composite data type specification of the *Quotation_Query* input parameter occurring in the *SQL_Quotation_Insert(In Quotation_Query)* operation formula.

Parameter	*Quotation_Query*

Data Type	TABLE of
	  Date : Text
	  SupplierName: Text
	  ProductNo : Text
	  Quantity : Integer
	  UnitPrice : Real
	  Total : Real
	End TABLE ;

Instances	<table><tr><th>Date</th><th>SupplierName</th><th>Total</th></tr><tr><td>20111025</td><td>Johnson Corp.</td><td>260,000.00</td></tr></table> <table><tr><th>ProductNo</th><th>Quantity</th><th>UnitPrice</th></tr><tr><td>A00001(Pen)</td><td>300</td><td>100.00</td></tr><tr><td>A00002(Mouse)</td><td>400</td><td>200.00</td></tr><tr><td>A00003(Camera)</td><td>500</td><td>300.00</td></tr></table>

Figure 20-251   Composite Data Type Specification

Figure 20-252 shows the composite data type specification of the *Purchase_Requisition_Query* input parameter occurring in the *SQL_Purchase_Requisition_Insert(In Purchase_Requisition_Query)* operation formula.

Parameter	*Purchase_Requisition_Query*

Data Type	TABLE of
	  Date : Text
	  OD : Text
	  ProductNo : Text
	  Quantity : Integer
	End TABLE ;

Instances	

Instances table:

Date	Originating_Department :
20111017	Sales Dept.

ProductNo	Quantity
A00001(Pen)	300
A00002(Mouse)	400
A00003(Camera)	500

Figure 20-252   Composite Data Type Specification

Figure 20-253 shows the composite data type specification of the *Purchase_Order_Query* input parameter occurring in the *SQL_Purchase_Order_Insert(In Purchase_Order_Query)* operation formula.

Parameter	*Purchase_Order_Query*

Data Type	TABLE of
	  Date : Text
	  SupplierName: Text
	  ProductNo : Text
	  Quantity : Integer
	End TABLE ;

Instances	 | Date	SupplierName
20111118	Johnson Corp. | | ProductNo	Quantity
A00001(Pen)	300
A00002(Mouse)	400
A00003(Camera)	500	

Figure 20-253   Composite Data Type Specification

Figure 20-254 shows the composite data type specification of the *Purchase_Query* input parameter occurring in the *SQL_Purchase_Insert (In Purchase_Query)* operation formula.

Parameter	*Purchase_Query*

Data Type	TABLE of
	  Date : Text
	  SupplierName: Text
	  ProductNo : Text
	  Quantity : Integer
	  UnitPrice : Real
	  ReturnQuantity : Integer
	  Total : Real
	End TABLE ;

Instances	(see tables below)

Date	SupplierName	Total
20111212	Johnson Corp.	258,000.00

ProductNo	Quantity	UnitPrice	ReturnQuantity
A00001(Pen)	300	100.00	0
A00002(Mouse)	390	200.00	10
A00003(Camera)	500	300.00	0

Figure 20-254   Composite Data Type Specification

Figure 20-255 shows the composite data type specification of the *Supplier_Data_Query* input parameter occurring in the *SQL_Supplier_Data_Insert(In Supplier_Data_Query)* operation formula.

Parameter	*Supplier_Data_Query*
Data Type	TABLE of SupplierName:Text Address :Text PhoneNumber:Text FaxNumber:Text E-mail : Text Rank : Text End TABLE ;
Instances	

SupplierName	Johnson Corp.
Address	1232 Fair Circle, Austin, TX
PhoneNumber	512-463-8472
FaxNumber	512-463-8499
E-mail	Johnson1122@gmail.com
Rank	B

Figure 20-255   Composite Data Type Specification

Figure 20-256 shows the implementation's CCD of the *purchasing management* systems definition version 3. In the figure, actor *Originating_Department* has a connection with the *Purchase_Order_Coordinator* component; actor *Supplier* has four connections with the *Purchasing_Department* component; component *Purchase_Order_Coordinator* has a connection with the *Purchase_Requisition_Processing_GUI* component; component *Purchasing_Department* has a connection with each one of the *Quotation_Processing_GUI*, *Purchase_Order_Processing_GUI*, *Purchase_Processing_GUI* and *Supplier_Data_Processing_GUI* components; each one of the *Purchase_Requisition_Processing_GUI*, *Quotation_Processing_GUI*, *Purchase_Order_Processing_GUI*, *Purchase_Processing_GUI* and *Supplier_Data_Processing_GUI* components has a connection with the *Purchasing_Database* component; component *Purchasing_Database* has a connection with the *Network_Operating_System* component.

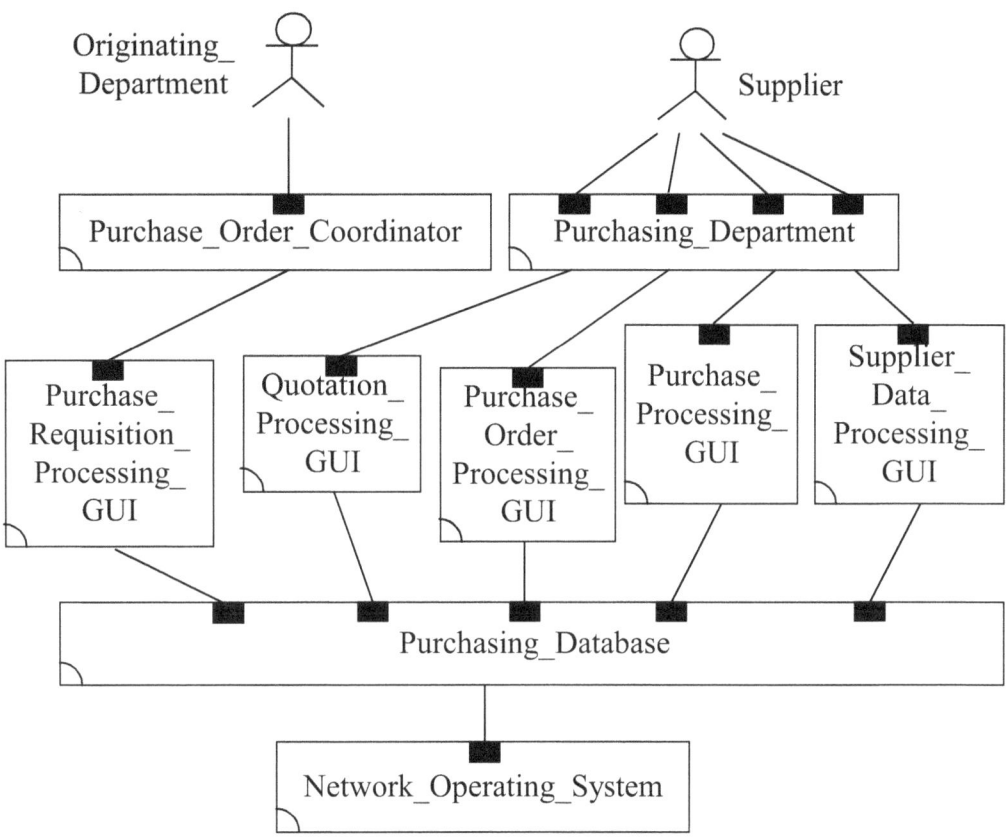

Figure 20-256   Implementation's CCD of the *Purchasing Management*
Systems Definition Version 3

20-3-2-4-2 Implementation's Systems Behavior of the Purchasing Management
Systems Definition Version 3

The entire implementation's systems behavior of the *purchasing management*
systems definition version 3 includes: a) *Implementation's SBCD* and b)
Implementation's IFD of the *purchasing management* systems definition version 3.

Figure 20-257 shows the implementation's SBCD of the *purchasing
management* systems definition version 3 in which interactions among the
Originating_Department, *Supplier* actors and the *Purchase_Order_Coordinator*,
Purchasing_Department,   *Purchase_Requisition_Processing_GUI*,
Quotation_Processing_GUI,   *Purchase_Order_Processing_GUI*,
Purchase_Processing_GUI, *Supplier_Data_Processing_GUI*, *Purchasing_Database*,
Network_Operating_System components shall draw forth the *Purchase_Requisition*,
Quotation, *Purchase_Order*, *Purchase* and *Collect_Supplier_Data* behaviors.

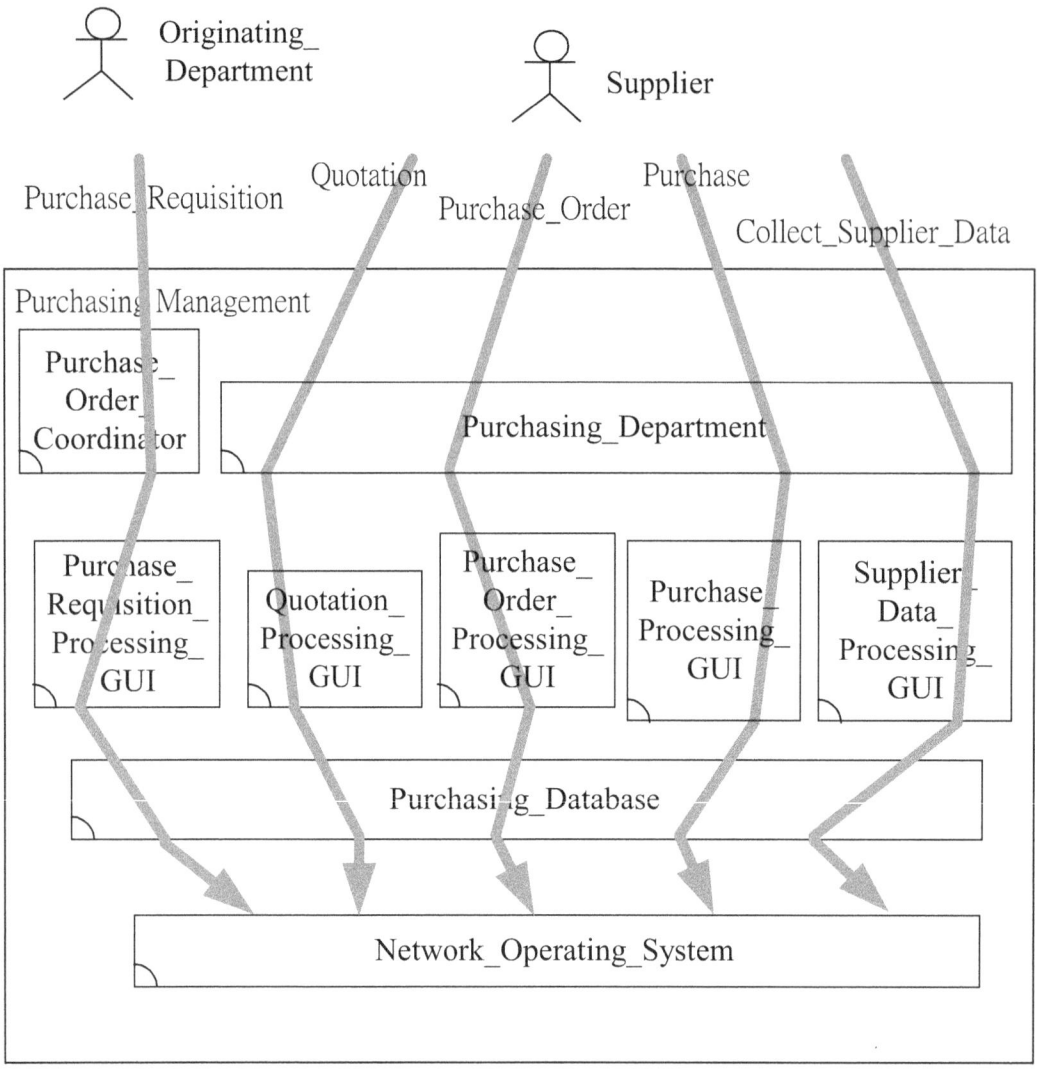

Figure 20-257   Implementation's SBCD of the *Purchasing Management*
Systems Definition Version 3

The overall implementation's behavior of the *purchasing management* systems definition version 3 includes the *Purchase_Requisition*, *Quotation*, *Purchase_Order*, *Purchase* and *Collect_Supplier_Data* behaviors. In other words, the *Purchase_Requisition*, *Quotation*, *Purchase_Order*, *Purchase* and *Collect_Supplier_Data* behaviors together provide the overall implementation's behavior of the *purchasing management* systems definition version 3.

Be noticed that the *Purchase_Requisition*, *Quotation*, *Purchase_Order*, *Purchase* and *Collect_Supplier_Data* behaviors are mutually independent of each other. They tend to be executed concurrently [Hoar85, Miln89, Miln99].

The overall implementation's behavior of the *purchasing management* systems definition version 3 includes five individual behaviors: *Purchase_Requisition*,

Quotation, *Purchase_Order*, *Purchase* and *Collect_Supplier_Data*. Each individual behavior is represented by an execution path. We use an interaction flow diagram (IFD) to define each one of these execution paths. Figure 20-258 shows the implementation's IFD of the *purchasing management* systems definition version 3 *Purchase_Requisition* behavior. First, actor *Supplier* interacts with the *Purchasing_Department* component through the *Quotation_Verify* operation call interaction, carrying the *Quotation_Form* input parameter. Next, component *Purchasing_Department* interacts with the *Quotation_Processing_GUI* component through the *Quotation_Processing_Button_Click* operation call interaction, carrying the *Quotation_Form* input parameter. Continuingly, component *Quotation_Processing_GUI* interacts with the *Purchasing_Database* component through the *SQL_Quotation_Insert* operation call interaction, carrying the *Quotation_Query* input parameter. Finally, component *Purchasing_Database* interacts with the *Network_Operating_System* component through the *Infrastructure_Resources_Share* operation call interaction.

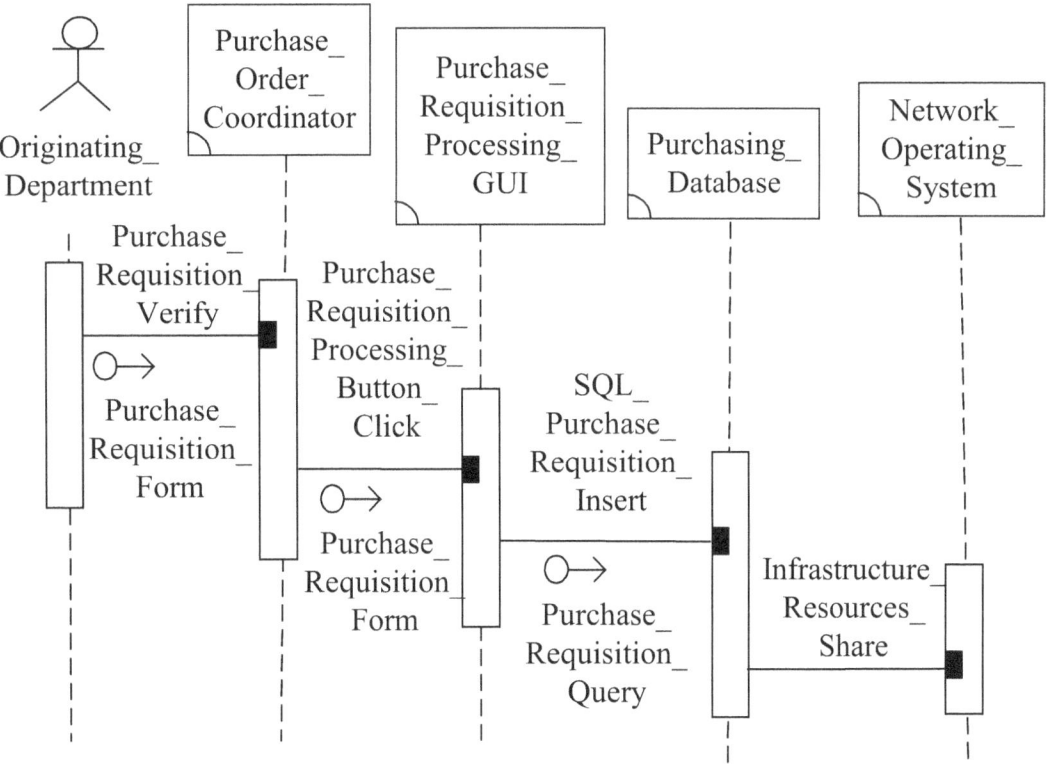

Figure 20-258   Implementation's IFD of the *Purchasing Management* Systems Definition Version 3 *Purchase_Requisition* Behavior

Figure 20-259 shows the implementation's IFD of the *purchasing management* systems definition version 3 *Quotation* behavior. First, actor *Supplier* interacts with the *Purchasing_Department* component through the *Quotation_Verify* operation call interaction, carrying the *Quotation_Form* input parameter. Next, component *Purchasing_Department* interacts with the *Quotation_Processing_GUI* component through the *Quotation_Processing_Button_Click* operation call interaction, carrying the *Quotation_Form* input parameter. Continuingly, component *Quotation_Processing_GUI* interacts with the *Purchasing_Database* component through the *SQL_Quotation_Insert* operation call interaction, carrying the *Quotation_Query* input parameter. Finally, component *Purchasing_Database* interacts with the *Network_Operating_System* component through the *Infrastructure_Resources_Share* operation call interaction.

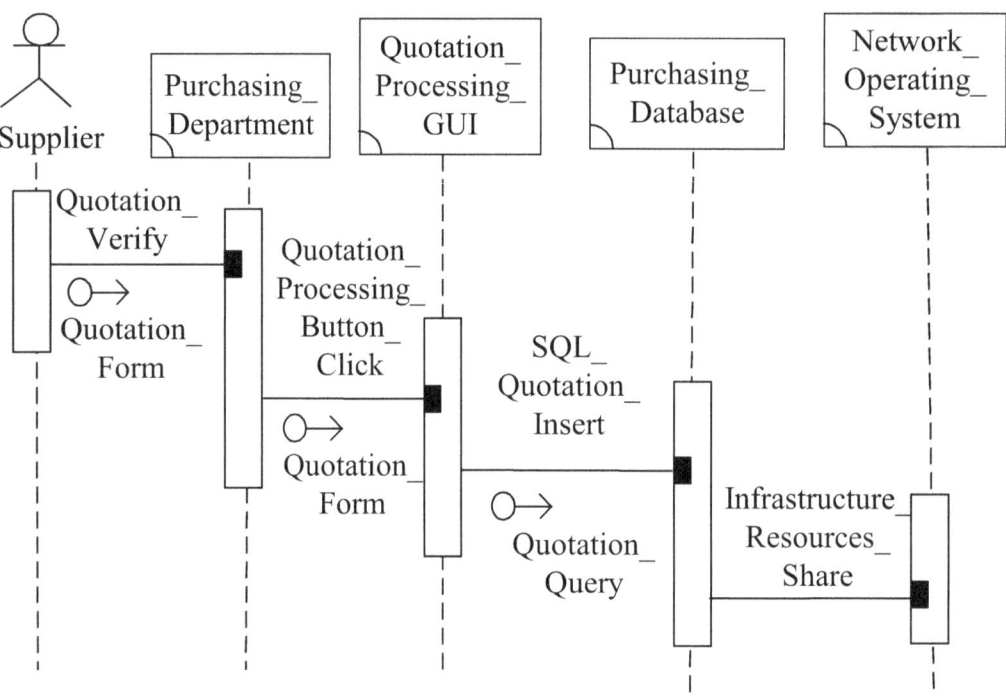

Figure 20-259   Implementation's IFD of the *Purchasing Management*
Systems Definition Version 3 *Quotation* Behavior

Figure 20-260 shows the implementation's IFD of the *purchasing management* systems definition version 3 *Purchase_Order* behavior. First, actor *Supplier* interacts with the *Purchasing_Department* component through the *Purchase_Order_Verify* operation call interaction. Next, component *Purchasing_Department* interacts with the *Purchase_Order_Processing_GUI* component through the *Purchase_Order_Processing_Button_Click* operation call interaction, carrying the *Purchase_Order_Form* input parameter. Continuingly, component *Purchase_Order_Processing_GUI* interacts with the *Purchasing_Database* component through the *SQL_Purchase_Order_Insert* operation call interaction, carrying the *Purchase_Order_Query* input parameter. Continuingly, component *Purchasing_Database* interacts with the *Network_Operating_System* component through the *Infrastructure_Resources_Share* operation call interaction. Continuingly, component *Purchasing_Department* interacts with the *Purchase_Order_Processing_GUI* component through the *Purchase_Order_Processing_Button_Click* operation return interaction, carrying the *Purchase_Order_Report* output parameter. Finally, actor *Supplier* interacts with the *Purchasing_Department* component through the *Purchase_Order_Verify* operation return interaction, carrying the *Purchase_Order_Report* output parameter.

474

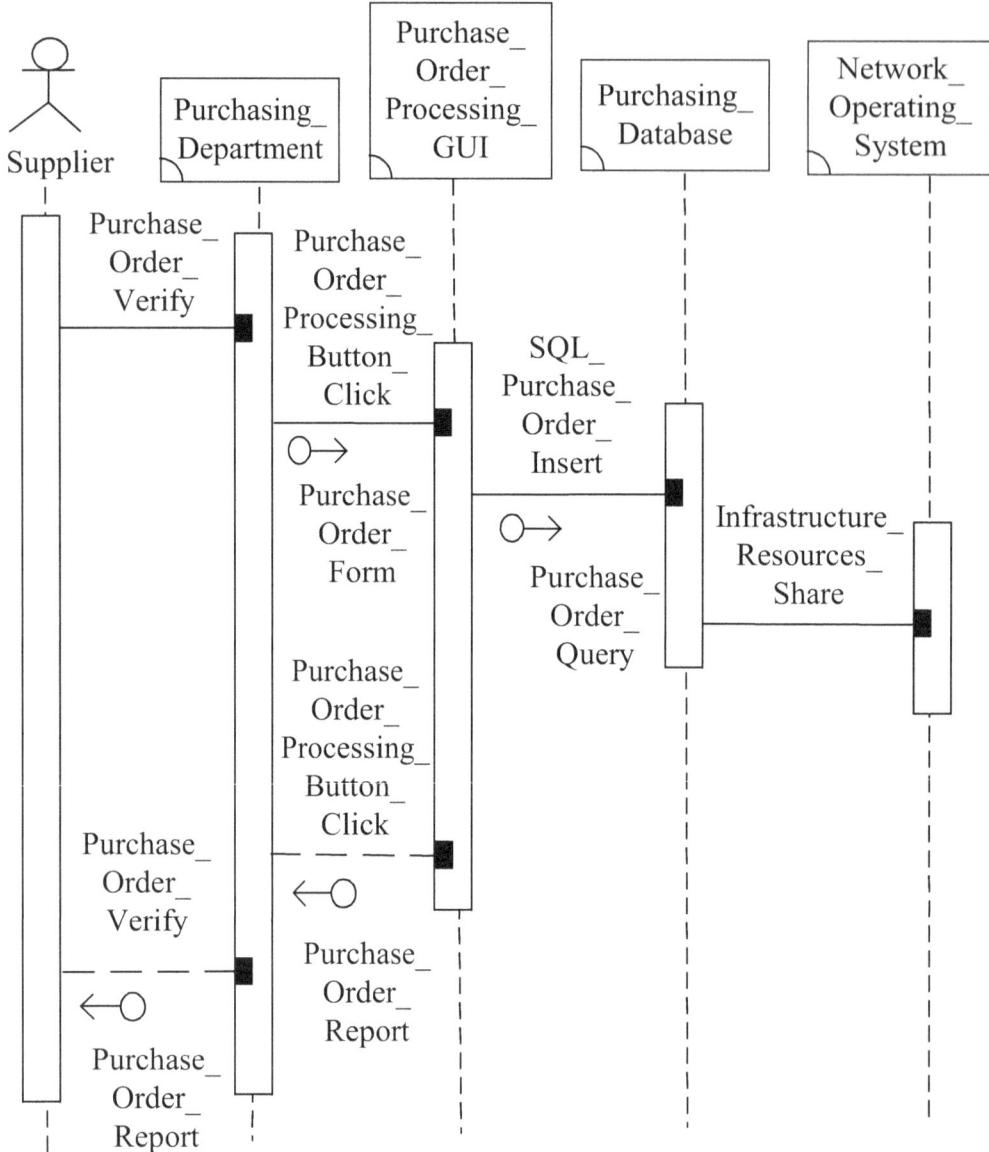

Figure 20-260   Implementation's IFD of the *Purchasing Management Systems Definition Version 3 Purchase_Order* Behavior

Figure 20-261 shows the implementation's IFD of the *purchasing management* systems definition version 3 *Purchase* behavior. First, actor *Supplier* interacts with the *Purchasing_Department* component through the *Purchase_Verify* operation call interaction, carrying the *Products_In* and *Invoice* input parameters. Next, component *Purchasing_Department* interacts with the *Purchase_Processing_GUI* component through the *Purchase_Processing_Button_Click* operation call interaction, carrying the *Purchase_Form* input parameter. Continuingly, component *Purchase_Processing_GUI* interacts with the *Purchasing_Database* component through the *SQL_Purchase_Insert* operation call interaction, carrying the

Purchase_Query input parameter. Continuingly, component *Purchasing_Database* interacts with the *Network_Operating_System* component through the *Infrastructure_Resources_Share* operation call interaction. Finally, actor *Supplier* interacts with the *Purchasing_Department* component through the *Purchase_Verify* operation return interaction, carrying the *Products_Out* output parameter.

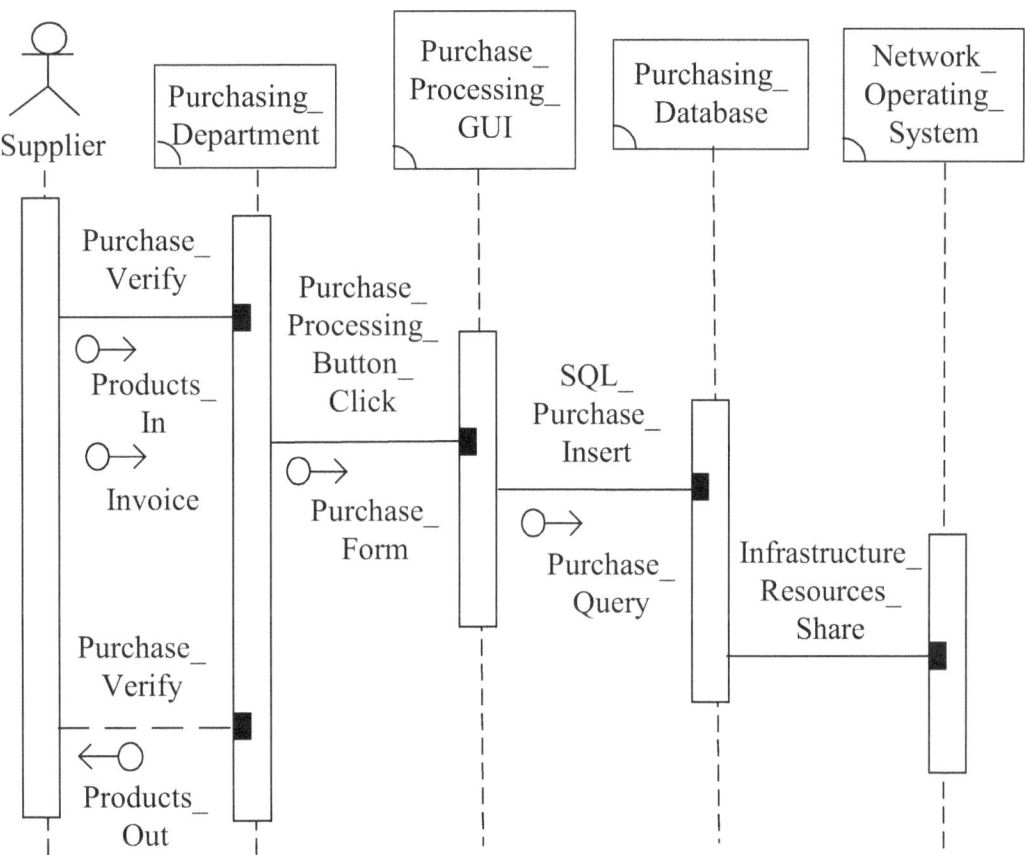

Figure 20-261   Implementation's IFD of the *Purchasing Management* Systems Definition Version 3 *Purchase* Behavior

Figure 20-262 shows the implementation's IFD of the *purchasing management* systems definition version 3 *Collect_Supplier_Data* behavior. First, actor *Supplier* interacts with the *Purchasing_Department* component through the *Interview* operation call interaction. Next, component *Purchasing_Department* interacts with the *Supplier_Data_Processing_GUI* component through the *Supplier_Data_Processing_Button_Click* operation call interaction, carrying the *Supplier_Data_Form* input parameter. Continuingly, component *Supplier_Data_Processing_GUI* interacts with the *Purchasing_Database* component through the *SQL_Supplier_Data_Insert* operation call interaction, carrying the

476

Supplier_Data_Query input parameter. Finally, component *Purchasing_Database* interacts with the *Network_Operating_System* component through the *Infrastructure_Resources_Share* operation call interaction.

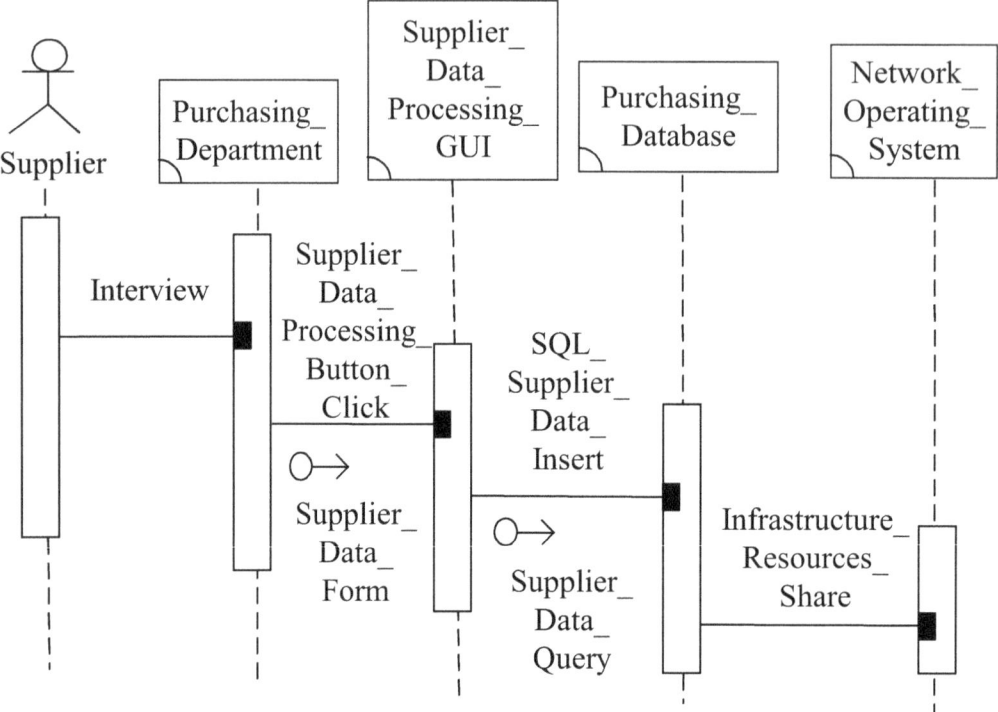

Figure 20-262   Implementation's IFD of the *Purchasing Management*
Systems Definition Version 3 *Collect_Supplier_Data* Behavior

APPENDIX A: SBC ARCHITECTURE DESCRIPTION LANGUAGE

(1) Architecture Hierarchy Diagram

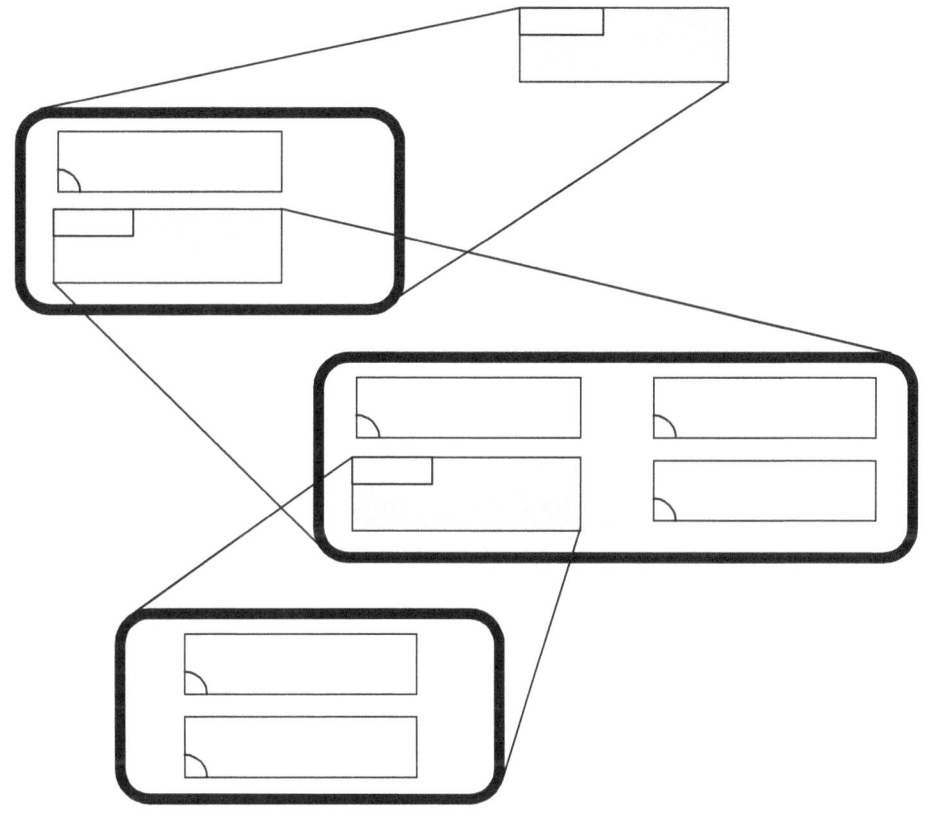

: Aggregated System

: Non-Aggregated System,　Component

(2) Framework Diagram

Business Layer		
Application Layer		
Data Layer		
Technology Layer		

: Component

(3) Component Operation Diagram

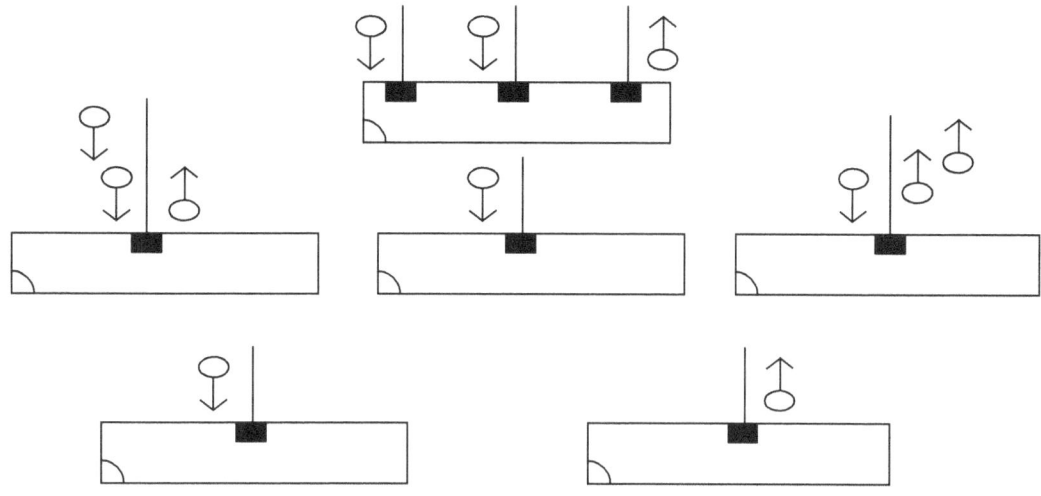

         ■   : Operation

          : Input Data

          : Output Data

          : Component

(4) Component Connection Diagram

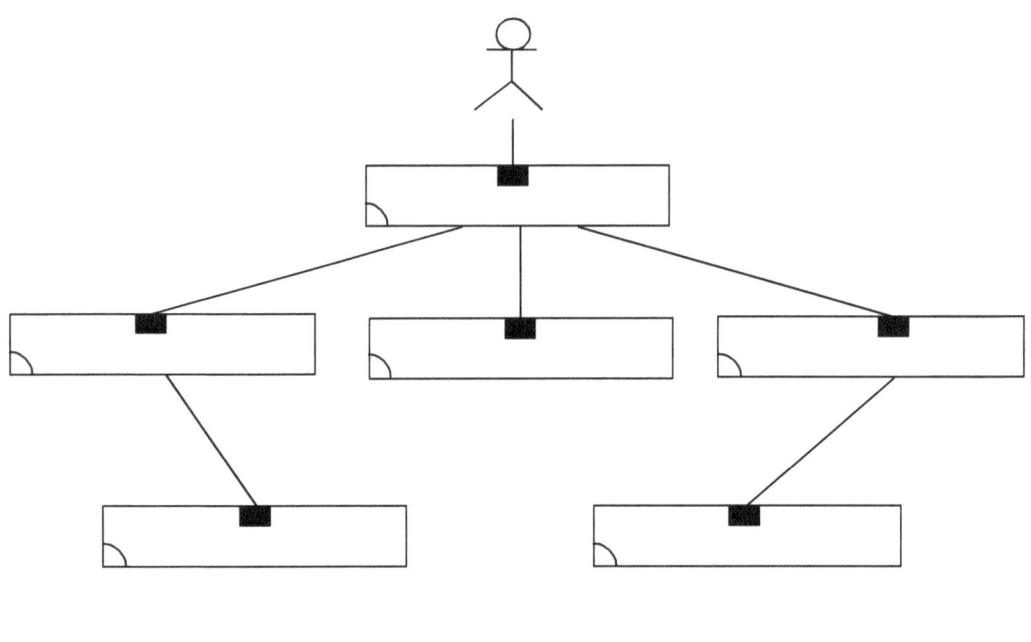

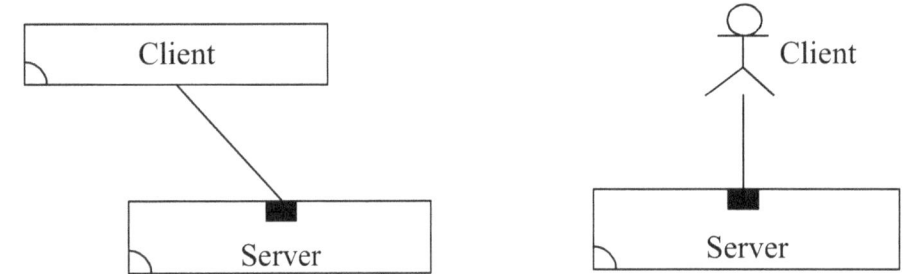

(5) Structure-Behavior Coalescence Diagram

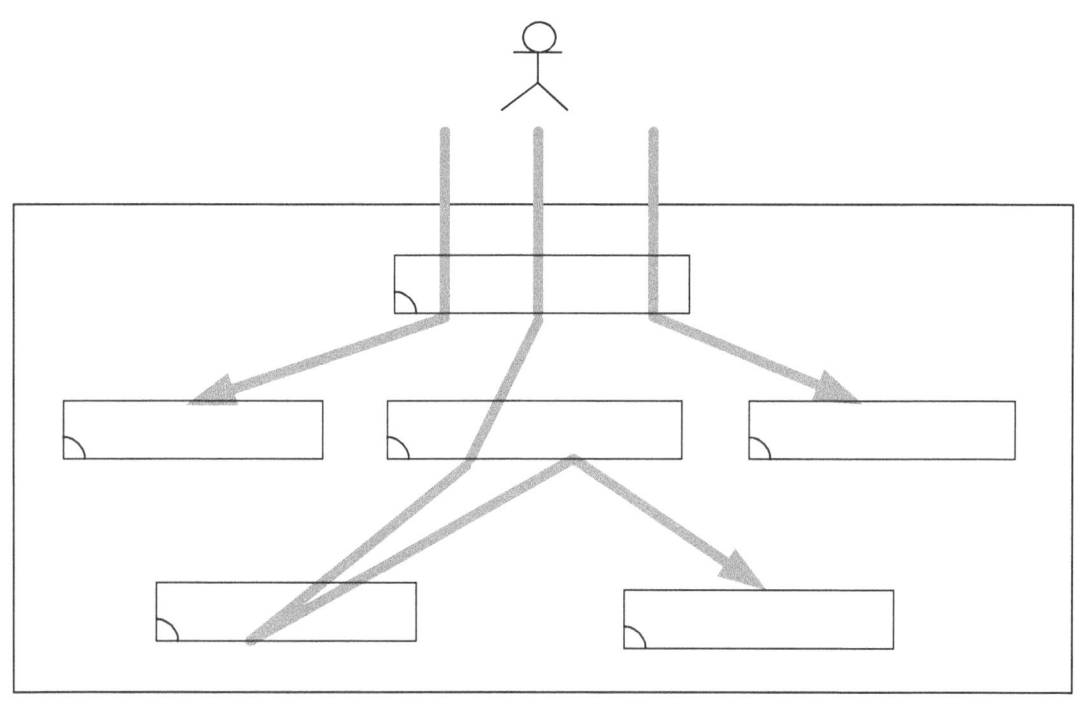

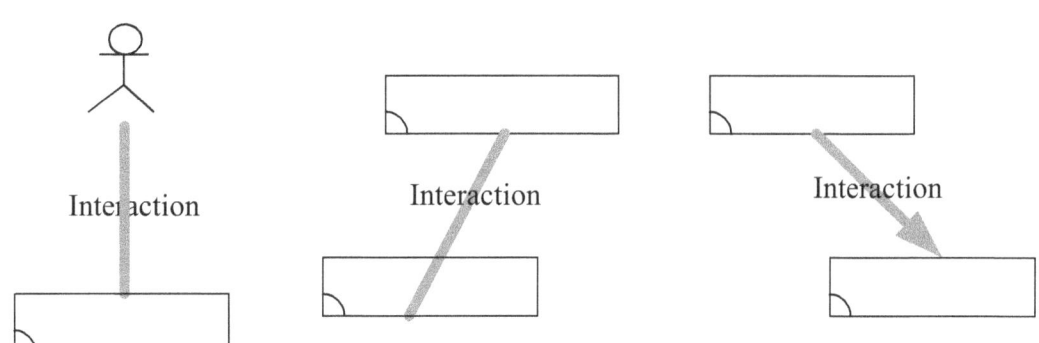

(6) Interaction Flow Diagram

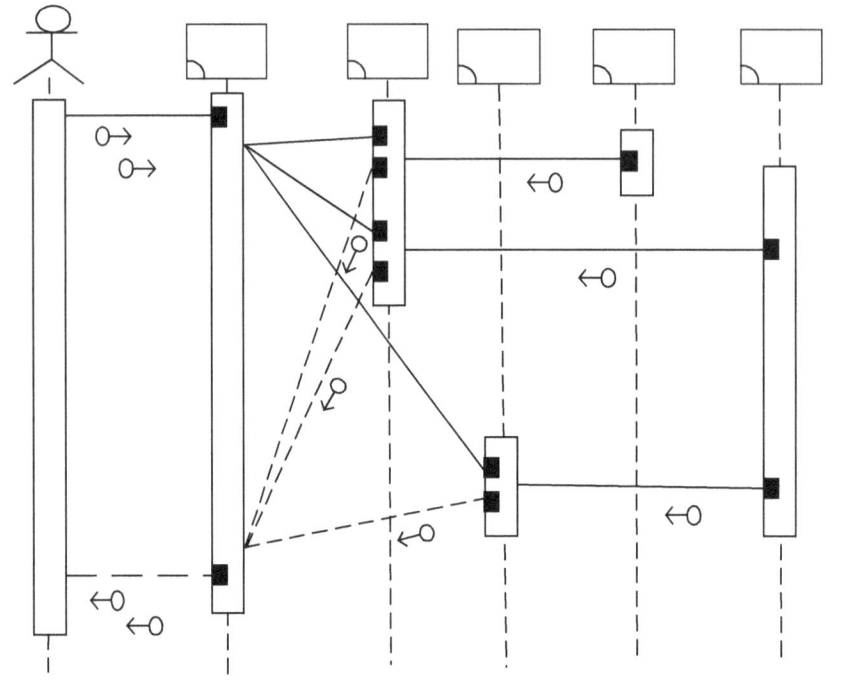

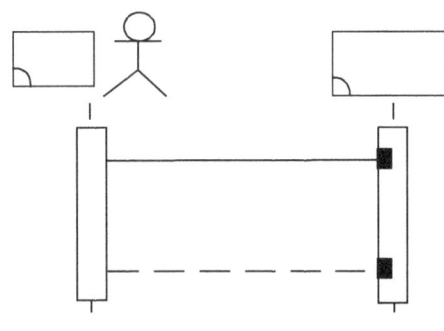

    : Operation Call Interaction

     : Operation Return Interaction

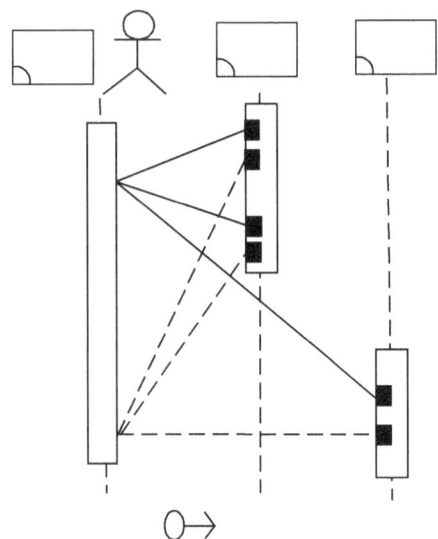

    : Conditional
Operation Call Interaction

     : Conditional
Operation Return Interaction

O→   : Input Data

←O   : Output Data

APPENDIX B: SBC ARCHITECTURE DEVELOPMENT METHOD (SBC-ADM)

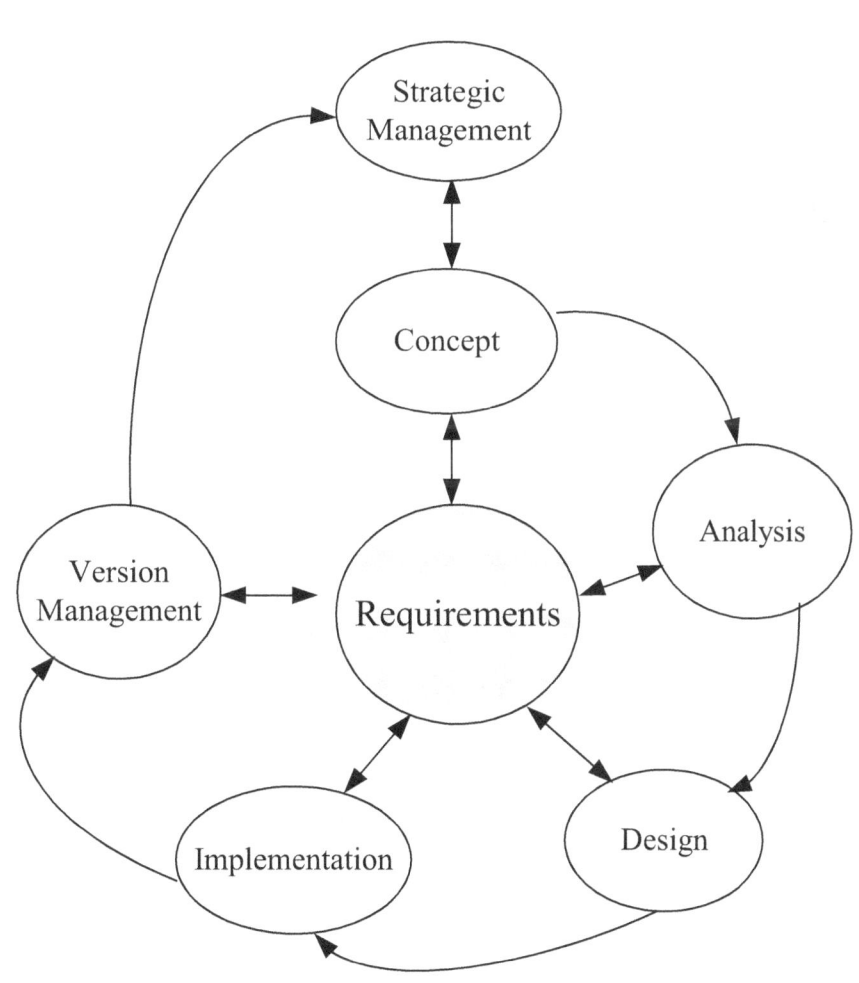

APPENDIX C: SBC VIEW MODEL (SBC-VM)

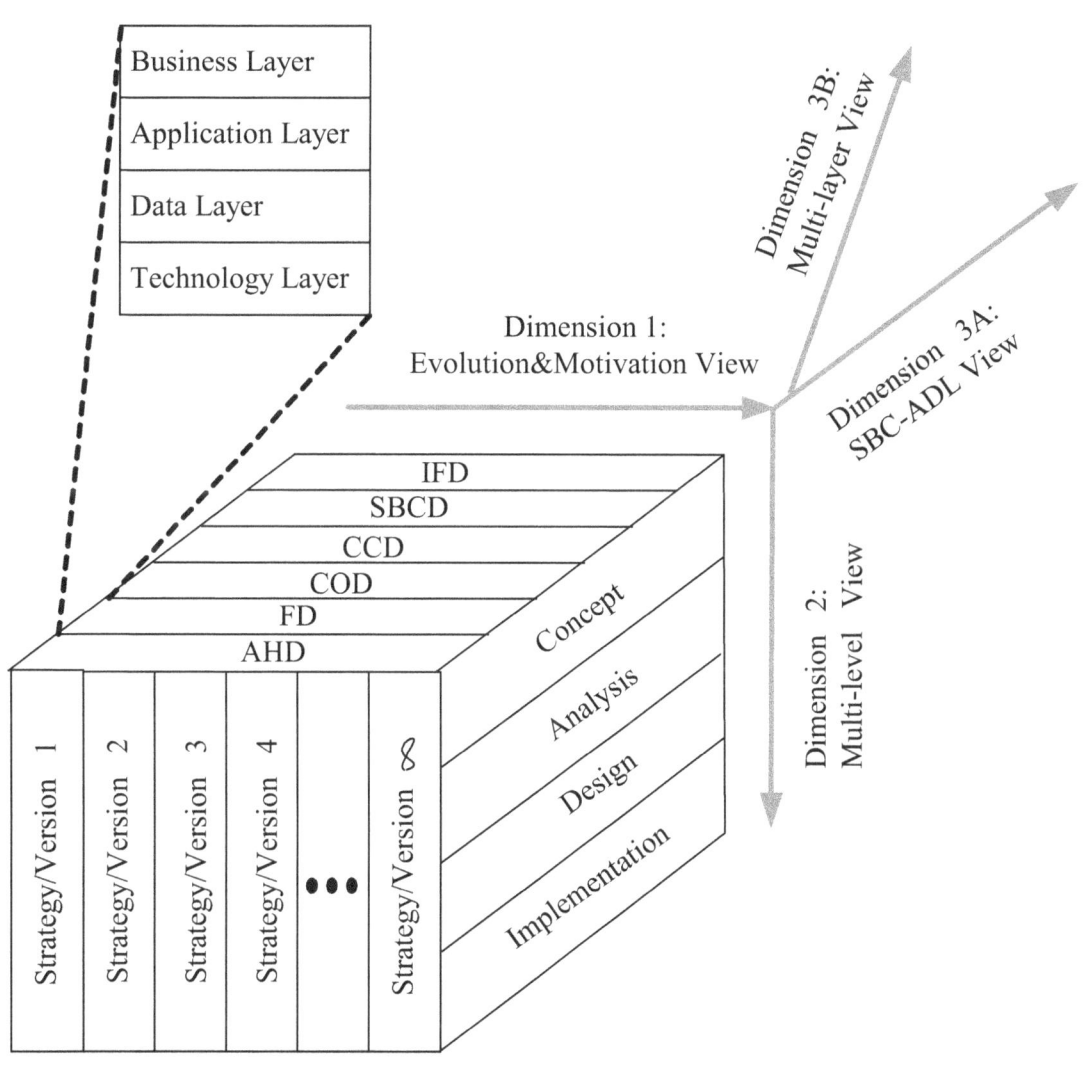

486

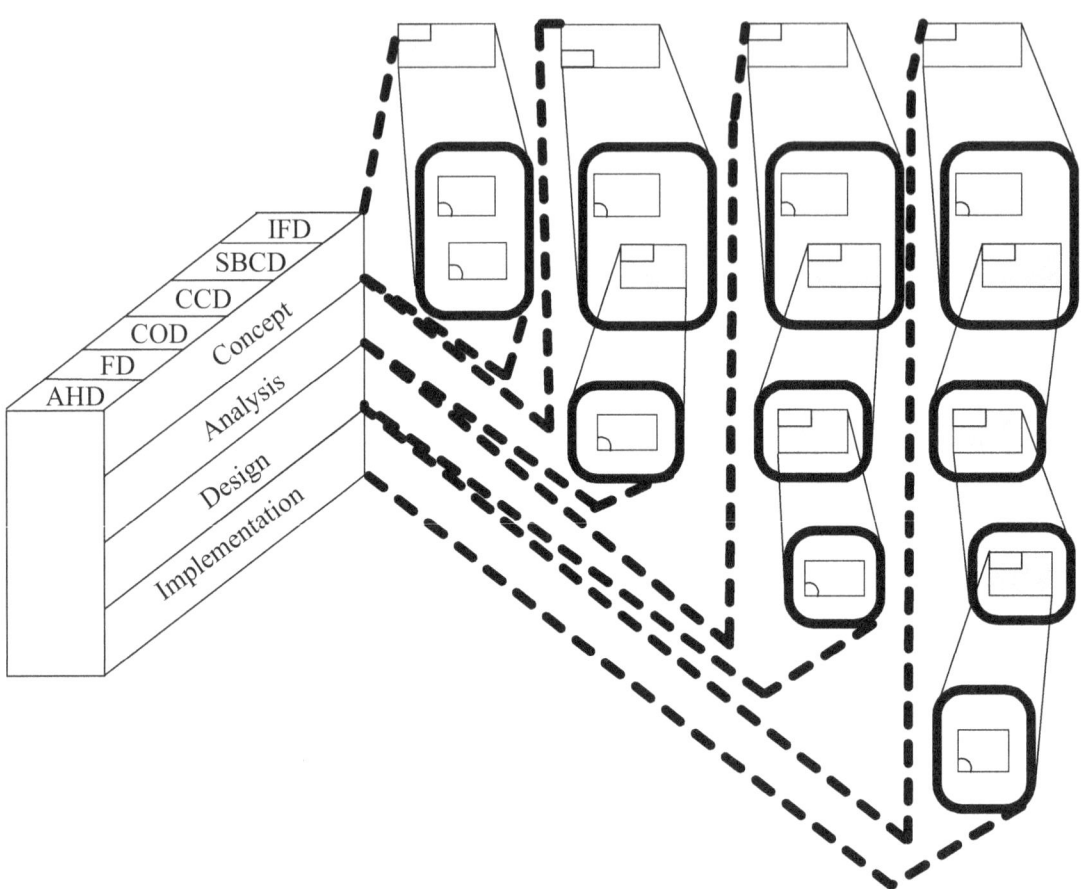

BIBLIOGRAPHY

[Acko68] Ackoff, R., "Toward a System of Systems Concepts," *Modern Systems Research for the Behavioral Scientist: A Sourcebook*, Aldine Publishing Company, 1968.

[Bare84] Barendregt, H. P., *The Lambda Calculus: Its Syntax and Semantics*, Elsevier Science Publishers, 1984.

[Beam90] Beam, W. R., *Systems Engineering: Architecture and Design*, McGraw-Hill, 1990.

[Bere09] Berenbach, B. et al., *Software & Systems Requirements Engineering: In Practice*, 1st Edition, McGraw-Hill Osborne Media, 2009.

[Berk08] Berkem, B., "From the Business Motivation Model to Service Oriented Architecture," *Journal of Object Technology*, Vol.7, No.8, 2008.

[Bert69] Von Bertalanffy, L., *General System Theory: Foundations, Development, Applications*, Revised Edition, George Braziller Inc., 1969.

[Bert81] Von Bertalanffy, L. et al., *Systems View of Man: Collected Essays*, Westview Pr, 1981.

[Burd10] Burd, S. D., *Systems Architecture*, 6th Edition, Cengage Learning, 2010.

[Chao09] Chao, W. S. et al., *System Analysis and Design: SBC Software Architecture in Practice*, LAP Lambert Academic Publishing, 2009.

[Chao12] Chao, W. S., *Systems Architecture: SBC Architecture at Work*, LAP Lambert Academic Publishing, 2012.

[Chao14a] Chao, W. S., *Systems Thingking 2.0: Architectural Thinking Using the SBC Architecture Description Language*, CreateSpace Independent Publishing Platform, 2014.

[Chao14b] Chao, W. S., *Systems Modeling and Architecting: Structure-Behavior Coalescence for Systems Architecture*, CreateSpace Independent Publishing Platform, 2014.

[Chao15a] Chao, W. S., *A Process Algebra For Systems Architecture: The Structure-Behavior Coalescence Approach*, CreateSpace Independent Publishing Platform, 2015.

[Chao15b] Chao, W. S., *An Observation Congruence Model For Systems Architecture: The Structure-Behavior Coalescence Approach*, CreateSpace Independent Publishing Platform, 2015.

[Chao15c] Chao, W. S., *Variants of SBC Process Algebra: The Structure-Behavior Coalescence Approach*, CreateSpace Independent Publishing Platform, 2015.

[Chao15d] Chao, W. S., *Single-Queue SBC Observation Congruence Model For Systems Architecture: The Structure-Behavior Coalescence Approach*, CreateSpace Independent Publishing Platform, 2015.

[Chao15e] Chao, W. S., *Multi-Queue SBC Observation Congruence Model For Systems Architecture: The Structure-Behavior Coalescence Approach*, CreateSpace Independent Publishing Platform, 2015.

[Chao17a] Chao, W. S., *Channel-Based Single-Queue SBC Process Algebra For Systems Definition: General Architectural Theory at Work*, CreateSpace Independent Publishing Platform, 2017.

[Chao17b] Chao, W. S., *Channel-Based Multi-Queue SBC Process Algebra For Systems Definition: General Architectural Theory at Work*, CreateSpace Independent Publishing Platform, 2017.

[Chao17c] Chao, W. S., *Channel-Based Infinite-Queue SBC Process Algebra For Systems Definition: General Architectural Theory at Work*, CreateSpace Independent Publishing Platform, 2017.

[Chao17d] Chao, W. S., *Operation-Based Single-Queue SBC Process Algebra For Systems Definition: General Architectural Theory at Work*, CreateSpace Independent Publishing Platform, 2017.

[Chao17e] Chao, W. S., *Operation-Based Multi-Queue SBC Process Algebra For Systems Definition: Unification of Systems Structure and Systems Behavior*, CreateSpace Independent Publishing Platform, 2017.

[Chao17f] Chao, W. S., *Operation-Based Infinite-Queue SBC Process Algebra For Systems Definition: Unification of Systems Structure and Systems Behavior*, CreateSpace Independent Publishing Platform, 2017.

[Chec99] Checkland, P., *Systems Thinking, Systems Practice: Includes a 30-Year Retrospective*, 1st Edition, Wiley, 1999.

[Cohe63] Cohen, P. J., "The Independence of the Continuum Hypothesis," *Proceedings of the National Academy of Sciences of the United States of America*, 50 (6), 1963, pp. 1143–1148.

[Date03] Date, C. J., *An Introduction to Database Systems*, 8th Edition, Addison Wesley, 2003.

[Dori95] Dori, D., "Object-Process Analysis: Maintaining the Balance between System Structure and Behavior," *Journal of Logic and Computation* 5(2), pp.227-249, 1995.

[Dori02] Dori, D., *Object-Process Methodology: A Holistic Systems Paradigm*, Springer Verlag, New York, 2002.

[Dori16] Dori, D., *Model-Based Systems Engineering with OPM and SysML*, Springer Verlag, New York, 2016.

[Elma10] Elmasri, R., *Fundamentals of Database Systems*, 6th Edition, Addison Wesley, 2010.

[Forr61] Forrester, J. W., *Industrial Dynamics*, Pegasus Communications, 1961.

[Free13] Freeman, L., *Strategy: A History*, Oxford University Press, 2013.

[Frie11] Friedenthal, S., et al., *A Practical Guide to SysML, Second Edition: The Systems Modeling Language*, 2nd Edition, Morgan Kaufmann, 2011.

[Gall03] Gall, J., *The Systems Bible: The Beginner's Guide to Systems Large and Small*, General Systemantics Pr/Liberty, 2003.

[Ghar11] Gharajedaghi, J., *Systems Thinking: Managing Chaos and Complexity: A Platform for Designing Business Architecture*, Morgan Kaufmann, 2011.

[Grad06] Grady, J. O., *System Requirements Analysis*, 1st Edition, Academic Press, 2006.

[Hend80] Henderson, P., *Functional Programming: Application and Implementation*, Prentice-Hall, 1980.

[Hoar85] Hoare, C. A. R., *Communicating Sequential Processes*, Prentice-Hall, 1985.

[Hoff10] Hoffer, J. A., et al., *Modern Systems Analysis and Design*, 6th Edition, Prentice Hall, 2010.

[Jorg12] Jorgensen, S. E., *Introduction to Systems Ecology (Applied Ecology and Environmental Management)*, CRC Press, 2012.

[Kapo94] Kaposi, A., et al., *Systems, Models and Measure*, Springer-Verlag London Limited, 1994.

[Kass07] Kasser, J. E., *A Framework for Understanding Systems Engineering*, BookSurge Publishing, 2007.

[Kill09] Killoran, D. M., *LSAT Logical Reasoning Bible: A Comprehensive System for Attacking the Logical Reasoning Section of the LSAT*, PowerScore Publishing, 2009.

491

[Klip09] Klipp, E. et al., *Systems Biology: A Textbook*, 1st Edition, Wiley-VCH, 2009.

[Koss11] Kossiakoff, A. et al., *Systems Engineering Principles and Practice*, 2nd Edition, Wiley-Interscience, 2011.

[Lasz96] Laszlo, E., *The Systems View of the World: A Holistic Vision for Our Time*, 2nd Edition, Hampton Pr, 1996.

[Luhm12] Luhmann, N., *Introduction to Systems Theory*, 1st Edition, Polity, 2012.

[Mann74] Manna, Z., *Mathematical Theory of Computation*, McGraw-Hill, 1974.

[Maie09] Maier, M. W., *The Art of Systems Architecting*, 3rd Edition, CRC Press, 2009.

[Mcke12] Mckeown, M., *The Strategy Book: How To Think and Act Strategically to Deliver Outstanding Results*, 1st Edition, FT Press, 2012.

[Mead08] Meadows, D. H., *Thinking in Systems: A Primer*, Chelsea Green Publishing, 2008.

[Miln89] Milner, R., *Communication and Concurrency*, Prentice-Hall, 1989.

[Miln99] Milner, R., *Communicating and Mobile Systems: the π-Calculus*, 1st Edition, Cambridge University Press, 1999.

[Mull11] Muller, G., *Systems Architecting: A Business Perspective*, CRC Press, 2011.

[Odum94] Odum, H. T., *Ecological and General Systems: An Introduction to Systems Ecology*, Rev Sub Edition, University Press of Colorado, 1994.

[Ogat03] Ogata, K., *System Dynamics*, 4th Edition, Prentice Hall, 2003.

[O'Rou03] O'Rourke, C. et al, *Enterprise Architecture Using the Zachman Framework*, 1st Edition, Course Technology, 2003.

[Palm09] Palm, W. III, *System Dynamics*, 2nd Edition, McGraw-Hill Science/Engineering/Math, 2009.

[Pele00] Peleg, M. et al., "The Model Multiplicity Problem: Experimenting with Real-Time Specification Methods". *IEEE Tran. on Software Engineering*. 26

(8), pp. 742–759, 2000.

[Pork78] Porkert, M., *Theoretical Foundations of Chinese Medicine: Systems of Correspondence*, The MIT Press, 1978.

[Prat00] Pratt, T. W. et al., *Programming Languages: Design and Implementation*, 4th Edition, Prentice Hall 2000.

[Pres09] Pressman, R. S., *Software Engineering: A Practitioner's Approach*, 7th Edition, McGraw-Hill, 2009.

[Raff11] Raff, H. et al., *Medical Physiology: A Systems Approach*, 1st Edition, McGraw-Hill Professional, 2011.

[Rayn09] Raynard, B., *TOGAF The Open Group Architecture Framework 100 Success Secrets*, Emereo Pty Ltd, 2009.

[Roza11] Rozanski, N. et al., *Software Systems Architecture: Working With Stakeholders Using Viewpoints and Perspectives*, 2nd Edition, Addison-Wesley Professional, 2011.

[Rumb91] Rumbaugh, J. et al., *Object-Oriented Modeling and Design*, Prentice-Hall, 1991.

[Sang03] Sangiorgi, D. et al., *The Pi-Calculus: A Theory of Mobile Processes*, Cambridge University Press, 2003.

[Scho10] Scholl, C., *Functional Decomposition with Applications to FPGA Synthesis*, Springer, 2010.

[Scot67] Scott, D. S., "A Proof of the Independence of the Continuum Hypothesis," *Mathematical Systems Theory*, Volume 1, 1967, pp. 89–111.

[Sebe12] Sebesta, R. W., *Programming the World Wide Web*, 7th Edition, Addison-Wesley, 2012.

[Seth96] Sethi, R., *Programming Languages: Concepts and Constructs*, 2nd Edition, Addison-Wesley, 1996.

[Shap00] Shapiro. S., *Foundations without Foundationalism: A Case for Second-order Logic*, Oxford University Press, 2000.

[Shel11] Shelly, G. B., et al., *Systems Analysis and Design*, 9th Edition, Course Technology, 2011.

[Sher09] Sherwood, L., *Human Physiology: From Cells to Systems*, 7th Edition, Brooks Cole, 2009.

[Sode03] Soderborg, N.R. et al., "OPM-based Definitions and Operational Templates," *Communications of the ACM* 46(10), pp. 67-72, 2003.

[Somm06] Sommerville, I., *Software Engineering*, 8th Edition, Addison-Wesley, 2006.

[Voit12] Voit, E., *A First Course in Systems Biology*, 1st Edition, Garland Science, 2012.

[Wall04] Wall, D., *Multi-Tier Application Programming with PHP: Practical Guide for Architects and Programmers*, Morgan Kaufmann, 2004.

[Warf06] Warfield, J. N., *An Introduction to Systems Science*, World Scientific Publishing Company, 2006.

[Weil00] Weil, A., *Spontaneous Healing: How to Discover and Embrace Your Body's Natural Ability to Maintain and Heal Itself*, Ballantine Books, 2000.

[Weil04] Weil, A., *Health and Healing: The Philosophy of Integrative Medicine and Optimum Health*, Revised Edition, Mariner Books, 2004.

INDEX

A

abstract system. *See* virtual system

actor, 59, 60

ADL. *See* architecture description language

ADM. *See* architecture development method

AF. *See* architecture framework

aggregated system, 58, 76, 88, 89, 93

AHD. *See* architecture hierarchy diagram

architecture description language, 38

architecture development method, 41

architecture framework. *See* view model

architecture hierarchy diagram, 57, 58

B

baseline architecture, 41

building block. *See* component

C

CCD. *See* component connection diagram

COD. *See* component operation diagram

component, 30, 36, 54, 57, 58

component connection diagram, 60

component operation diagram, 59

concrete system. *See* physical system

connection

    client, 102

    operation provider, 102

    operation user, 102

    server, 102

D

data flow diagram, 48

data type

    composite, 96, 97, 99, 100

    primitive, 96, 97, 99

DFD. *See* data flow diagram

E

entity. *See* component

evolution&motivation view, 29, 44, 53, 66, 68, 72

    strategy/version n view, 29

    strategy/version n+1 view, 29

external environment. *See* actor

F

FD. *See* framework diagram

FD view. *See* multi-layer view

first-order function, 165

first-order logic, 166

first-order system, 167

framework diagram, 58

functional decomposition, 18

G

general architectural theory. *See* general systems theory 2.0

general systems theory 1.0, 17

general systems theory 2.0, 71, 181, 221

498